Inorganic substances

Inorganic substances

A prelude to the study of descriptive inorganic chemistry

DEREK W. SMITH

Associate Professor and Chairperson of Chemistry Department
University of Waikato, Hamilton, New Zealand

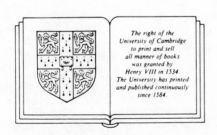

The right of the
University of Cambridge
to print and sell
all manner of books
was granted by
Henry VIII in 1534.
The University has printed
and published continuously
since 1584.

Cambridge University Press

Cambridge
New York Port Chester
Melbourne Sydney

8

Published by the Press Syndicate of the University of Cambridge
The Pitt Building, Trumpington Street, Cambridge CB2 1RP
40 West 20th Street, New York NY 10011, USA
10 Stamford Road, Oakleigh, Melbourne 3166, Australia

© Cambridge University Press 1990

First published 1990

Printed in Great Britain by
Redwood Burn Limited, Trowbridge, Wiltshire

British Library cataloguing in publication data

Smith, Derek W.
Inorganic substances.
1. Inorganic chemistry
I. Title
546

Library of Congress cataloging in publication data

Smith, Derek W. (Derek William), 1943–
Inorganic substances: a prelude to the study of descriptive
inorganic chemistry/Derek W. Smith.
    p.   cm.
Includes bibliographies and index.
ISBN 0 521 33136 6 – ISBN 0 521 33738 0 (paperback)
1. Chemistry, Inorganic.   I. Title.
QD151.5.S65   1990
546—dc19   88-38329 CIP

ISBN 0 521 33136 6 hard covers
ISBN 0 521 33738 0 paperback

Contents

Preface

This book was conceived in response to the disquiet which many chemists have expressed concerning the direction taken by courses and textbooks of inorganic chemistry in the last 20–30 years. The traditional inorganic text was divided into two parts. The book began with an outline of the underlying theory: atomic structure, bonding theory, structural chemistry and relevant aspects of thermodynamics etc. The bulk of the traditional text was taken up by descriptive inorganic chemistry: a systematic survey of the chemistry of the elements, according to the Periodic classification. Here the reader was supposed to acquire a sound factual knowledge of inorganic substances – what compounds have been prepared, how they are prepared, their structures, chemical reactions and physical properties. The purpose of the theoretical part was mainly to make the study of descriptive inorganic chemistry more palatable and intellectually-satisfying; it provided a framework around which substances and properties can be classified, and which enabled the reader to 'understand' and 'explain' experimentally-observed facts, perhaps even to 'predict' facts which were not explicitly set out.

Over the years, the theoretical part tended to grow at the expense of the descriptive material. This reflects the increasingly diverse preoccupations of inorganic chemists. Since the 1950s, they have invaded territory which was formerly within the domain of the physical chemists, e.g. structural chemistry, spectroscopy, magnetochemistry, quantum chemistry and reaction mechanisms. These preoccupations naturally influenced the structure of the inorganic course and associated texts. The theoretical part tended to become the end rather than the means, and the more selective coverage of factual chemistry often seemed to be chosen to illustrate the theoretical principles and vindicate them. By the 1970s, many teachers had abandoned any attempt to cover descriptive inorganic chemistry in the traditional sense. Thus we can encounter the student who can write an erudite account of structural minutiae in copper(II) chemistry, ligand field spectra and the bonding in carbonyl complexes, but who knows little about the more mundane compounds of the transition elements and would be hard pressed to locate indium in the Periodic Table, let alone venture anything about its chemistry. Or, we might find the student who, if asked to write on cadmium chloride – its structure and properties – would begin by presenting an immaculate molecular orbital diagram for a linear triatomic molecule. Topics like MO theory, ligand field spectra etc. are fascinating and worthy of serious academic study, as are many of the currently-fashionable areas of structural and preparative chemistry. But there has been a growing awareness of the need to retain descriptive inorganic chemistry, to make a serious study of the Periodic Table as a whole, and to come to terms with its beauties and paradoxes.

The appearance in 1984 of Greenwood and Earnshaw's *Chemistry of the Elements* was a landmark. Here was a modern textbook of essentially descriptive inorganic chemistry (although these authors might object to such a title). Such a comprehensive work is necessarily hefty (1500+ pages, 2 kg) and perhaps not very practical for classroom purposes, let alone for browsing on the bus or at the breakfast table.

I have tried to write a book which might serve as a supplement or as a prelude to the study of Greenwood and Earnshaw, or of other advanced inorganic texts such as Cotton and Wilkinson's *Advanced Inorganic Chemistry*. It is not a textbook of descriptive inorganic chemistry, nor is it really a textbook of theoretical inorganic chemistry. It is intended to bridge the gap between the 'general chemistry' taken by students in their first year at university and the more advanced texts which make a serious effort to cover systematic, descriptive inorganic chemistry. It may also be useful for students who will not be studying inorganic chemistry at an advanced level but who need rather more than is usually covered in first-year service courses.

It is expected that the reader will have some acquaintance with:

> atomic orbitals and electron configurations;
> elements of valence bond and molecular orbital theories;
> molecular shapes (VSEPR theory);
> simple ionic and other three-dimensional solid structures;
> elementary coordination chemistry and crystal field theory;
> elementary thermodynamics, kinetics and mechanisms;
> solution chemistry, redox potentials, acid/base equilibria.

Few students nowadays are likely to encounter these topics for the first time in the pages of an inorganic text. Accordingly they are not treated *ab initio* in this book. However, only a very elementary knowledge – such as the reader might be expected to acquire at the beginning of a degree course, or even in the last year at school – should be sufficient. Some readers will, of course, find it necessary to revise their existing understanding of, e.g., thermodynamics, and others may feel the need to seek more comprehensive expositions of some topics here presented. A good book should motivate its readers to read further, to challenge their grasp of subjects they think they have 'done', and help to break down the seemingly watertight compartments in the mind which divide closely-related and interdependent modules of knowledge.

The subject matter of the book is the physical and theoretical background to descriptive inorganic chemistry. Its purpose is the same as that of the first part of the traditional inorganic text, before the theory came to be seen as an end in itself and to engulf the factual chemistry. Much of the same ground is covered in contemporary inorganic texts – but I have tried to keep one eye fixed firmly on descriptive chemistry. The theory is thus that which is directly relevant to an appreciation of what substances can be prepared, and how; the structures adopted; and the more striking physical and chemical properties. I have tried – perhaps not always with success – to steer away from 'explaining' details of structures, reaction mechanisms, spectra etc. which, although undeniably fascinating, tend to divert attention from the study of 'real' inorganic chemistry. The ultimate aim is to prepare the reader for a study of the Periodic Table which will be agreeable and intellectually-satisfying, leaving the student with a well-ordered factual knowledge of the elements and a feeling that most of it makes sense.

I will not prolong this preface by trying to explain and justify my choice of material and arrangement. Apart from Chapter 2 (which could usefully be read at a later stage) and Chapter 7 (a cursory perusal of which may be desirable before Chapter 2 for a student having little or no previous acquaintance with molecular orbital theory), the book is best read starting with Chapter 1 and finishing with Chapter 10. Please do not skip over sections you feel will be 'old hat' because you have 'done' that topic before. I have tried to retexture some of the more well-worn garments in the inorganic chemist's wardrobe; you might find something thought-provoking or surprising in the least likely places.

I have to acknowledge a host of mentors, colleagues and students who have inspired and maintained my enthusiasm for inorganic chemistry over the years. For secretarial assistance in writing this book, I thank Karen Beamish-White, Pauline Blithe, Kim Hewlett, Elaine Norton and Raewyn Oulton. My wife, Mandy, has given her full support throughout, at the expense of much sacrifice of family life. The work is ultimately dedicated to my little daughter, Lucy, whose complete indifference to chemistry reminds me that there are other things in life.

1

The theoretical framework of inorganic chemistry

1.1 The scope of inorganic chemistry

Inorganic chemistry is what inorganic chemists do. The perpetuation of the traditional divisions of chemistry, and of a number of subdivisions, is related to the tribal loyalties of chemists. Just as most people are proud of their nationality, most chemists, in common with other scientists, feel the need to belong to a particular group. An inspection of the political map of the world will reveal many countries whose emergence and survival as sovereign states are difficult to understand, since they are not bounded by any obvious geographical features. Boundaries between nations are often determined, and persist, for historical reasons which have faded into relative obscurity. Likewise, chemistry has been divided for reasons thought proper at one time but which are no longer appropriate. Members of a tribe commune with each other via their writings in journals (many of which are devoted to the specialised interests of the tribe) and by meeting at conferences. The members share a common folklore and hagiology, which, together with the distinctive jargon of their subject, engender a feeling of kinship. There are movements afoot to abolish terms such as 'inorganic chemistry'; but these are not likely to prevail in the near future.

What, then, do inorganic chemists do? An inorganic chemist is likely to be engaged in at least one of the following:

(1) The preparation and characterisation of substances other than those which fall within the domain of organic chemistry (the extent of this domain will be discussed later in this section). Characterisation of a product involves finding out what it is, and usually stops somewhere between an elemental analysis – which determines the empirical formula – and a full structure determination.

(2) The determination of the structures of inorganic substances. This requires a full description of the geometrical arrangement of

1

atoms in the substance, and the determination of all relevant bond lengths and interbond angles. Such information is usually yielded by X-ray or neutron diffraction studies of crystalline solids, and by electron diffraction or microwave spectroscopy in the case of gases. Limited, qualitative structural information can often be obtained by other physical methods, such as infra-red or NMR spectroscopy. Some inorganic chemists are completely devoted to the application of physical techniques for structure determination; but as long as their aim is to obtain structural information and to set it within the context of inorganic chemistry as a whole, they may be admitted to full membership of the tribe.

(3)   The investigation of the chemical reactions of inorganic substances. This may involve detailed studies of reaction mechanisms and bring the inorganic chemist into the realms of physical chemistry and molecular physics.

(4)   The investigation of the physical properties of inorganic substances. This may be undertaken with a view to exploiting practical applications, or to obtain chemically-relevant information from the physical measurements. Magnetic measurements, for example, may tell us something about the electronic structure and bonding in a coordination compound.

(5)   The establishment of the theoretical principles which underlie any of the above experimental studies. Some people whose work is wholly devoted to theoretical studies still regard themselves as inorganic chemists if they feel that their work is largely of interest to other inorganic chemists, and the theory is not seen as an end in itself.

What distinguishes inorganic from organic chemistry? Organic chemistry may be defined as that area of chemistry which is directly consequent upon the unique properties of the carbon atom, viz.:

(i)   Its (nearly) constant valency of 4, having four valence orbitals and four valence electrons.

(ii)   The restricted number of bonding situations in which a C atom is usually found.

(iii) The relatively rigid bonding about a carbon atom, whether it be in sp, sp^2 or sp^3 hybridisation, which entails high activation barriers to substitution, addition or elimination reactions.

(iv)   The propensity of the carbon atom towards catenation, the formation of chains and rings.

There are, of course, important areas of overlap between the domains of organic and inorganic chemistry; organometallic chemistry and bio-inorganic chemistry are two which have grown very rapidly in recent decades. Even in these, however, it is often possible to distinguish between the approaches and preoccupations of organic and inorganic chemists, although the two are steadily coming together.

1.2 The role and status of theory in inorganic chemistry

According to the *Oxford English Dictionary*, a theory is a 'supposition or system of ideas explaining something, especially one based on general principles independent of the facts, phenomena etc. to be explained'. The contemporary theoretical chemist seeks to reduce a chemical problem to applied mathematics. The ultimate aim is to calculate, from first principles (the laws of physics, and universal physical constants), any experimental result which can be expressed numerically. Such *ab initio* calculations involve no simplifying assumptions, although some approximations have to be made for all but the most trivially simple systems. The mathematics may look fearsome, but 'black box' computer packages are accessible to most chemists. At the other end of the scale, we have simple, qualitative models which are based upon more rigorous, quantitative theories. Between these extremes, we have a range of methods from 'back-of-the-envelope' calculations to quite elaborate quantum-mechanical treatments in which we make some simplifying assumptions in order to economise on computing time.

A theory is rather like a Delphic oracle, an omniscient being to whom we address questions and who gives intellectually-satisfying answers; the answers may not be wholly unambiguous, and 'predictions' may be more impressive in hindsight. However, as the great theoretical chemist C. A. Coulson was wont to say, 'we like to think that the molecules are our friends'. You should be able to write vivid pen-portraits of your best friends, describing their habits and personalities; you may not always predict accurately their behaviour in certain circumstances, but you should usually be able to rationalise such behaviour after the event. So it is in theoretical chemistry.

The most elaborate and rigorous theories do not necessarily lead to the most satisfying 'explanations'. Suppose, for example, you ask a theoretical chemist the question: 'why is the H_2O molecule bent?' The theoretician will attack the problem by calculating the H–O–H angle which will minimise the total energy of the molecule. A quantum-mechanical calculation of the total energy can be performed for any assumed value of this angle, and the theoretician will be well satisfied if a plot of the total energy against the bond angle shows a minimum close to the experimental value of $104.5°$. Such good agreement will satisfy the theoretician that the calculation is sound, and that any underlying approximations and assumptions are valid. But has the bent structure of the H_2O molecule been 'explained'? The total energy of the H_2O molecule (expressed in units appropriate to the gaseous substance) is $-200\,758\,kJ\,mol^{-1}$; this is the energy relative to the separated nuclei and free electrons. The energy is the sum of the kinetic energies of the electrons ($+200\,758\,kJ\,mol^{-1}$) and the potential energy ($-401\,516\,kJ\,mol^{-1}$), arising from electron–nucleus attraction, electron–electron repulsion and nucleus–nucleus repulsion. The total molecular binding energy – the energy of the substance $H_2O(g)$ relative to $2H(g)$ and $O(g)$ – is $917\,kJ\,mol^{-1}$, or less than 0.5% of the total energy. The difference in energy between linear H_2O and bent (104.5°) H_2O is only about $50\,kJ\,mol^{-1}$. The potential energy term can be expressed as the sum of several thousand terms (mostly involving interelectron repulsion, and with many equalities due to symmetry). To try to 'explain' why the molecule is bent is as daunting a task as that of an accountant who is asked to explain why a company having a turnover of £600 000 per year, involving thousands of transactions large and small, has made £50 more profit in one year compared with another. Obviously, no simple explanation can be offered from such an exercise.

The reader will probably be familiar with at least one simple, qualitative 'explanation' for the shape of the water molecule (there are several). The VSEPR approach (see Section 1.4) 'predicts' that the bond angle should be somewhere between 90° and 109.5°, which is good enough for the purposes of most inorganic chemists. The fact that some of the underlying assumptions in this and other simple theories can be challenged does not necessarily vitiate the theory.

Simple, qualitative arguments are often unable to explain a single observation in isolation. They may, however, rationalise (i.e. make sense of) a large body of observations. For example, you may pose the question: 'why is H_2O a liquid at room temperature and atmospheric pressure, while H_2S is a gas?' It is a pertinent question, because the weak London attraction between molecules in condensed phases – which is often almost entirely responsible for holding them together – should increase with increasing atomic/molecular weight (see Section 3.3). A complete answer to the question would require a calculation of the enthalpies of vaporisation for both liquid water and liquid H_2S from first principles. If the calculated values are in good agreement with experiment, we can then enquire into the underlying reasons for the higher boiling point of water. Much more simply, however, we can rationalise the boiling points of binary hydrides in a qualitative way, based on considerations of hydrogen bonding as well as London forces.

To most chemists, theory means the theory of the chemical bond, or valence theory. A satisfactory theory should:

(1) Describe the physical origin of chemical bonding, and explain why molecules, and non-molecular crystals, are stable with respect to the separated atoms.

(2) Explain (or at least rationalise) the stoichiometries of compounds, and thus guide us in the study of known compounds and in the quest for new ones.

(3) Explain the structures of molecules and crystals, e.g. why is the H_2O molecule bent, or why is SiO_2 a non-molecular crystal rather than a molecular gas like CO_2?

(4) Explain the reactivities of substances; why is X a strong acid, or why is Y a powerful oxidising agent?

(5) Provide a framework for the interpretation of, e.g., spectroscopic and magnetic data, so that such data can provide information about structure and bonding.

None of the various theories which the reader will encounter is perfect; all have their strengths and weaknesses. 'Horses for courses' should be the maxim of the inorganic chemist. Some theories are more firmly underpinned by experimental data and by fundamental physical laws than others. The fact that a theory 'works' to a useful extent does not prove that it is literally true, i.e. that it presents a physically-realistic description of the system. Conversely, the fact that some of the underlying assumptions in a theory can be shown to be unsound does not, *per se*, require that the theory be discarded. A simplistic approach can serve us well provided that we understand its limitations and do not take it too literally.

It is quite permissible to skip from one theory to another while discussing the same problem, just as an English-speaking person may occasionally use words and phrases from other languages in order to find *le mot juste*. For example, the term 'sp² hybridisation' properly belongs to the language of valence bond theory. You may see it appear, however, in a discussion of the molecular orbitals of a molecule such as benzene or ethylene; it provides a convenient shorthand notation to describe the σ bonding. These points should be borne in mind in reading Section 1.4 and subsequent chapters in which various theories are invoked.

1.3 Valency and oxidation numbers: a historical sketch of bonding theory prior to quantum mechanics

The historical perspective in chemistry is, alas, too often neglected in contemporary teaching. This fact is not unrelated to the confusion and infelicities which bedevil much of our nomenclature and notation (see Chapter 3). The historical reasons for the introduction of certain terminology are but dimly remembered and their validity is now highly questionable.

For some 50 years after Dalton's atomic theory was published in the early 1800s the theory of chemical bonding made little progress. In the previous section, it was noted that one of the primary functions of bonding theory is to rationalise stoichiometry. But the stoichiometric formulae of chemical substances were often uncertain until about 1860, despite rapid progress in analytical techniques. The determination of stoichiometry demanded a reliable table of atomic weights, the establishment of which depended, in part, on a knowledge of stoichiometry. This vicious circle was broken by the application of Avogadro's Hypothesis – first promulgated in 1808 but not generally accepted until the 1860s – which made possible the determination of the molecular weights of gaseous substances by vapour density measurements. As the problems of stoichiometry were clarified between 1850 and 1860, the theory of valency was developed, largely by Frankland (English, 1825–99), Couper (Scottish, 1831–92) and Kekulé (Alsatian, 1829–96). The valency number of an atom (sometimes constant, but often variable) provided a numerical measure of its combining power. By about 1870, chemists were becoming familiar with 'graphic formulae', which depicted molecular structures with the atoms joined by straight lines; a substance would be stable if it could be represented by a molecular structure in which the usual valencies of the constituent atoms were satisfied. Some of the formulae which might have appeared in the lecture notes of a chemistry student of the 1870s would be acceptable today, e.g.:

$$
\begin{array}{cccc}
& & & Cl \\
& & & | \\
H\!-\!N\!-\!H & & O\!=\!C\!=\!O & O\!=\!S\!=\!O \\
| & & & | \\
H & & & Cl
\end{array}
$$

Others would, however, be unacceptable, e.g.:

$$
F\!-\!Ca\!-\!F \qquad Ca\underset{O}{\overset{O}{\diagdown\diagup}}S\underset{O}{\overset{O}{\diagup\diagdown}} \qquad HO\!-\!N\underset{O}{\overset{O}{\diagdown}}
$$

Calcium fluoride and calcium sulphate are now known not to contain discrete molecules, and are regarded as ionic, $Ca^{2+}(F^-)_2$ and $Ca^{2+}SO_4^{2-}$. In the case of the nitric acid molecule, we now view pentavalent nitrogen atoms as objectionable because there are sound reasons for believing that the N atom cannot form more than four bonds, and the bonding in HNO_3 can be otherwise described without violation of this rule. The experimental evidence in favour of ionic structures for solids such as CaF_2 and $CaSO_4$ did not emerge until the 1920s; and pentavalent nitrogen lingered on for a few years more.

It is important to understand that the valency number concept is based upon a *molecular* view of chemistry. Frankland was careful to warn his students against the notion that his structural formulae were to be viewed as actual representations of molecular structures. Kekulé, near the end of his life, was sceptical about the real existence of atoms. These attitudes were strongly supported by the great positivist scientist-philosophers Mach (1838–1916) and Ostwald (1853–1932), who saw the atomic theory as a convenience, providing a metaphorical language in which chemistry could be expressed. However, the belief in molecules as real entities gained ground in the 1870s. Maxwell's kinetic theory of gases depended on an explicit recognition of molecules, as did Van't Hoff and Le Bel's postulate of the tetrahedral carbon atom which forced organic chemists to think in terms of three-dimensional molecules.

The valency or valence of an atom in a substance is equal to the number of bonds it forms in the most satisfactory molecular formulation of the substance. This is appropriate for substances which do consist of discrete molecules; it is not without value (although liable to cause confusion) for substances which, though not molecular, can be given a plausible molecular formulation. However, the valency number concept must fail for substances which cannot be given a plausible molecular formulation. Such compounds began to attract attention in the 1870s and 1880s, just as valency theory was attaining general acceptance (the Periodic Law of Mendeleev had given it a considerable boost, since valency numbers were related to the Groups of the Periodic Table). An example of a troublesome compound was the mineral cryolite, then formulated as $AlF_3.3NaF$ and of great technological importance because of its role in the Hall process for the extraction of aluminium (1886). It was definitely a compound and not a mixture. Given the usual valencies of 3 for Al and 1 for Na and F, it is impossible to devise a molecular structure. Such compounds were often known as 'molecular compounds' – the implication being that three NaF molecules were somehow bound to one AlF_3 molecule – or as 'complexes'. We now formulate cryolite as $Na_3[AlF_6]$, i.e. it consists of Na^+ ions and AlF_6^{3-} ions. A hundred years ago, however, the existence of ions in solids was not suspected (although their presence in solutions was gaining acceptance); and in any case the notion that an Al atom could bind six F atoms would have been dismissed by most chemists. Another class of 'complex' compounds were amines such as $NiCl_2.6NH_3$. Their molecular structures were at one time rendered by postulating chains of pentavalent nitrogen atoms, e.g. $Ni(NH_3-NH_3-NH_3-Cl)_2$, which preserved the usual valency of 2 for nickel. This had been superseded by about 1900 by Werner's Coordination Theory, one of the milestones in the history of chemistry. Werner was still wedded to the molecular approach – he died in 1917, a few years before the first X-ray structures revealed the absence of discrete molecules in many inorganic compounds. However, he accepted Arrhenius's theory of ions in solution, and made much use of it in the characterisation of complex compounds. Werner postulated that many atoms, especially those of the transition elements, had to be assigned two valencies – primary and secondary. The former was essentially the same as the classical valency, as postulated for simple compounds (e.g. 2 for nickel). The secondary valency (commonly 6, for atoms of the 3d elements) was satisfied by groups (ligands) which were arranged in a definite geometrical figure about the central atom (an octahedron, for a secondary valency of 6). We now write $[Ni(NH_3)_6]Cl_2$ instead of $NiCl_2.6NH_3$, i.e. the compound contains octahedral $Ni(NH_3)_6^{2+}$ ions and Cl^- anions.

Primary and secondary valencies were later replaced by *oxidation number* and *coordination number* respectively. The oxidation number is based upon an extreme ionic view, being the charge on an atom in the most plausible ionic formulation. For many inorganic compounds this is far more satisfactory than the valency number; very often, an ionic description is possible where no molecular formulation can be devised. Indeed, coordination compounds may be historically defined as compounds whose stoichiometric formulae cannot be rationalised within the classical valency number concept. This definition is still valid. For example, it would be generally agreed that Cs_2CuCl_4 falls within the purview of coordination chemistry, while the isomorphous compound Cs_2SO_4 does not. The latter would have been given a molecular formulation 100 years ago:

$$Cs - O \diagdown \atop Cs - O \diagup \enspace S \diagup\diagdown \atop \diagdown\diagup O \atop O$$

We now regard this formulation as inappropriate; the implication that each Cs atom is uniquely bonded to one O atom is inconsistent with the crystallographic evidence which shows that each Cs is about equidistant from eight O atoms. We prefer to regard the crystal as an assembly of Cs^+ and SO_4^{2-} ions. In the case of Cs_2CuCl_4, we similarly recognise the presence of discrete $CuCl_4^{2-}$ ions. In contrast to Cs_2SO_4, however, this compound could not have been given a plausible molecular structure 100 years ago, without invoking some highly unlikely valencies. This is the reason for its present-day classification as a coordination compound. Many chemists would prefer to write $Cs_2[CuCl_4]$, to stress the presence of a discrete complex anion; but few would insist on the formulation $Cs_2[SO_4]$ for caesium sulphate. The use of square brackets in such cases is further discussed in Section 3.4.

It should be apparent that valency number and oxidation number are two quite different concepts, not to be confused. Where it is possible to assign either a valency or an oxidation number to an atom, the two are often the same, but this is not invariably so. For example, in the N_2

molecule, the oxidation number of the N atom is zero, but its valency is 3. Unfortunately, there persists in the literature a tendency to view the two as virtually synonymous. Thus we still hear of 'the divalent (or bivalent?) state of tin', 'low-valent compounds of chromium' and 'iron in the III valence state'. (The use of 'valence state' where 'oxidation state' or 'oxidation number' is meant can be confusing, since 'valence state' has a distinct meaning in valence bond theory – see Chapter 6.) Most execrable of all is the term 'zerovalent', applied to compounds of (e.g.) nickel in the 0 oxidation state. It makes no more sense to describe the Ni atom in the molecular compound $Ni(CO)_4$ as being zerovalent than it would be to so describe the N atom in N_2 or C in CH_2Cl_2.

The molecular view of chemistry and its associated terminology which flourished in the second half of the nineteenth century has other legacies. If you look at almost any jar on a laboratory shelf, you will see printed on the label the 'molecular weight' of the substance therein, whether the substance is molecular or not. One of the most common sources of frustration among chemistry teachers is the deeply-ingrained belief among many students in the molecular formulation of all substances, and the tendency to use the terms 'molecule' and 'substance' as synonymous and interchangeable. This causes little difficulty in organic chemistry, nearly all of which is concerned with molecular substances. But it can cause serious problems in the teaching and study of inorganic chemistry.

The oxidation number/coordination number terminology is widely (but not universally) applicable in inorganic chemistry, and is not restricted to coordination chemistry. Valency numbers are more useful, however, for the organic chemist; it is rarely profitable to attempt the assignment of an oxidation number to a carbon atom but its (almost) constant valency of 4 is one of the cornerstones of organic chemistry. In this book, we will use valency numbers and associated terminology rather sparingly, and only in circumstances where it is clearly more appropriate than the alternative terminology. There are, of course, many situations where neither valency nor oxidation numbers can be unequivocally assigned to particular atoms. What, for example, is the oxidation number of the Rh atom in $Rh_{13}(CO)_{24}H_3^{2-}$?

1.4 Contemporary theories of structure and bonding in inorganic chemistry

In this section, we survey the various theories of bonding and electronic structure which will be invoked elsewhere in the book and which will be regularly encountered by any serious student of inorganic chemistry. As pointed out in Section 1.2, it must be stressed that no one theory is wholly adequate, and the inorganic chemist must have at least a passing knowledge of several. The 'all-purpose' theory would probably be about as useful as an all-purpose motor vehicle, Jack-of-all-trades and master of none. This section should convey to the reader the strengths and weaknesses of the various theories, and help choose the correct horse for the course.

The ionic model

In 1916, the German physicist Kossel proposed that many compounds could be described as consisting of cations and anions, so that the chemical bond was electrostatic in origin. This idea was not entirely new. Davy and Berzelius in the early nineteenth century had proposed such a theory, and its popularity for a while held back the progress of chemistry; since it could not readily accommodate molecules such as H_2, N_2 and Cl_2, followers of this 'dualistic theory' were reluctant to accept the diatomic description of these gases which followed from Avogadro's hypothesis, and this led to problems in atomic weight determination. What was new in Kossel's theory was the explicit recognition of electrons as being central to chemical bonding, and the notion that atoms tend to exchange electrons in order to attain the supposedly stable noble gas configuration. The atomic numbers of the elements had been largely determined by about 1914 (via X-ray spectroscopy), and the lack of chemical reactivity of the noble gases was established by about 1900. The Rutherford/Bohr model of the atom had gained wide acceptance by 1916, and the location of electrons at the periphery of the atom clearly suggested that the chemical bond was of electronic origin. Thus Kossel had all the information he needed to rationalise the compositions of many (although by no means all) inorganic substances. Compounds such as $NaCl$, CaF_2, LaI_3 etc. could all be described as containing ions having noble gas configurations. Calcium phosphate $Ca_3(PO_4)_2$ could be rendered as $3Ca^{2+}2P^{5+}8O^{2-}$. Kossel's theory led directly to the introduction of oxidation numbers.

Almost simultaneous with the publication of Kossel's paper there appeared a rival electronic theory. The American chemist Lewis introduced the idea of the covalent electron-pair bond. Like Kossel, he was impressed by the apparent stability of the noble gas configuration. He was also impressed by the fact that, apart from many compounds of the transition elements, most compounds when rendered as molecules have even numbers of electrons, suggesting that electrons are usually found in pairs. Lewis devised the familiar representations of molecules and polyatomic ions (Lewis structures, or Lewis diagrams) in which electrons are shown as dots (or as noughts and crosses) to show how atoms can attain noble gas configurations by the sharing of electrons in pairs, as opposed to complete transfer as in Kossel's theory. It was soon apparent from the earliest X-ray studies that Kossel's theory was more appropriate for some compounds, and Lewis's for others. Further elaboration of Kossel's electrovalent theory led to the ionic theory in use today.

The chief merit of the ionic theory is the ease with which quantitative calculations can be performed within a simple electrostatic approach. These will be discussed in Chapter 5. Even if it could be conclusively proven that crystals such as NaCl do not contain ions, chemists would be reluctant to discard the ionic description. The model even has successes – and this should arouse suspicion – in situations where few chemists would seriously entertain the real existence of ions. For example, the 4d and 5d elements tend to exhibit high oxidation states more readily than their counterparts in the 3d series. This can be explained if we note (see Section 4.3) that the energies required to attain highly ionised states are much smaller for the heavier atoms, provided that we are prepared to believe that compounds like WF_6, ReO_3 and OsO_4 are ionic!

Valence shell electron pair repulsion theory (VSEPR)

The VSEPR theory has its roots in the observation prior to 1940 that isoelectronic molecules or polyatomic ions usually adopt the same shape. Thus BF_3, BO_3^{3-}, CO_3^{2-}, COF_2 and NO_3^- are all isoelectronic, and they all have planar triangular structures. As developed in more recent years, the VSEPR theory rationalises molecular shapes in terms of repulsions between electron pairs, bonding and nonbonding. It is assumed that the reader is familiar with the rudiments of the theory; excellent expositions are to be found in most inorganic texts.

An advantage of VSEPR is its foundation upon Lewis electron-pair bond theory. No mention need be made of orbitals and overlap. If you can write down a Lewis structure for the molecule or polyatomic ion in question, with all valence electrons accounted for in bonding or nonbonding pairs, there should be no difficulty in arriving at the VSEPR prediction of its likely shape. Even when there may be some ambiguity as to the most appropriate Lewis structure, the VSEPR approach leads to the same result. For example, the molecule HIO_3 could be rendered, in terms of Lewis theory as:

(a)　　　　　　　(b)

Structure (a) violates the 'octet rule' of Lewis, but 'octet expansion' became recognised as common and acceptable for a number of heavier atoms such as I. In either case, five of the I atom's valence electrons are used for bonding, and there is one lone pair on the central atom. The VSEPR prediction – that the molecule will be a trigonal pyramid – is the same for either.

VSEPR theory can successfully account for many of the fine details in a structure, especially bond angles. However, we will be mainly concerned with the gross geometries of molecules and polyatomic ions. Structural minutiae are of considerable interest to most inorganic chemists, but they are important in the study of descriptive inorganic chemistry only to the extent that they may illuminate details of bonding which are relevant to the very existence of a substance, and to its reactions.

The VSEPR approach is largely restricted to Main Group species (as is Lewis theory). It can be applied to compounds of the transition elements where the nd subshell is either empty or filled, but a partly-filled nd subshell exerts an influence on stereochemistry which can often be interpreted satisfactorily by means of crystal field theory. Even in Main Group chemistry, VSEPR is by no means infallible. It remains, however, the simplest means of rationalising molecular shapes. In the absence of experimental data, it makes a reasonably reliable prediction of molecular geometry, an essential preliminary to a detailed description of bonding within a more elaborate, quantum-mechanical model such as valence bond or molecular orbital theory.

Valence bond (VB) theory

VB theory was developed in the 1930s – mainly by Linus Pauling – as an attempt to invest the successful Lewis electron-pair bond theory with quantum-mechanical validity. An 'ordinary' single covalent bond between atoms A and B can be formed if a singly-occupied orbital on A overlaps with a singly-occupied orbital on B. A dative or coordinate covalent bond is formed by the overlap of a doubly-occupied orbital on A with an empty orbital on B. (This distinction can be removed by considering a dative bond $A \rightarrow B$ to involve the overlap between singly-occupied orbitals on A^+ and B^-.) The origin of the bond is best thought of as arising from the enhancement of electron–nucleus attraction in the overlap region (where electrons are within the field of both positive nuclei), overcoming the increase in interelectron and internuclear repulsion which must accompany the close approach of the atoms. 'One-electron' bonds are found only in cations such as H_2^+, Li_2^+ and Cl_2^+, none of which occurs in stable substances although they can be identified spectroscopically. The pairing of electron spins in a covalent bond is a consequence of the Pauli Principle, which prohibits the occupancy of overlapping orbitals by electrons of the same spin.

The strategy for devising a VB description of the bonding in a molecule or polyatomic ion may be summarised as follows:

(1) Write down a Lewis-type structural formula, identifying the bonds as being single, double etc., and accounting for all valence electrons in bonding or nonbonding pairs. (Two or more Lewis structures may have to be considered.)

(2) Place the atoms in the correct geometrical arrangement, as determined by experimental data or the prediction of VSEPR theory.

(3) Write down the *valence states* of the atoms (not to be confused with oxidation states!). An atom is in the appropriate valence state if the occupancies and orientations of its valence orbitals permit the formation of the required bonds, to match the Lewis structure; valence states are discussed in Section 6.1. A single bond requires one σ bond, a double bond one σ and one π, and a triple bond one σ and two π bonds. To form the necessary σ bonds, it is usually necessary to invoke hybridisation in order that the valence orbitals have the correct orientations. For the formation of π bonds, pure p or d orbitals, or sometimes pd hybrids, are required; hybridisation optimises σ overlap, but hybrid orbitals having an appreciable amount of s character yield poor π overlaps because the s components are quite useless in this respect. The attainment of the valence state will require some investment in promotion energy if it corresponds to a configuration different from that of the ground state. This investment must be recouped by the formation of reasonably strong bonds if the molecule is to be stable.

(4) The molecule can now be fitted together. If more than one Lewis structural formula can be written down and can be plausibly described in terms of VB theory, resonance is invoked.

The concepts of hybridisation and resonance are the cornerstones of VB theory. Unfortunately, they are often misunderstood and have consequently suffered from much unjust criticism. Hybridisation is not a phenomenon, nor a physical process. It is essentially a mathematical manipulation of atomic wave functions which is often necessary if we are to describe electron-pair bonds in terms of orbital overlap. This manipulation is justified by a theorem of quantum mechanics which states that, given a set of n 'respectable' wave functions for a chemical system which turn out to be inconvenient or unsuitable, it is permissible to transform these into a new set of n functions which are linear combinations of the old ones, subject to the constraint that the functions are all mutually orthogonal, i.e. the overlap integral $\int \psi_i \psi_j d\tau$ between any pair of functions ψ_i and ψ_j ($i \neq j$) is always zero. This theorem is exploited in a great many theoretical arguments; it forms the basis for the construction of molecular orbitals as linear combinations of atomic orbitals (see below and Section 7.1).

Resonance is closely related to hybridisation. It depends upon the same quantum-mechanical theorem, applied now to the total wave functions for complete molecules, as opposed to atomic orbitals. For each of the n plausible Lewis structures, there will be a VB wave function. The 'true' ground state wave function will be a linear combination of these, and it can be shown that of the n permissible (i.e. orthogonal) linear combinations, the one which is lowest in energy – and which will therefore be the ground state – will always be more stable than any of the n contributing functions. This is an application of the Variation Principle, also used in molecular orbital theory. An individual resonance structure has no more real existence than does an individual atomic orbital in a molecule. VB theory gained wide acceptance between about 1930 and 1950; this was accelerated by its close relationship to the earlier and familiar Lewis theory. VB theory represented an evolutionary stage in the development of bonding theory. Its rival, which was developed more quietly during the same period, was revolutionary rather than evolutionary.

Molecular orbital (MO) theory

Here, we seek to obtain wave functions – molecular orbitals – in a manner analogous to atomic orbital (AO) theory. We harbour no preconceptions about the chemical bond except that, as in VB theory, the atomic orbitals of the constituent atoms are used as a basis. A naive, 'zeroth-order' approximation might be to regard each AO as an MO, so that the distribution of electron density in a molecule is simply obtained by superimposing the constituent atoms whose AOs remain essentially unaltered. But since there is inevitably an appreciable amount of orbital overlap between atoms in any stable molecule – without it there would be no bonding! – we must find a set of orthogonal linear combinations of the constituent atomic orbitals. These are the MOs, and their number must be equal to the number of AOs being combined.

Thus an MO treatment involves the division of the AOs into sets; within each set, all the AOs overlap with each other and must be combined into a new set of orthogonal MOs. This division is most readily accomplished by symmetry arguments, using the powerful and elegant methods of group theory. The resulting MOs (whose energies are obtained by calculations invoking the Variation Principle) are classified as bonding, nonbonding or antibonding, according to how their occupancy will contribute to the stability of the molecule relative to the separated atoms. The final step – in a qualitative treatment – is to insert the appropriate number of electrons into the lowest-energy MOs to give the ground state configuration of the species under scrutiny. (See Chapter 7 for further details.)

Comparison of VB and MO theories

The ascendancy of MO over VB theory since the 1950s has come about for several reasons:

(1) The application (even at the qualitative level) of MO theory to systems other than the very simplest – e.g. diatomic molecules – demands some knowledge of group theory. Until the 1960s, few chemists had been taught group theory and those who wanted to learn it had difficulty in finding comprehensible expositions.

(2) By the 1960s, most chemists had access to an electronic computer, and quantitative MO calculations (albeit often at a fairly low level of approximation) became feasible. VB theory is less amenable to such computations.

(3) In the 1950s and 1960s, inorganic chemists became increasingly interested in phenomena such as paramagnetism, electron spin resonance, electronic spectroscopy and photoelectron spectroscopy. VB theory is ill-adapted to the interpretation of these properties, and inorganic chemists were forced to look elsewhere (see Sections 2.6–2.8).

Point (3) above requires some amplification. At the quantitative level, the ultimate aim of either a VB or an MO calculation is to obtain the total molecular wave function. Such a function will lead to an electron density map for the molecule which should yield information about its bonding and insights into its reactivity. The function may also be manipulated in order to calculate various molecular constants whose theoretical values can be compared with experimental ones, if available. The kind of function we are talking about is a many-electron function; it contains the coordinates of all the electrons in the molecule, and is usually expressed as a product of one-electron functions (i.e. orbitals). In MO theory, these are the MOs. The constraints of symmetry and orthogonality ensure that these MOs are amenable in themselves to quantum-mechanical manipulations. In VB theory, however, the one-electron functions are localised bond orbitals which are not quite 'respectable' and are not immediately amenable to manipulation. The total molecular wave function obtained from a VB calculation is not necessarily inferior to its MO counterpart; however, its factorisation into one-electron functions is designed to preserve the useful and successful notion of the localised electron-pair bond. This has the disadvantage that the one-electron functions are less useful for quantum-mechanical purposes.

Suppose that we seek a detailed theoretical interpretation of the electron spin resonance spectrum of a paramagnetic molecule having one unpaired electron. We need a realistic, one-electron wave function representing the orbital in which the odd electron resides. MO theory will yield a more suitable function, because it is able to focus our attention on individual electrons, while VB theory looks at a whole whose parts are of dubious real significance. This applies to all properties in which individual electrons and their orbitals have to be scrutinised.

However, provided that we are chiefly interested in ground state properties of diamagnetic species having closed-shell configurations, VB theory retains some advantages over MO theory. The idea of the localised chemical bond (invoking resonance where needed) is still useful. Within the above limitations, it is far easier to describe the bonding in a molecule and to account for its existence by VB arguments than by MO calculations. In this book, we will use VB theory in those situations to which it is well adapted; the heavier artillery of MO theory will be spared for those occasions when it is really needed.

Crystal field (CF) and ligand field (LF) theories

CF theory dates back to 1929, and takes its name from an intellectual exercise performed by Hans Bethe (better known as one of the pioneers of nuclear weapons). He discussed, using group theoretical techniques, 'the splitting of terms in crystals', i.e. the splitting of the electronic states of ions under the perturbation of surrounding ions of opposite charge. His fellow-physicists were very interested at the time in the quantum-mechanical theory of magnetism, and found Bethe's approach to be of great value when dealing with compounds of the transition elements. Their fascinating optical and magnetic properties can be explained by considering the splitting of the central atom d orbitals – which are presumed to be uncontaminated by overlap with other orbitals – under the perturbation of the ligands, i.e. the 'crystal field'. The ligands are viewed as point charges or point dipoles.

CF theory made little impact on inorganic chemists until the early 1950s. They were mostly quite content with the VB theory of bonding in coordination compounds. Perhaps the most important factor which seduced coordination chemists towards CF theory was the appearance (c. 1950) of the relatively cheap commercially-manufactured visible/ ultraviolet spectrophotometer. It soon became a matter of routine among inorganic chemists to record the spectra which gave rise to the colours exhibited by most compounds of the transition elements. These d–d spectra were useful in the characterisation of complexes, and their theoretical interpretation aroused much activity. VB theory was found wanting, for the reasons explained above. The timely 'rediscovery' of CF theory was a great stimulus to coordination chemistry. It proved to be superior to VB theory in the analysis of magnetic properties, whose value to coordination chemists was being increasingly recognised in the 1950s. The concept of CFSE (crystal field stabilisation energy) helped to rationalise many thermodynamic and kinetic properties of complexes.

Before long, however, it became apparent to some of the leading practitioners that the fundamental assumptions at the roots of CF theory – the electrostatic description of the bonding and the purity of the d orbitals – were highly questionable, and an impressive array of experimental evidence, drawn from several independent sources, could be cited in favour of this view. For example, electron spin resonance experiments showed convincingly that the unpaired electrons in complexes – which, according to CF theory, lay in pure central atom d orbitals – were often appreciably delocalised over the ligand atoms. This could be explained by appeal to MO theory.

Faced with this familiar situation – a simple and useful theory whose foundations have become somewhat shaky – the attitudes of inorganic chemists fell into three categories:

(1) For many purposes, CF theory works perfectly well. Let us continue to use it. The fact that its underlying assumptions are unsound is insufficient reason to discard it.

(2) CF theory is unsound and should be discarded. It is liable to lead us into misunderstandings about electronic structure and bonding. It should be possible – with the aid of the electronic computers now available (c. 1960) – to perform quantitative MO calculations, and the results should tell us everything we need to know.

(3) CF theory can be refurbished. There are great advantages in having to look at only five orbitals, instead of dozens which come out of a full MO calculation. For quantitative calculations of d–d spectra, magnetic susceptibilities, electron spin resonance parameters etc., all that has to be done is to introduce one or two adjustable parameters – 'fudge factors', if you will – which implicitly take care of the effects of covalency. We can largely retain the language and the spirit of CF theory, while recognising that some of its assumptions should not be taken too literally.

The force of argument (1) was such that CF theory, even in its simplest form, is alive and well today. The proponents of (2) suffered a number of setbacks in the 1960s. It was shown that the simplified, semi-empirical MO treatments then in vogue were quite inappropriate for the quantitative interpretation of such quantities as d–d transition energies. It can fairly be said that argument (3) won the day. This 'adjusted CF theory' is sometimes called 'ligand field theory' to distinguish it from CF theory. (Other writers prefer to view LF theory as embracing all theoretical approaches which focus attention on the partly-filled subshell in complexes of the transition elements.)

It is possible to reformulate the CF approach by considering the ligand field in terms of explicit σ and π interactions. This approach, the *angular overlap model*, is as simple as the old CF approach and can yield results of real chemical significance. For example, analysis of a d–d spectrum can provide information relevant to the question of whether π bonding is significant in a particular complex.

VB and MO approaches will be discussed in more detail in Chapters 6, 7 and 8. Because simple CF arguments will be used – where appropriate – throughout this book, we conclude this section with a brief summary of the essential concepts and terminology.

The primary task of CF/LF theory is to provide a straightforward description of the electronic states of d block ions in chemical situations. These are determined by the allocation of electrons to nd orbitals, having realised that the fivefold degeneracy of a d subshell in a free atom/ion is lost in a real chemical environment. CF splitting diagrams can be deduced from simple electrostatic considerations; a d orbital is destabilised relative to its energy in the free ion to the extent that it interacts with the crystal field set up by an arrangement of point charges representing the ligand environment. The orbital energies are usually shown relative to the fivefold degenerate level produced by an equivalent electrostatic field of spherical symmetry. The most important CF diagrams are shown in Fig. 1.1.

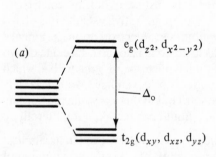

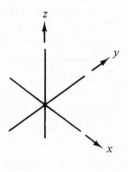

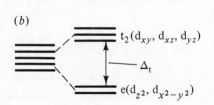

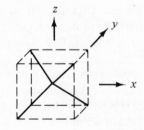

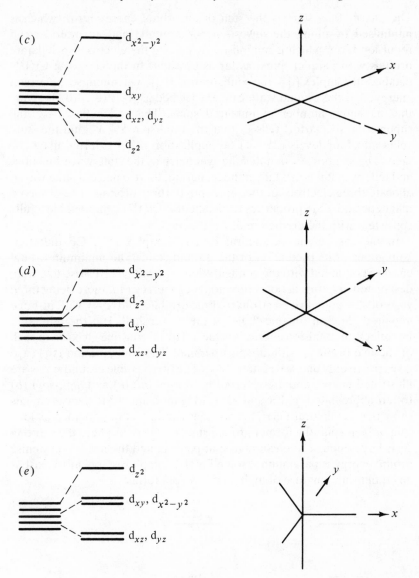

Fig. 1.1. Crystal field splitting diagrams for (*a*) octahedral ML_6, (*b*) tetrahedral ML_4, (*c*) square planar ML_4, (*d*) square pyramidal ML_5 and (*e*) trigonal bipyramidal ML_5.

Experimental information about ground states comes from magnetic and ESR studies (Section 2.8). These tell us about unpaired electrons – how many there are per formula unit, and which orbitals they inhabit.

The energy of a state is the sum of an orbital energy term (which is minimised by filling the lowest-energy orbitals) and an interelectron repulsion term, which is minimised by placing the electrons in separate orbitals with parallel spins, as far as possible. In the case of a Cr(III) octahedral complex (d^3), the configuration $(t_{2g})^3$ will minimise the orbital energy, and the ground state can be specified as $(d_{xy})^1(d_{xz})^1(d_{yz})^1$ giving the maximum number of unpaired spins. This is labelled $^4A_{2g}$; the superscript 4 (quartet) tells us that the total spin $S = 3/2$ and the state splits into four levels $(2S + 1)$ on application of an external magnetic field. The symbol A_{2g} denotes the symmetry of the state wave function, and tells us that it is orbitally nondegenerate, there being only one way to allocate three electrons of the same spin to three orbitals. The observed magnetic and ESR properties of octahedral Cr(III) complexes are fully consistent with the prediction of CF theory.

In the case of an octahedral Fe(III) complex (d^5), CF theory is ambiguous in its prediction of the ground state. The minimum orbital energy is obtained with the configuration $(t_{2g})^5$ which gives rise to a state designated $^2T_{2g}$; this denotes threefold degeneracy (T), there being three ways of allocating five electrons to three orbitals, and $S = \frac{1}{2}$, with just one unpaired electron, or one 'hole' in the t_{2g} subshell. But the minimum interelectron repulsion energy is achieved by placing one electron in each of the five orbitals, with all spins parallel. The configuration $(t_{2g})^3(e_g)^2$ gives rise to only one sextet state, $^6A_{1g}$. The two possible ground states are illustrated below, and are referred to as *low-spin* (a) and *high-spin* (b) (often abbreviated as LS and HS). Magnetic and ESR measurements show that octahedral Fe(III) complexes can in fact be classified as low-spin or high-spin; the former ground state is observed where the splitting Δ_o is large enough to encourage spin-pairing, and the need to minimise orbital energy is paramount. Some Fe(III) complexes appear to exist as an equilibrium mixture of high- and low-spin forms.

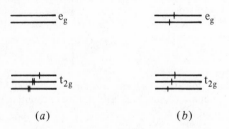

(a)   (b)

The interpretation of d–d spectra requires the description of excited states arising from the reallocation of electrons to orbitals, relative to the ground state. CF theory makes it possible – using well-established methods that were developed *c.* 1930 for the classification of atomic states

– to determine the appropriate symmetry designations, spin states and orbital wave functions for all the states arising from the d^n configuration in any chemical environment. The relative energies of the states – and hence the spectroscopic transition energies measured by running the d–d spectrum – can be expressed in terms of orbital splitting parameters and interelectron repulsion parameters.

The concept of crystal field stabilisation energy (CFSE) is very useful in coordination chemistry, and in understanding the thermochemistry of ionic solids MO, MF_2 etc., and hydrated ions $M^{n+}(aq)$. The CFSE of a particular M^{n+} ion in its chemical environment is equal to the lowering in orbital energy it enjoys relative to its orbital energy in the equivalent, spherically-symmetric crystal field whose energy is the 'centre-of-gravity' or baricentre about which the nd orbitals are split. Thus in an octahedral complex the three t_{2g} orbitals are each stabilised by $(2/5)\Delta_o$, and the e_g orbitals destabilised by $(3/5)\Delta_o$, relative to the baricentre. An octahedral Cr(III) complex thus has a CFSE of $1.2\Delta_o$, but a high-spin Fe(III) complex has no CFSE because the stabilisation of three t_{2g} electrons is exactly balanced by the destabilisation of two e_g electrons. For octahedral complexes – excluding low-spin systems where the term is appropriate, i.e. d^{4-7} – maximum CFSE is obtained for d^3 and d^8; for low-spin systems, we have a maximum at d^6. Maximum CFSE for tetrahedral complexes is found at d^2 and d^7, while the d^8 configuration has maximum CFSE for square planar and trigonal bipyramidal systems. All these maxima are associated, to varying degrees, with complexes having some 'extra' thermodynamic and kinetic stability; for example, the great majority of square planar complexes have the d^8 configuration, and Co(II) (d^7) forms more tetrahedral complexes than any other central atom.

However, the CFSE concept has been the butt of some well-founded objections. In Section 1.5, and elsewhere in this book, it is stressed that stability is not an absolute quality to be associated with certain configurations like a talisman that protects them from all evil. It is usually desirable to specify exactly what changes our 'stable' system might undergo spontaneously, and analyse the reasons why these do not occur. For example, in Sections 9.5 and 10.4 we will be looking at the kinetic inertness of octahedral Cr(III) complexes, i.e. their very slow rates of ligand exchange. This is often attributed to the high CFSE associated with the d^3 configuration in octahedral coordination. More relevant is the change in CFSE in going to a suitable transition state, e.g. by loss of one ligand to give a five-coordinate complex. As it happens, this is greater for d^3 complexes than for any others, but it is not immediately obvious that it should be so. We will make much use of CFSE arguments in this book, but it is well to bear in mind that such considerations may well be little more than mnemonics and that far more complicated factors are really at work.

Another much-venerated totem among the sub-tribe of coordination chemists is the Jahn–Teller theorem. The two physicists after whom the phenomenon is named proved a theorem in 1937 to the effect that a non-linear molecule (or complex ion) in an orbitally-degenerate state would be unstable with respect to a distortion which would remove the degeneracy. The ground state of a complex will possess orbital degeneracy if any degenerate set of orbitals is unevenly occupied, i.e. if it is other than empty, half-filled or filled. The stronger the interaction between the orbitals concerned and the ligand field – i.e. the higher in energy are the orbitals – the more pronounced will be the effect. Thus among octahedral complexes we expect a large distortion for high-spin d^4 $(t_{2g})^3(e_g)^1$, low-spin d^7 $(t_{2g})^6(e_g)^1$ and d^9 $(t_{2g})^6(e_g)^3$, and a smaller effect for d^1, d^2, low-spin d^4 and low-spin d^5. As experimental evidence for the validity of the theorem, we may cite the structural chemistry of copper(II) (d^9), which includes few – if any – authentic cases of regular octahedral six-coordination, but many cases of an axially-elongated octahedron (see Fig. 8.3), i.e. an octahedron in which two opposite ligands have been withdrawn to give bond lengths some 20–30% greater than expected. Jahn and Teller considered vibronic interactions – i.e. vibrational/electronic coupling – in their analysis but the effect can usually be understood in terms of CFSE arguments. Thus if the axial elongation is defined along the z axis, it is easy to see that d_{z^2} is stabilised relative to $d_{x^2y^2}$ with a gain in CFSE:

(See also Sections 8.2 and 8.3.)

1.5   Thermodynamic arguments in inorganic chemistry

Stability and instability

Thermodynamic arguments rarely explain anything in an absolute sense. They can, however, provide that essential first step towards understanding a chemical (or any other) problem – a rational restatement of the problem, and the factors which have to be considered in its solution. For example, consider the question: 'Why is HF a weak acid in aqueous solution?' and the answer: 'Because its pK_a is 3.17 at 25 °C, which means that ΔG^0 for its dissociation is 18 kJ mol^{-1}; the dissociation is thermodynamically unfavourable'. This answer does not immediately explain why HF is a weak acid; but it does set out the considerations which must be involved in any explanation. Further thermodynamic analysis of the problem requires the breakdown of the acid dissociation into steps. These are not intended to represent a mechanism, a sequence of actual events; they are intended to be read like a balance sheet, which can be drawn up in several ways and in which many transactions may be lumped together under one heading. A chemist may be able to identify the decisive factors which determine the signs and magnitudes of the essential thermodynamic functions for the process under scrutiny, just as an accountant can read a balance sheet to assess the reasons for a company's profitability or otherwise. The accountant has difficulty, however, if the overall profit or loss is very small compared with the total turnover, and compared with some individual transactions. The task is easier if the accountant is asked to compare two balance sheets – for two similar companies, or for the same company in different years. We have already alluded to this situation in a qualitative way in Section 1.2. Here we illustrate the problem by looking at ΔH^0 values at 25 °C for processes relevant to the acid dissociation of HF and HCl in aqueous solution, with a view to answering the question: 'Why is aqueous HF a weak acid, while HCl is a strong acid?' The relevant data are collected in Table 1.1. The ΔH^0 values for the overall dissociation are both relatively small compared with some of the other quantities involved, and it is difficult to identify any decisive factor which accounts for the fact that HF is a weak acid, while HCl is a strong acid. Entropy must, of course, be considered. The $T\Delta S^0$ values at 25 °C for the overall dissociation reactions are respectively -34 and $-16 \, \text{kJ mol}^{-1}$ for HF and HCl, so that entropy is as important as enthalpy in determining their relative strengths, and tends in both cases to oppose dissociation. The most important single factor would appear to be the much greater dissociation

Table 1.1 *Enthalpy data $\Delta H°$ for the dissociation of HX (X = F, Cl) in kJ mol^{-1} at 25°C*

Process	HF	HCl
$HX(aq) \rightarrow HX(g)$	48	18
$HX(g) \rightarrow H)g) + X(g)$	567	431
$X(g) + e \rightarrow X^-(g)$	-328	-348
$X^-(g) \rightarrow X^-(aq)$	-524	-378
$H(g) \rightarrow H^+(g) + e$	1312	1312
$H^+(g) \rightarrow H^+(aq)$	-1091	-1091
$HX(aq) \rightarrow H^+(aq) + X^-(aq)$	-16	-56

energy of the HF molecule compared with HCl, set alongside the smaller difference between the hydration enthalpies of the gaseous anions, i.e. the ΔH^0 values for the processes:

$$X^-(g) \rightarrow X^-(aq)$$

The question of stability is central to the study of descriptive inorganic chemistry. A substance, or any chemical system, is stable if it does not undergo spontaneously any perceptible change under specified conditions. If a substance has been prepared and characterised, it may be supposed that it has some degree of stability. If a substance has not been prepared, it is pertinent to ask whether there is any reason to suppose that it is unstable. When we talk of stability, we must always consider the question: stable (or unstable) with respect to what? A compound is unstable if it spontaneously undergoes chemical change to give other products, which must be specified before we make further enquiry into the reasons for instability. Likewise, in order to understand why a compound is stable, we have to ask ourselves what other product or products could conceivably be formed preferentially, and into which the compound under scrutiny might be transformed spontaneously. These points are best illustrated by means of some examples.

Hydrogen peroxide

Some thermodynamic data relevant to the stability of hydrogen peroxide are summarised below (all energies in $kJ\,mol^{-1}$ at 25 °C):

	ΔH^0	ΔG^0
$H_2O_2(l) \rightarrow H_2(g) + O_2(g)$	188	120
$H_2O_2(l) \rightarrow 2OH(g)$	266	188
$H_2O_2(g) \rightarrow 2OH(g)$	214	174
$H_2O_2(g) \rightarrow HO_2(g) + H(g)$	365	332
$H_2O_2(l) \rightarrow H_2O(l) + \frac{1}{2}O_2(g)$	−98	−117

Evidently, hydrogen peroxide is stable at room temperature with respect to decomposition to the elemental substances hydrogen and oxygen. But it is unstable with respect to water and oxygen. Yet, pure hydrogen peroxide can be prepared, and its properties have been exhaustively studied. On heating, however, it rapidly decomposes to water and oxygen. This reaction can occur at room temperature if a suitable catalyst is present; many inorganic d-block cations can accomplish this. Here we have an example of a compound that is kinetically stable, though thermodynamically unstable. There is no convenient reaction pathway under standard conditions whereby H_2O_2 can undergo decomposition without prohibitive activation barriers. What is the likely mechanism for the decomposition (or disproportionation)? From the data shown above, the rupture of the O–O or O–H bonds in a unimolecular rate-determining step involves high activation energies and will be immeasurably slow at room temperature. Obviously some bimolecular mechanism is needed, and some readers may be able to suggest one in which electron pairs are shifted around between a pair of H_2O_2 molecules in a concerted manner, so that bonds are forming at the same time as others are breaking. The most effective catalyst for the decomposition of H_2O_2 is the enzyme catalase, an iron(III) complex which occurs in various mammalian organs. This appears to be able to bind two H_2O_2 molecules in exactly the correct attitudes, and to activate them towards the bimolecular rearrangement of electrons.

Arsenic pentachloride

This elusive compound was the subject of much theoretical speculation prior to its preparation in 1976. Repeated failures to obtain it by the most obvious methods led to the general conclusion that the reaction:

$$AsCl_5 \rightarrow AsCl_3 + Cl_2$$

must be thermodynamically favourable in all circumstances. However, this reaction should have a positive entropy change, which means that it becomes less favourable thermodynamically (less negative ΔG^0) as the temperature is lowered. Thus $AsCl_5$ is most likely to be stable at low temperatures. It was eventually prepared by ultraviolet irradiation of a mixture of $AsCl_3$ and Cl_2 at $-105\,°C$; the Cl_2 molecules undergo photolysis to give the reactive Cl atoms, and the formation of molecular $AsCl_5$ is facilitated. The compound decomposes on heating to about $-50\,°C$. The mechanism of its decomposition is almost certainly unimolecular. Trigonal bipyramidal molecules like $AsCl_5$ are very flexible, and it is not difficult to envisage a smooth pathway which may be depicted as:

In Chapter 6 we will have much to say about a closely-related substance, PH_3F_2, which is thermodynamically stable to reactions such as:

$$PH_3F_2 \rightarrow PH_3 + F_2$$
$$\rightarrow PHF_2 + H_2$$
$$\rightarrow PH_2F + HF$$

but is unstable with respect to disproportionation:

$$2PH_3F_2 \rightarrow PH_2F_3 + PH_3 + HF$$

The latter reaction clearly demands a bimolecular rate-determining step, and will be slow at low temperatures and low pressures. Thus PH_3F_2 can be obtained as a reasonably stable gaseous substance at room temperature.

Calcium(I) chloride

No such substance as CaCl has been isolated; however, the CaCl molecule has been observed spectroscopically in gaseous mixtures at high temperatures, and a number of its molecular constants have been measured. Before asking whether CaCl 'exists' or not, it is necessary to specify whether you are talking about the molecule or the substance. We may presume that the substance calcium(I) chloride would, if isolated, take the form of an ionic solid at room temperature. Thermochemical calculations of the kind described in Chapter 5 show that this substance should be stable with respect to decomposition to the elemental substances. It is definitely unstable, however, towards the disproportionation reaction:

$$2CaCl(s) \rightarrow Ca(s) + CaCl_2(s)$$

If calcium metal is treated with a deficiency of chlorine gas, we obtain only $CaCl_2$, and there is no reaction between Ca and $CaCl_2$ to yield CaCl. Could CaCl be stabilised kinetically? Probably not. It is impossible to conceive of any other thermodynamically-favourable reaction whereby it might be produced (except as a molecular gas at high temperatures) and one can easily envisage a facile electron-transfer process in the solid state whereby disproportionation would take place before a crystal of CaCl reached macroscopic proportions.

Other cases of unstable and 'nonexistent' species will be explored in Chapters 5 and 6.

Thermodynamics looms much larger in the affairs of inorganic chemists than it does among their organic colleagues. This fact is one cogent reason for retaining a distinction between the disciplines. Many organic compounds are thermodynamically unstable with respect to the elemental substances, and many more are unstable with respect to other simpler products, e.g.

$$CH_3COOH(l) \rightarrow CH_4(g) + CO_2(g)$$
$$\Delta G^0 = -55 \, kJ \, mol^{-1}$$
$$2N(CH_3)_3(l) \rightarrow N_2(g) + 4CH_4(g) + 2C(s) + H_2(g)$$
$$\Delta G^0 = -389 \, kJ \, mol^{-1}$$

But these facts are rarely of great concern to organic chemists, for whom kinetic considerations are usually more important. The unique bonding capabilities of the C atom confer upon organic compounds a degree of rigidity which inorganic compounds tend to lack, and reactions in which a C atom undergoes substitution, or a change in hybridisation, tend to be relatively slow. This can be illustrated by comparing analogous compounds of carbon and silicon. Ethylene C_2H_4 is unstable with respect to its polymer $(CH_2)_n$ (and much else besides) but the polymerisation requires a catalyst. Si_2H_4 has never been isolated as a pure substance, and apparently polymerises rapidly under all conditions. (As discussed further in Section 10.6, the trick in preparing the elusive $R_2Si=SiR_2$ is to use a bulky R group which inhibits polymerisation.) Both CH_4 and SiH_4 are unstable in air, with respect to the dioxide and water. Methane requires a flame or spark to ignite, but SiH_4 does not. Carbon tetrachloride and silicon tetrachloride are both unstable with respect to hydrolysis; but CCl_4 is kinetically inert to water.

Enthalpy- and entropy-driven processes

A reaction is said to be enthalpy-driven if it involves a large, negative ΔH^0 with a smaller and usually unfavourable $T\Delta S^0$ at all accessible temperatures. In a thermochemical analysis of such a reaction, and in comparing several such related reactions, only the enthalpy terms need normally be considered. Most redox reactions and acid–base reactions come into this category. The latter term can be interpreted liberally to include many instances of complex formation, e.g.:

$$ZnCl_2(s) + 2NH_3(g) \rightarrow ZnCl_2(NH_3)_2(s)$$

for which the thermodynamic functions are:

$\Delta H^0 = -170 \, \text{kJ mol}^{-1}$
$\Delta G^0 = -111 \, \text{kJ mol}^{-1}$
$T\Delta S^0 = -59 \, \text{kJ mol}^{-1}$ (at 25 °C)

That the entropy change is unfavourable could be confidently predicted, given the presence of two moles of gas on the left-hand side. As the temperature is increased, the $T\Delta S^0$ term becomes more important; neglecting the small temperature dependence of ΔH^0 and ΔS^0, it can be easily shown that ΔG^0 will become zero at about 850 K, at which temperature the decomposition of the complex should be complete. Such decomposition can be achieved at lower temperatures if the partial pressure of ammonia is kept low, by pumping. Most thermal decompositions – which are often the reverse of acid–base reactions (see Section 9.2) – are entropy-driven. All substances containing chemical bonds can be decomposed by heating to a sufficiently high temperature.

The mixing of two gaseous substances, or of two non-polar liquids, are further examples of entropy-driven processes. These involve negligible enthalpy changes (no strong chemical bonds are formed or broken) but the increased randomness and disorder in the system lead to a positive entropy change.

The dissolution of an ionic solid in a polar solvent, or the dissociation of an acid, are examples of processes for which ΔH^0 and ΔS^0 are both relatively small and are often of equal importance in determining the sign and magnitude of ΔG^0. This was seen to be the case for HF as discussed above, and the thermodynamic functions for the several steps into which its dissociation can be analysed are much greater in magnitude. As discussed further in Section 3.3, the same is true in the analysis of the solubilities of ionic solids.

All values of the thermodynamic quantities ΔH^0, ΔS^0 and ΔG^0 given in this book refer to the standard states at 25 °C and one atmosphere pressure. This is not explicitly stated unless it is felt important to remind the reader of the restriction.

1.6   Further reading

For the theoretical approach to inorganic chemistry, see the books listed in Section A.7 of the Appendix. See also the books listed in Section 4.8, especially Chapter 1 of Johnson (1982). The historical development of bonding theory is thoroughly treated by Palmer, W. G. (1965). *A History of the Concept of Valency to 1930*. Cambridge University Press.

2

Physical methods in the characterisation of inorganic substances

2.1 Introduction

This chapter presents a brief survey of the more important physical techniques to which reference will be found in more comprehensive texts. It was noted in Chapter 1 that inorganic chemists have become increasingly preoccupied with physical methods since about 1950, often to the extent that the development of the technique and the underlying theory overshadow the inorganic chemistry. Inorganic chemists have indeed invaded large parts of the territories traditionally associated with physical chemistry. It was pointed out in the Preface that the neglect of factual, descriptive chemistry in contemporary inorganic courses and texts has been a direct consequence of the emphasis placed upon physical and theoretical matters. It is proper, however, that an account of descriptive inorganic chemistry should make some mention of the methods used in determinations of structures etc. A scientist should always be prepared to give a rational and convincing answer to the question: 'How do you know that?' In order that the student who asks such a question receives a meaningful reply, it is necessary to have some knowledge of the kind of information conveyed by a given method, its strengths and its limitations.

A great number of physical properties have been used by inorganic chemists, and new physical methods appear regularly, some to join existing methods as indispensable weapons in the arsenal and others to sink into obscurity after a short period of popularity. At one extreme, we have a technique such as X-ray crystallography, which can yield accurate atomic positional parameters in a crystal and hence bond lengths, interbond angles etc. At the other, we have a more mundane physical property such as the melting point of a substance. There is no straightforward, quantitative relationship between the constitution of a substance and its melting point. Nevertheless, it is useful to measure the melting point of a newly-prepared substance; this will assist its identification in the future, and some structural information can be inferred. Thus (as discussed in Chapter 3) substances which consist of discrete molecules in the solid state tend to have lower melting points than those which do not. Between these extremes, we have a considerable number of physical techniques which yield limited structural information, often sufficient to characterise the substance without providing quantitative bond lengths etc.

Inorganic chemists are interested in chemical reactions as well as the static properties of substances. The measurement of thermodynamic quantities for chemical reactions will not concern us, although we will make extensive use of the experimental results elsewhere in this book. In Chapter 9 we will look in more detail at inorganic reactions and their mechanisms: blow-by-blow accounts of what actually happens at the atomic level as the reaction proceeds. Some of the spectroscopic methods described in this chapter are important in mechanistic studies; they may be used to follow the rate of a reaction or to identify short-lived intermediates. Other techniques (such as isotopic labelling) are useful in the determination of reaction mechanisms.

Many readers will find this chapter uneven and ill-balanced. The contemporary chemistry student will doubtless be exposed to a great deal more about nuclear magnetic resonance spectroscopy, and probably other techniques, than can be covered in this book; the treatment will often be from the perspective of the organic or physical chemist, however. On the other hand, some readers will never have heard of nuclear quadrupole resonance, and may never hear of it again except in passing. This chapter is largely restricted to diffraction and spectroscopic methods which are useful in the characterisation of inorganic substances, which may yield information relevant to the bonding therein, or which may otherwise be encountered in the study of descriptive inorganic chemistry.

This material has been placed early in the book because structural and spectroscopic data are invoked in practically every other chapter. However, the reader whose previous acquaintance with molecular orbital (MO) theory and molecular symmetry is negligible might find a cursory reading of Chapter 7 useful, especially before tackling Sections 2.6–2.8.

2.2   Diffraction methods

These embrace X-ray diffraction, neutron diffraction and electron diffraction. The first two of these are almost entirely used in the study of crystalline solids, while electron diffraction is of most value (to inorganic chemists at least) for structure determinations of gaseous substances. X-ray diffraction has been used to obtain structural information for species in solution, and electron diffraction has applications in the study of molecules adsorbed on surfaces. However, the term 'crystallography' is almost synonymous nowadays with X-ray diffraction analysis.

X-ray diffraction

It was observed in 1912 that a beam of X-rays is scattered when it interacts with a crystal, in a manner similar to the diffraction of light by a grating. Within a few years, this phenomenon was being actively exploited in the determination of structure in crystalline solids. The wavelength of the X-rays used is of the same order of magnitude as the shortest internuclear distances in crystals, c. 100 pm. It is not the nuclei that diffract the X-rays, however; rather it is the regularly-repeating pattern of electron density in the crystal which acts as a three-dimensional grating. An X-ray diffraction pattern (as recorded on a photographic film or plate) appears as an array of spots; the pattern is qualitatively determined by the size and symmetry of the unit cell, i.e. the repeating unit which, when multiplied in three dimensions, gives the whole crystal. The intensity of each spot is a complicated function of the electron density distribution in the unit cell. A mathematical analysis of the intensity data leads to a three-dimensional electron density map; the nuclei are located at the maxima. Relatively heavy atoms can be located with great precision, so that internuclear distances can be determined with estimated errors of less than 1 pm; the accuracy is largely dependent on the ratio of the number of intensity measurements made to the number of independent structural parameters which need to be determined. The more high-quality data, and the fewer structural parameters, the better.

In the 1950s, X-ray crystallography was a highly-specialised and laborious business. But the introduction in the 1960s of the automated diffractometer and efficient programs for processing the data with modern computers have revolutionised the technique. A crystal structure determination which might have taken a professional crystallographer six months to solve c. 1960 can now be completed in a couple of weeks by a chemist having no specialist training. The solution of the structure of Vitamin B_{12} (see Section 9.8), which has a molecular weight of 1356, occupied a team of about 10 crystallographers for 10 years (1948–59). Today, the solution of a comparable crystal structure might occupy one Ph.D. student for a year or so.

The reader may wonder why – given the completeness with which an X-ray analysis should reveal the structure of a crystalline solid and the extent to which the method has been automated – inorganic chemists should bother with any other method of characterisation for solid compounds. It is true that many inorganic substances are nowadays characterised by X-ray crystallography alone. There are, however, cases where the technique is impracticable; and even where an X-ray analysis can be performed, it often pays to find out as much as possible about the product by other means before embarking upon an X-ray analysis. These remarks require some amplification.

In the first place, it is not practicable to perform a complete X-ray analysis on every newly-prepared compound. A modern crystallographic laboratory, whose diffractometer is dedicated to routine data collection, might be able to deal with 100–200 crystals in a year. An energetic synthetic group – and there may be several in the same institution – should be able to produce far more new compounds than that. Crystallographers and their clients have to be selective. Reasonably large single crystals of good quality are required for a full X-ray analysis. Many crystalline materials refuse obstinately to produce suitable crystals, and can only be obtained as powders. Such powders do give diffraction patterns, and in very simple cases these can lead to a full structure analysis. With more complex structures, powder diffraction patterns provide useful 'fingerprints' for future identification. They may also show that two or more compounds are isomorphous and hence (*mutatis mutandis*) isostructural. For example, the fact that Cs_2ZnCl_4 is isomorphous with Cs_2SO_4 suggests very strongly that it contains tetrahedral $ZnCl_4^{2-}$ ions. But for most powdered samples, structural information must be sought by means of other techniques.

If the substance under scrutiny contains hydrogen atoms whose location is important, X-ray analysis will usually have to be supplemented by other methods. Very light atoms contribute very little to the scattering of X-rays and – especially if much heavier atoms are present – may be effectively invisible to the crystallographer. There are a number of other circumstances where, even if good crystals are available, X-ray crystallography may fail to yield the desired structural information.

Nor is the technique infallible. Hearken the following cautionary tale. In 1980, a paper presented at a conference created a sensation, particularly among inorganic chemists interested in fluorine compounds. It was reported that one of the products of the shock-wave compression of a mixture of CuF_2 and $CuCl_2$ was $ClF_6^+ CuF_4^-$. X-ray analysis revealed octahedral ClF_6^+ cations and square planar CuF_4^- anions. Such a product was unexpected, and the CuF_4^- ion had not previously been identified. This report met with a degree of scepticism, however, and the matter was not cleared up until 1983 when it was shown that the product was in fact a rather more prosaic compound $[Cu(H_2O)_4][SiF_6]$. The original workers had mistaken $Cu(H_2O)_4^{2+}$ for CuF_4^- because O and F atoms, differing by only one unit in atomic number, are difficult to distinguish crystallographically, and H atoms are invisible unless very carefully looked for. Likewise, SiF_6^{2-} was mistaken for ClF_6^+ since Si and Cl scatter X-rays to approximately the same extent; the authors did not suspect the presence of silicon in the sample (doubtless arising from etching of a quartz vessel by moist CuF_2), and the technique does not distinguish between cations and anions, or between 1+ and 2+ or 1− and 2− ions. The original authors can be excused their error on the grounds that only a tiny sample was available. Otherwise, they might have performed an elemental analysis (which would have shown the absence of Cl and the presence of Si), magnetic measurements (which would have confirmed that the copper was in the II state and not III) or taken an infra-red spectrum (revealing the presence of water molecules). However, the dangers inherent in the characterisation of a substance by X-ray crystallography alone should be apparent.

It might be thought that the electron density maps which arise from an X-ray analysis should provide detailed information about electron distribution and hence bonding. The electron densities obtained from routine X-ray studies are not sufficiently accurate for such purposes. However, the finer details of electron density distributions in crystals can be obtained in favourable cases by very careful and accurate X-ray diffraction studies. Neutron diffraction (see below) can also provide such information.

Neutron diffraction

According to the familiar De Broglie equation:

$$\lambda = h/mv$$

which relates the momentum mv of a moving particle to the wavelength λ of the associated wave, h being Planck's constant, a beam of neutrons travelling at a velocity of $3970\,m\,s^{-1}$ corresponds to a wavelength of 100 pm. Such a beam of neutrons does undergo diffraction on interaction with a crystal, and it is indeed the nuclei (rather than, as in the case of X-rays, the electrons) which are responsible. The theory and practice of neutron diffraction is very similar to that of X-ray diffraction. However, neutron scattering factors do not depend on atomic number as do X-ray scattering factors; within a factor of about 3, they are the same for all nuclei. Thus the proton scatters neutrons nearly as well as many much heavier nuclei, and a deuterium nucleus is even better. Neutron diffraction is thus the method of choice where H atoms in a crystal have to be located with precision.

The neutron possesses spin, and hence a magnetic moment. The interaction of neutrons with substances having unpaired electrons leads to magnetic scattering in addition to the usual scattering, and its detailed analysis can be of great value in the study of properties such as ferromagnetism. Of particular interest is the interaction of polarised neutron beams with crystals containing paramagnetic molecules or ions.

Accurate maps depicting the unpaired electron density can be obtained, and the results compared with theoretical calculations.

Neutron diffraction is subject to a number of experimental limitations. Sources of neutrons of suitable velocity are relatively few, and are mostly associated with large nuclear reactors. Large crystals are needed, the data collection process is slow and a structure determination is very expensive. Thus inorganic chemists have to be even more selective in their choice of samples for neutron diffraction analysis than for X-ray diffraction. It is only worthwhile to examine crystals containing features which cannot be elucidated by any other technique, i.e. the precise location of H atoms or the study of magnetic structure. For example, if you were to characterise (using spectroscopic methods and chemical analysis) a compound as $[N(CH_3)_4][Fe_2Co(CO)_{11}]$, you would be wasting time and money by embarking upon a neutron diffraction study. With X-rays, you could neglect the hydrogen atoms leaving 30 atomic positions to be determined. With neutrons, you should be able to locate the hydrogens but with rather more trouble, since 42 atoms now have to be located. But the location of methyl hydrogens is rarely of importance. Interest in this compound is focussed upon the anion, whose structure can be obtained much more cheaply (and possibly more accurately) by means of X-rays. However, in the case of $[N(CH_3)_4][Fe_3(CO)_{11}H]$, a neutron diffraction study would be worthwhile.

Electron diffraction

Electrons cannot penetrate far into a solid (unlike X-rays or neutrons) and electron diffraction studies of solids are restricted to the examination of surfaces. The first observations of electron diffraction – confirming the 'wave nature' of the electron – in 1927 were in fact obtained by reflection of an electron beam from the surface of a metal. However, electrons which have been accelerated through about 40 kV undergo diffraction by gases. As with X-rays, the scattering of electrons by an atom increases with its atomic number, so that hydrogen atoms can be difficult to locate with precision. The diffraction pattern is processed mathematically to yield a radial distribution function; this when plotted against the internuclear distance shows a number of peaks, each of which corresponds to a pair of atoms. The intensity of the peak – measured by the area under its curve – is proportional to the product of the atomic scattering factors and inversely proportional to the internuclear distance. If a sufficient number of well-resolved peaks can be observed, it should be possible to devise a unique structure which fits the data. As an example, Fig. 2.1 shows the radial distribution function for the trigonal bipyramidal molecule PCl_4F (see also Section 2.3).

The technique is subject to serious limitations. Significant distortions from an idealised, regular symmetry may not be readily discernible from

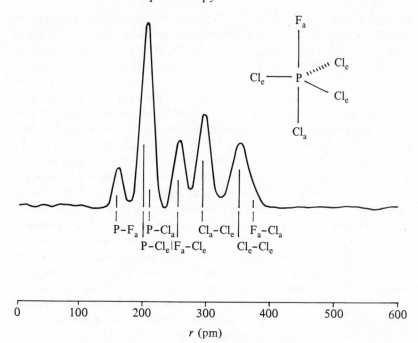

Fig. 2.1. Radial distribution function obtained from electron diffraction of PCl_4F (redrawn from Macho, C. *et al.* (1986), *Inorganic Chemistry*, **25**, 2828–35, with permission of the American Chemical Society).

the radial distribution function. Sometimes the data are consistent with quite a number of equally plausible structures, and other techniques must be invoked in order to choose among these. In any case, the number of parameters which can be obtained is small; if the molecule is fairly complex and lacking in symmetry, some assumptions have to be made (e.g. that the coordination of a saturated carbon atom is exactly tetrahedral). Another problem is the 'shrinkage effect', which occurs if linear or nearly linear atomic groupings undergo bending vibrations of large amplitude. Within these limitations, electron diffraction has been one of the mainstays of structural inorganic chemistry since about 1930, and new volatile compounds suitable for electron diffraction analysis are being synthesised in numbers large enough to warrant continued interest in the technique.

2.3 Vibrational spectroscopy

The motion of atoms relative to one another in molecules and crystals involves changes in energy. The frequencies of these vibrations are mostly in the range 10^{12}–10^{14} s^{-1}, which corresponds to wavelengths of the order of 10^{-6}–10^{-4} m. Vibrational frequencies are usually rendered in units of cm^{-1} (wavenumbers). Thus the 'frequency' of radiation whose wavelength is 5×10^{-6} m (i.e. 5×10^{-4} cm) is the reciprocal of the wavelength, i.e. 2000 cm^{-1}. The true frequency is equal to $(2 \times 10^5 \, \text{m}^{-1}) \times (3 \times 10^8 \, \text{m s}^{-1}) = 6 \times 10^{13}$ s^{-1}, multiplying the 'frequency' in reciprocal metres by the velocity of light. Electromagnetic radiation in the infra-red region can excite vibrations in molecules and crystals. Other methods, of which the Raman effect is particularly important, can be used in the study of such vibrations. Infra-red (IR) spectroscopy and Raman spectroscopy are complementary tools, although the former is much more commonly encountered in chemical laboratories. Both techniques are relatively versatile, in that they can tackle solids, liquids and gases with minimal sample preparation, and they can readily be adapted for experiments under high pressure, or at temperatures much higher or lower than ambient. Although they are much used as sources of 'fingerprints' for the routine identification of known compounds, they can provide more detailed structural information. Most of the rest of this section will be couched in language appropriate for the study of discrete molecules/ions, but the same principles, with some elaboration, are applicable to non-molecular crystals as well.

Infra-red (IR) spectroscopy

Here we are studying the absorption of infra-red radiation from a continuous source. The spectrum is a plot of the absorbance of the sample as a function of the frequency of the incident radiation. Each peak (or band) in the spectrum corresponds to the absorption of a photon with the concomitant excitation of a *normal mode of vibration*; some relatively weak bands (overtones) involve the double excitation of a vibrational mode, while the simultaneous excitation of two different modes may appear as a combination band, but these need not concern us further.

A molecule containing N atoms has $3N - 6$ normal modes of vibration (or $3N - 5$ if it is linear). Each can be represented by a set of arrows, one on each atom, giving the direction along which the atom moves back and forth as the mode is executed. For example, the three normal modes of a bent AB_2 molecule can be represented as:

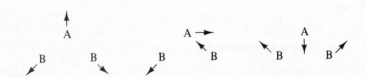

These are referred to as the symmetric stretching, antisymmetric stretching and bending modes respectively. The bending mode appears in the spectrum at a much lower frequency than the stretching modes. Modes of vibration are more conveniently rendered mathematically in terms of *internal coordinates*. If we number the atoms 1, 2, 3, . . ., N, each bond stretch can be labelled by the coordinate Δr_{ij}, and each interbond angle deformation by the coordinate $\Delta \theta_{ijk}$. Any mode of vibration can be written as a linear combination of coordinates. The permissible combinations are found in much the same way as molecular orbitals are constructed as linear combinations of atomic orbitals (see Section 1.4 and Chapter 7). By means of a fairly straightforward group-theoretical procedure, we obtain combinations of equivalent coordinates and assign to each a symmetry label. If two or more of these combinations have the same symmetry, they become mixed. After taking account of such mixing, we obtain *normal coordinates*, one for each normal mode. Thus in the case of the bent AB_2 molecule, the symmetric stretch and the bending mode have the same symmetry. All three modes occur in the IR; in the case of gaseous H_2O, they appear at $3756\,cm^{-1}$ (antisymmetric stretch), $3657\,cm^{-1}$ (symmetric stretch) and $1595\,cm^{-1}$ (bend). But, strictly speaking, the $3657\,cm^{-1}$ band involves some H—O—H bending as well as O—H stretching, while the $1595\,cm^{-1}$ band involves some stretching although it is predominantly a bending mode.

The mechanism whereby IR radiation is absorbed by a molecule involves the interaction between the oscillating electric field associated with electromagnetic radiation and the oscillating dipole moment of the molecule. A vibrational mode whose execution does not involve a fluctuating dipole moment will be inactive and will not appear in the IR spectrum. This *selection rule* is of great value in determining the symmetry of the molecule. A bent AB_2 molecule such as H_2O has a non-zero dipole moment in its equilibrium configuration, and it should be apparent that this will fluctuate as any of the three modes is executed. A linear AB_2 molecule such as CO_2 has no dipole moment in its equilibrium configuration. As it executes the symmetric stretching mode, at no time does it have even temporarily a dipole moment. This mode is *forbidden* in the IR. But as the CO_2 molecule performs the antisymmetric stretch, the two C–O distances are not the same and the centre of symmetry is lost. We now have a fluctuating dipole moment which is time-averaged to zero. This mode, and the bending mode, are IR-active. Obviously, we can determine whether an AB_2 molecule is linear or bent by inspection of its IR spectrum.

At the most empirical level, IR spectra provide valuable fingerprints of substances since the spectrum is so sensitive to structural details. At a slightly higher level of elaboration, we can identify specific bonds or groupings of atoms whose stretching and deformation frequencies fall within fairly narrow ranges according to their chemical environments. More detailed structural information about the environment of a particular bond or group can be obtained from a host of empirical correlations and relationships. At a more sophisticated level still, we can exploit the selection rules to determine the molecular symmetry. This is best done in conjunction with Raman spectroscopy, where the selection rules are different.

One of the most active areas of research in inorganic chemistry in the last few decades has been the carbonyl complexes of the transition elements, where CO molecules are bonded to d block atoms. IR spectroscopy has been a particularly valuable tool in the characterisation of these compounds. The triply-bonded CO molecule has a stretching frequency of $2170\,cm^{-1}$; this appears in carbonyl complexes (usually known as 'metal carbonyls') at rather lower frequencies, between 1800 and $2100\,cm^{-1}$, since the C–O bond is somewhat weakened when the molecule functions as a ligand (see Section 8.4). The carbonyl stretching bands are strong and sharp, and are usually well differentiated from other vibrational bands. Apart from fingerprinting, it is usually possible to identify specific $M(CO)_n$ groups and their geometrical disposition. Bridging and terminal CO ligands can be readily distinguished. The calculation of CO stretching force constants and their value is mentioned later in this section.

IR spectroscopy is not confined to stable substances. In recent years, matrix isolation IR spectroscopy has become important in the investigation of short-lived, unstable molecular species. A gas containing such highly-reactive molecules – produced by photolysis of a reaction mixture, or in a high-temperature furnace – is suddenly cooled by contact with an inert solid (e.g. argon at $c.\,40\,K$). The matrix-isolated molecules are protected by the low temperature from unimolecular decomposition, and – by sheer isolation, if the dilution is sufficient – from bimolecular processes such as dimerisation or disproportionation. For example, the photolysis of $Mn(CO)_5H$ by a laser produces the otherwise unstable $Mn(CO)_5$ and $Mn(CO)_4H$ molecules whose IR spectra can be measured in an argon matrix. Because of the low temperature, the lack of intermolecular interactions and the rigidity with which the molecules are trapped in the matrix, such spectra are often very well resolved, better than can be achieved by conventional methods. Thus matrix isolation spectroscopy is widely used in the study of stable species, in preference to conventional techniques.

Essentially two types of IR spectrometer are commercially available: dispersive or interferometric. The former is the more familiar to most chemists. The source of continuous IR radiation is dispersed, by means of a prism or grating, and the spectrum is scanned within the required limits of frequency; the slower the scanning speed, the better the resolution.

The interferometer splits the beam of incident radiation, then recombines the beams to produce interference usually before impinging upon the sample. The result is an interference pattern which can be analysed, via a Fourier transformation, into an absorption spectrum. The interferometer has some advantages over the dispersive instrument. The whole spectrum is measured instantaneously, and a single instrument can be used over a larger range of frequency than is possible with a dispersive spectrometer. The disadvantage is that the Fourier transform analysis requires a computer, but with the decreasing cost and size of suitable computers, this is less of a problem than it used to be. The availability of the computer can be exploited, for example, to store a large number of spectra so that an unknown sample can be compared and perhaps identified from its fingerprint, providing an automated method of chemical analysis in favourable cases.

Raman spectroscopy

Raman spectroscopy is more expensive and more time-consuming than IR spectroscopy. Thus although practically every chemical laboratory has IR facilities, relatively few have Raman spectrometers. But the technique does have its advantages, and for some purposes both the IR and Raman spectra are required.

When visible or near ultraviolet radiation is shone upon a sample, it may undergo transmission, absorption, reflection or scattering, usually some combination of these. Most of the scattered light has the same wavelength as the incident beam; but a small fraction of it is of longer wavelength (lower frequency) than the incident light and an even smaller proportion is scattered with an increase in frequency. This is the Raman effect. The difference in frequency corresponds to a vibrational mode; in the case of light scattered at a lower frequency than the incident beam (or *exciting line*), we can imagine that part of the energy of the incident photon has been absorbed by the sample in the form of vibrational energy. The spectrum of the scattered light shows a very intense peak coincident with the exciting line, and a series of weak peaks at lower frequencies; the differences in frequency between these Raman lines and the exciting line are vibrational frequencies.

The selection rules for the Raman effect are quite different from those for IR spectroscopy. The mechanism involves interaction between the incident radiation and the fluctuating polarisability of the molecule, in contrast to the fluctuating dipole moment in IR absorption. The dipole moment is a vector quantity, and can be resolved into components along three Cartesian axes. The polarisability is a tensor quantity, whose components can be written as products of Cartesian axes. For a molecule having no symmetry at all, or having only a plane of symmetry, all vibrational modes are allowed in both the Raman and IR. If the symmetry is higher, some modes which are forbidden in the IR may appear in the Raman, and vice versa. If the molecule has a centre of symmetry, no vibrational mode can appear in both the IR and Raman spectra. Thus the combination of the two spectra provides a more powerful means of determining molecular symmetry than either in isolation. Straightforward group-theoretical procedures enable us to classify vibrational modes according to their symmetry properties, and according to whether they are IR-active, Raman-active or (as happens in a few situations) inactive (forbidden) in both types of spectra.

One advantage of Raman spectroscopy over IR is the help given by polarisation data. If the exciting line is plane-polarised, the Raman lines may be depolarised to varying degrees, and it is possible to make some definite assignments of frequencies to vibrational modes. A totally-symmetric vibration – one which preserves the symmetry of the molecule in its equilibrium condition as it is executed, e.g. the symmetric stretching of an AB_2 molecule – appears as a polarised line in the Raman spectrum. Comparable information from IR spectra can be obtained only from measurements on single crystals, which involve considerable experimental difficulty, but polarisation/depolarisation in the Raman effect can be observed for gases or liquids.

Raman spectroscopy is a good technique for studying species in aqueous solution. Liquid water has a relatively weak Raman effect, and solutes can readily be studied. In contrast, water is a very strong absorber of IR radiation, and the broad bands blot out most of the region. Quantitative measurements of equilibria in aqueous solution can be carried out by means of Raman spectroscopy.

Raman spectroscopy was revolutionised in the 1960s by the use of lasers as exciting lines. With the mercury lamps previously used, Raman spectroscopy was a tedious business, the scattering being very weak. Samples which absorb visible radiation (i.e. coloured materials) were particularly difficult to study. These problems have largely been overcome by the use of laser sources. Indeed, the presence of an absorption maximum close to the wavelength/frequency of the exciting line can lead to a great enhancement of part of the Raman spectrum, and can provide additional information. This is the *resonance Raman effect*. For example, suppose we are dealing with a coordination compound whose visible spectrum includes a charge transfer band (see Section 2.6) corresponding to an electron jump from an orbital localised on a ligand donor atom L to the central atom M. Such a transition is likely to be accompanied by a considerable change in the M–L bond. The Raman spectrum may then be dominated by a progression in the M–L stretching mode, i.e. a series of lines of frequencies v, $2v$, $3v$, ... up to perhaps 20 lines in all. Such observations can be of great value. They can assist materially in the assignment of electronic transitions in the visible and near ultraviolet regions (Section 2.6). Because the effect can be very strong, it can be used to probe the active sites of large and complex biological molecules. Suppose, for example, that a protein is known to contain molybdenum, and it has an absorption band in the visible part of the spectrum. If the resonance Raman spectrum reveals a progression in a frequency close to that of the symmetric stretch of the tetrahedral MoS_4^{2-} ion, it may be inferred that the absorption band involves a tetrahedral MoS_4 unit.

Force constants

A diatomic molecule has only one mode of vibration, represented by the coordinate Δr, the change in the internuclear distance r. The potential energy of the molecule $U(r)$, whose vibration is approximated as simple harmonic motion, is given by:

$$U(r) = \tfrac{1}{2}k(\Delta r)^2$$

The constant k is the *force constant*, which has units of $(\text{force})(\text{length})^{-1}$. In SI units, force constants are therefore expressed in $N\,m^{-1}$, although you will often see them quoted in millidynes per Ångström $(1\,md\,Å^{-1} = 100\,N\,m^{-1})$.

For a diatomic molecule whose atoms have masses m_1 and m_2 (in atomic mass units), the force constant k (in $N\,m^{-1}$) is related to the vibrational frequency (in cm^{-1}) by:

$$k = 5.94 \times 10^{-5}v^2m_1m_2/(m_1 + m_2)$$

A small correction is usually made to allow for the fact that the vibration is not strictly harmonic. It will be apparent that for a constant value of k, the vibrational frequency of a diatomic molecule will decrease as the atoms become heavier.

Force constants of diatomic molecules range from about $2000\,N\,m^{-1}$ for the triply-bonded molecules N_2 and CO to less than $10\,N\,m^{-1}$ for molecules such as Cs_2, with weak bonds between large atoms. There is no simple relationship between the force constant and the bond strength (as measured by the thermochemical bond energy; see Section 6.2) although correlations can be established for series of related molecules. The force constant is equal to the second derivative $d^2U(r)/dr^2$ of the molecular potential energy with respect to the internuclear distance. It may therefore be regarded as a measure of the 'stiffness' of the bond when the atoms are in their equilibrium positions. This will depend upon the sensitivities of many types of interactions – some attractive, some repulsive – to a small displacement of the atoms from their equilibrium positions. If stretching force constants can be defined for bonds in polyatomic molecules, there is reason to believe that these should be very sensitive to differences in the strengths of comparable bonds over a range of related molecules. After all, the force constants of diatomic molecules cover a wider range than internuclear distances or thermochemical dissociation energies. Compare, for example, CsH and HF:

	r (pm)	Dissociation energy (kJ mol^{-1})	Force constant (N m^{-1})
HF	92	569	966
CsH	249	176	47

The determination of force constants for a polyatomic molecule involves the procedure known as *normal coordinate analysis*. It was noted at the beginning of this section that a normal mode of vibration can be described as a combination of internal coordinates, each coordinate representing the stretching of a bond or the deformation of an interbond angle. A force constant is associated with each internal coordinate; we must also consider interaction constants between two coordinates if, for example, one bond is significantly affected by the stretching of another. If the molecular geometry and the force constants are known, it is a straightforward matter to calculate the vibrational frequencies, and to obtain explicit expressions for the normal modes of vibration as normal coordinates. From the resulting normal coordinates, other information – for example, the amplitudes of motion of atoms in a given vibrational mode – can be extracted. In practice, however, we have experimental values for the vibrational frequencies and we need to know the force constants. Even if a complete vibrational spectrum has been obtained – from IR and Raman data, and sometimes from other sources as well – and an unequivocal assignment of frequencies to normal modes has been accomplished, there are nearly always more unknowns (force constants) than knowns (frequencies) in the numerical analysis of the problem. It is sometimes possible to rectify the situation by isotopic substitution of some atoms in the molecule. For example, if ^1H is replaced by ^2H (deuterium) in a particular bond, the greater mass of the latter will affect a vibrational frequency to the extent that the H atom moves in the appropriate normal coordinate. However, the force constants should not be appreciably affected by isotopic substitution so that we obtain more experimental quantities without increasing the number of unknowns. In most cases it is necessary to simplify the problem by reducing the number of force constants. Some interaction constants can be set equal to zero. Relationships between force constants can sometimes be derived or postulated, and constraints imposed upon their relative or absolute magnitudes. Because of these approximations, the results of most normal coordinate analyses must be viewed with caution. But if a consistent set of approximations is made in a study of a related group of molecules, meaningful comparisons can be made.

The following example illustrates not only the complementary nature of IR and Raman spectroscopy, and the value of normal coordinate analysis, but also the complementarity between vibrational spectroscopy and other structural methods. In 1986, a group of West German chemists reported a study of the molecules PCl_nF_{5-n} ($n = 0$–5). It has been known for a long time that PF_5 and PCl_5 molecules are trigonal bipyramidal in shape, as predicted by VSEPR theory (although the substance PCl_5, at room temperature and atmospheric pressure, is a solid consisting of PCl_4^+ and PCl_6^- ions). Assuming that the mixed species are also trigonal bipyramidal, geometrical isomerism is possible, because the axial and equatorial positions of the trigonal bipyramid are not equivalent. For PXY_4, we could have:

The subscripts a and e denote the axial and equatorial positions respectively; the labels C_{3v} and C_{2v} convey to the initiated the molecular symmetry. For PX_2Y_3, three isomers are possible:

From the selection rules, the number of allowed IR and Raman bands for each isomer is as follows:

	IR-allowed	Raman-allowed
$PX_5(D_{3h})$	5	6
$PXY_4(C_{3v})$	6	7
$PXY_4(C_{2v})$	11	12
$PX_2Y_3(D_{3h})$	5	6
$PX_2Y_3(C_s)$	12	12
$PX_2Y_3(C_{2v})$	11	12

Inspection of the observed spectra showed beyond reasonable doubt that $PClF_4$ and PCl_4F have C_{2v} and C_{3v} symmetries respectively, while PCl_3F_2 is D_{3h}. It is difficult to decide between C_s and C_{2v} for PCl_2F_3 from the vibrational spectra alone; electron diffraction results (which were published in the same article) clearly indicated the latter, and confirmed the geometries of the other species. Normal coordinate analyses were performed; the stretching force constants thus obtained are given along with the bond lengths from the electron diffraction data in Table 2.1. These reveal an excellent correlation; the force constant increases steeply as the bond length decreases. It could be said that the force constant is a much more sensitive measure of bond strength than the internuclear distance. This is much exploited in the area of carbonyl complexes (see below).

The results summarised in Table 2.1 show a number of interesting trends and features. It will be noted that the axial bonds are apparently longer – and presumably weaker, as suggested by the force constants – than the equatorial bonds. This is true of the vast majority of trigonal bipyramidal molecules and polyatomic ions where the central atom belongs to the p block (i.e. Main Group atoms); it does not necessarily apply for d block central atoms. Note also that Cl atoms have a marked preference for equatorial positions about the central P atom while F atoms prefer axial positions. This observation is consistent with the broad generalisation, valid for a large number of trigonal bipyramidal molecules and ions, that the less electronegative substituents prefer equatorial positions. Note too that the P–Cl and P–F bond strengths decrease as the number of Cl atoms increases; it appears that the stabilities of compounds containing pentavalent phosphorus decrease with decreasing electronegativity of atoms bonded to P. The foregoing generalisations are all closely connected, and will be discussed further in Chapters 6 and 7.

The force constants which emerge from a normal coordinate analysis are strongly dependent on the assignment of vibrational frequencies to specific modes. An unequivocal assignment is not always possible from

Table 2.1 Internuclear Distances (in pm) and, in brackets, stretching force constants (in N m⁻¹) for
PCl_nF_{5-n} molecules

	PF_5	$PClF_4$	PCl_2F_3	PCl_3F_2	PCl_4F	PCl_5
P–F$_e$	153.4(646.7)	153.5(622.9)	153.8(605.7)	—	—	—
P–F$_a$	157.7(545.3)	159.1(510.7)	159.3(474.0)	159.6(439.4)	159.7(427.5)	—
P–Cl$_e$	—	200.0(337.1)	200.2(325.3)	200.5(317.7)	201.1(293.5)	202.3(278.5)
P–Cl$_a$	—	—	—	—	210.7(218.4)	212.7(195.9)

the experimental data alone; in such cases, the assignment which leads to the most sensible set of force constants is chosen. In the case of PCl_2F_3, the analysis was used to correct an earlier assignment which led to force constants inconsistent with the trends in bond lengths. This is not so much a case of question-begging as an illustration of the way in which theory influences our evaluation of experimental data. We must be careful, however, not to reject a clear-cut assignment solely on the grounds that it leads to unsatisfactory force constants.

It might be thought that the vibrational analysis for PCl_nF_{5-n} was redundant, since the electron diffraction data provided complete structural information. This is not quite true; the two studies were in fact complementary. In the radial distribution functions obtained from electron diffraction, some of the peaks were ill-resolved; their better resolution in order to obtain accurate structural parameters was assisted by the amplitudes of vibration which can be calculated by normal coordinate analysis. The vibrational study was also valuable when, in 1987, the same team tackled the structural characterisation of the analogous arsenic compounds. These presented some experimental difficulties, because they are thermally less stable than their phosphorus analogues; they tend to decompose to give As(III) species, e.g.

$$AsCl_3F_2(g) \rightarrow AsF_2Cl(g) + Cl_2(g)$$

At lower temperatures, molecular $AsCl_2F_3$ is unstable with respect to the non-molecular substance $AsCl_4^+AsF_6^-$. Complete electron diffraction analysis could not be performed for the whole range of molecules. However, full vibrational spectra were obtained; guided by the results for PCl_nF_{5-n}, the force constants thus extracted showed that the $AsCl_nF_{5-n}$ have the same structures. This illustrates the point that vibrational spectra can often be obtained in situations where diffraction data cannot.

Force constant calculations have been especially valuable in the important field of carbonyl complexes $M_xL_y(CO)_z$, where M is one of the d block elements and L represents some other ligand(s). It has already been noted that the C–O stretches around $2000\,cm^{-1}$ are very useful in characterising such compounds. The C–O stretching force constants, if obtainable, should provide additional information. The nature of the bond between M and CO is discussed in more detail in Chapter 8; for the moment, we may say that the stronger the M–C bond, the weaker will be the C–O bond, if the conventional description is valid. Furthermore, the theory predicts the relative strengths of C–O bonds in carbonyls where the CO ligands are not all equivalent. For example, in complexes such as these:

$$
\begin{array}{ccc}
& \overset{\displaystyle O}{\underset{\displaystyle |}{C}} & \\
R_3P & | & C^O \\
& \diagdown\!\!\diagup Mo \diagdown\!\!\diagup & \\
_O{}^C & | & PR_3 \\
& \underset{\displaystyle O}{\overset{\displaystyle |}{C}} &
\end{array}
$$

$$
\begin{array}{ccc}
& \overset{\displaystyle O}{\underset{\displaystyle |}{C}} & \\
^O{}C & | & SiCl_3 \\
& \diagdown\!\!\diagup Fe \diagdown\!\!\diagup & \\
_O{}^C & | & SiCl_3 \\
& \underset{\displaystyle O}{\overset{\displaystyle |}{C}} &
\end{array}
$$

$$
\begin{array}{ccc}
^O{}C & H & C^O \\
& \diagdown\!\!\diagup Mn \diagdown\!\!\diagup & \\
_O{}^C & | & C_O \\
& \underset{\displaystyle O}{\overset{\displaystyle |}{C}} &
\end{array}
$$

it can be predicted that a CO ligand which is *trans* to another CO (i.e. the two occupy opposite corners in the octahedral figure, as opposed to the *cis* arrangement where the same groups occupy adjacent positions) will have a weaker M–C bond and hence a stronger C–O bond than one *trans* to a ligand such as H^- or $SiCl_3^-$. These predictions are difficult to verify from crystallographically-determined bond lengths. The C–O distance in carbonyls varies by only a few pm over the many hundreds of X-ray determinations that have been performed, and cannot be measured with sufficient accuracy (given the proximity of much heavier d block atoms) for meaningful correlations to be established. However, as in the PCl_nF_{5-n} molecules (Table 2.1), the C–O force constant is much more sensitive than the bond length to the bonding within the ligand molecule. Complete normal coordinate analyses are rarely performed on carbonyl complexes. However, approximate methods which focus attention on the C–O stretches are available. The resulting force constants are of dubious absolute significance but they are useful in making comparisons over a range of related compounds. The C–O stretching force constant is $1847\,N\,m^{-1}$ in the free molecule; in complexes, it is in the range $1200–1800\,N\,m^{-1}$.

2.4   Nuclear magnetic resonance (NMR) spectroscopy

Around 1960, organic chemistry can be said to have entered a new era, thanks to the availability of commercial high-resolution NMR spectrometers. Since then, the technique has advanced far more rapidly than any other; there seems to be no limit to the development of new electronic wizardry in NMR, and many of the more severe restrictions on its applicability in inorganic chemistry have been swept away. The reader of this book is certain to encounter a good deal more about NMR than can be covered in this book. Here is presented a very brief discussion of the use of NMR in the characterisation of inorganic compounds, in following inorganic reactions and in studying molecular dynamics of interest to inorganic chemists.

NMR is concerned with the property of nuclear spin, which is expressed in terms of the quantum number I and, for a given nucleus, is a fixed integral multiple of $\frac{1}{2}$. Nuclei whose mass numbers and atomic numbers are both even, e.g. $^{12}_{6}C$, $^{16}_{8}O$, $^{40}_{16}Ca$ etc., have zero nuclear spin and are said to be NMR-inactive. A nucleus whose mass number is odd has a spin $n/2$ where n is odd, while a nucleus of even mass number and odd atomic number has a spin $n/2$ where n is non-zero and even. Under the perturbation of an applied magnetic field, the nucleus may exist in $(2I+1)$ energy states characterised by the magnetic quantum numbers m taking the values $I, I-1, I-2, \ldots, -I$. The magnetic states are equally spaced, by an amount of energy proportional to the product of the gyromagnetic (or magnetogyric) ratio g and the field strength H. In an NMR experiment, we are observing transitions from the magnetic quantum number m to $m-1$; for example, the ^{11}B nucleus has a spin of $\frac{3}{2}$; the energy states in a magnetic field (in order of increasing energy) have magnetic quantum numbers $m = \frac{3}{2}, \frac{1}{2}, -\frac{1}{2}, -\frac{3}{2}$. We can therefore observe three transitions, as shown below:

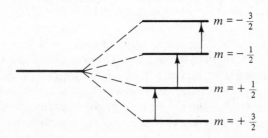

These all require exactly the same amount of energy, and the selection rule forbids any other transitions. At a typical magnetic field strength of $2\,T$ (i.e. 2 tesla, or 20000 gauss), the energy spacing between the states corresponds to electromagnetic radiation of frequency $2.7 \times 10^7\,s^{-1}$, or

27 MHz. This is in the radiofrequency range (wavelength c. 10 m). Thus NMR can – in principle at least – be observed if we have a source of radiofrequency radiation, a magnet of around 1–2 T, and the means of tuning either the field strength or the frequency of the radiation (or both) so that the energy of each photon matches exactly the spacing between the magnetic states.

The frequency required to induce resonance at a fixed applied magnetic field (or, the field required at a fixed frequency) varies over all the NMR-active nuclei within a factor of about 50. Thus by setting the field and frequency, we can choose to focus our attention on a particular nucleus. Until about 1970, NMR was restricted to a few nuclei of spin $\frac{1}{2}$, which are abundant for the elements concerned and of high sensitivity to NMR. The most widely studied were ^1H, ^{19}F and ^{31}P. Other nuclei having a spin of $\frac{1}{2}$ could be studied with isotopically-enriched samples, e.g. ^{13}C. Nuclei having spins of 1, $\frac{3}{2}$, 2, . . . etc. were less useful because they have quadrupole moments, leading to broadening of the NMR lines which obscures useful information. Nuclei of low natural abundance and/or low sensitivity were very difficult to study. These limitations are far less serious since the introduction of pulsed Fourier transform (FT) techniques of measuring NMR spectra (cf. the use of FT methods in IR spectroscopy in Section 2.3), together with a battery of cunning techniques for signal enhancement. Virtually the whole of the Periodic Table is now accessible. There have also been developments in the use of very strong magnetic fields, using superconducting magnets. These provide improved sensitivity as well as better resolution. The need for liquid helium – expensive and not always readily available – to cool the superconducting coil may disappear in the near future, with the expected advent of new, high-temperature superconductors (see Section 7.5).

The signal exhibited by a particular nucleus in an NMR experiment is dependent upon two perturbations: the chemical shift, and the effects of spin–spin coupling.

The chemical shift

If the frequency of the radiation source is fixed, the precise field at which resonance occurs depends upon the chemical environment of the nucleus under scrutiny. Under low resolution – where the effects of spin–spin interaction (see below) are not observable – the spectrum should show one peak for each chemically-distinct nucleus of the type under scrutiny; the intensity of a peak (measured by its area) is proportional to the number of equivalent nuclei encompassed by its envelope. We can thus count the number of nuclei of each chemically-distinct type in the sample. The chemical shift depends on the extent to which the nucleus is shielded from the applied field by electrons. Detailed, quantitative theories are available but chemists often rely upon empirical observations. The chemical shift is measured relative to a standard, e.g. $Si(CH_3)_4$ tetramethylsilane (TMS) for ^1H spectra. If, at a constant applied field, resonance is observed at a frequency v, while v_0 is the resonance frequency for the standard, the chemical shift is usually expressed in parts per million from the standard and is given the symbol δ:

$$\delta = 10^6(v - v_0)/v_0$$

From the chemical shift alone, useful structural information can be inferred. In ^1H NMR, for example, H atoms attached to benzene rings, aliphatic H atoms and H atoms directly bonded to atoms of the transition elements can all be readily distinguished and identified by their chemical shifts. At a more subtle level, the state of hybridisation of an atom can be specified by looking at the chemical shifts of nuclei bonded to it.

Spin–spin coupling

The resonance of a nucleus may be affected by the proximity of other nuclei in the molecule whose spins are non-zero: these need not be of the same atomic number as the nucleus under scrutiny. Suppose we have two nuclei A with a spin I_A and B with a spin I_B. The resonance of nucleus A will be split into $(2I_B + 1)$ peaks, equally spaced and of equal intensity; this happens because the precise frequency at which A absorbs depends upon the magnetic state of B. Because the $(2I_B + 1)$ states are so close in energy, they are for most practical purposes, equally occupied; hence the equal intensities of the peaks in the resonance of nucleus A. The spacing between the peaks is the *coupling constant J* between the nuclei. This is expressed in frequency units; the coupling constant is independent of the operating frequency of the spectrometer, in contrast to the chemical shift. If the nuclei A and B are chemically remote, the coupling may be so small that it cannot be observed. This is usually the case if the nuclei are separated by more than about three bonds in the molecule.

Spin–spin coupling between nuclei which are magnetically equivalent do not lead to such splittings. The term 'magnetic equivalence' requires proper definition. Two nuclei are magnetically equivalent if they have the same chemical shift, and have equal coupling constants with all other non-equivalent nuclei. Thus the ^1H spectrum of benzene or the ^{19}F spectrum of SF_6 (octahedral molecules) show only a single peak. The proton NMR spectrum of CH_3F consists of a doublet, i.e. two bands of equal intensity; the equivalent protons do not interact with each other, but the peak is split by the ^{19}F ($I = \frac{1}{2}$) nucleus. The ^{19}F spectrum of CH_3F is a quartet of four peaks, whose relative intensities are in the ratios $1:3:3:1$. If we label the sum of the magnetic quantum numbers of the three protons as M, this may take four possible values: $\frac{3}{2}, \frac{1}{2}, -\frac{1}{2}, -\frac{3}{2}$, arising from the ways in which the spins may be combined, viz.:

m_1	m_2	m_3	M
$\frac{1}{2}$	$\frac{1}{2}$	$\frac{1}{2}$	$\frac{3}{2}$
$\frac{1}{2}$	$\frac{1}{2}$	$-\frac{1}{2}$	$\frac{1}{2}$
$\frac{1}{2}$	$-\frac{1}{2}$	$\frac{1}{2}$	$\frac{1}{2}$
$-\frac{1}{2}$	$\frac{1}{2}$	$\frac{1}{2}$	$\frac{1}{2}$
$\frac{1}{2}$	$-\frac{1}{2}$	$-\frac{1}{2}$	$-\frac{1}{2}$
$-\frac{1}{2}$	$\frac{1}{2}$	$-\frac{1}{2}$	$-\frac{1}{2}$
$-\frac{1}{2}$	$-\frac{1}{2}$	$\frac{1}{2}$	$-\frac{1}{2}$
$-\frac{1}{2}$	$-\frac{1}{2}$	$-\frac{1}{2}$	$-\frac{3}{2}$

The fact that there are three ways of obtaining $m = \frac{1}{2}$ or $-\frac{1}{2}$ leads to the relative intensities.

It is convenient to classify molecules according to the NMR spin systems to which they belong. You may see reference made to AX, AMX, AB_2C systems etc. Each nucleus is assigned a capital letter; where their chemical shifts are very different, or if they are of different elements, they are given letters well apart in the alphabet. Thus a molecule is described as an AX system if it has two distinct NMR-active nuclei; the scheme is usually restricted to nuclei of $I = \frac{1}{2}$, so that the spectrum of either A or X will show a doublet. Examples include $PFCl_2$ (^{19}F and ^{31}P are the NMR-active nuclei), or Cl_3Si—OH (considering ^{29}Si and ^1H as A and X). An A_2X system has two equivalent nuclei of one sort and one of the other, e.g. SiH_2Cl_2 (A = ^1H, X = ^{29}Si). A molecule such as CHCl=CHBr would be classed as AB (referring to two not-very-different ^1H nuclei), while *cis*-WF_4ClBr is A_2BC:

cis-WF_4ClBr trans-WF_4ClBr

The two F atoms labelled F′ are mutually equivalent but the two labelled F″ and F‴ are not, one being opposite (or *trans* to) a Cl atom while the other is opposite a Br atom. Thus the ^{19}F spectrum is of A_2BC type (A = F′, B = F″, C = F‴). The experienced NMR user can quickly identify the type of system represented by a given spectrum, and hence obtain structural information. We could readily distinguish between the *cis* and *trans* geometrical isomers of WF_4ClBr; the latter, with four equivalent F atoms at the corners of a square, should have a ^{19}F spectrum with only a single peak while the former, as already noted, is of A_2BC type.

Where the molecule contains many non-equivalent magnetic nuclei, the spectrum may be so complex as to defy analysis. In these circumstances, it may be possible to simplify the spectrum by *spin decoupling*, which has the effect of eliminating all coupling effects of a particular type of nucleus. For example, suppose that we have prepared a complex which consists of octahedral molecules $Mo(PR_3)_4(L)(L')$, where R is an alkyl group and L,L' are other neutral ligands having no influence on the NMR spectrum. The molecule may have *cis* or *trans* geometry:

trans-   cis-

Looking at the ^{31}P spectrum, the *trans* isomer should give a fairly simple spectrum, essentially one resonance split by coupling with the protons of the R groups. But the *cis* isomer will have a very complex spectrum; the A_2BC system will be further split by coupling between ^{31}P and ^{1}H. By decoupling these nuclei, the spectrum of the *cis* isomer should become easily recognisable as A_2BC and the characterisation of the substance is largely complete. Spin decoupling is also valuable in order to remove the effects of nuclei such as ^{14}N whose quadrupole moments may broaden the resonance of other nuclei and obscure vital information.

NMR and molecular dynamics

One of the most important applications of NMR is in the study of *fluxional* or *stereochemically-nonrigid* molecules, whose configurations are rapidly changing under normal conditions with the breaking and making of covalent bonds. The classic example is found in the case of PF_5. This gaseous substance (b.p. $-85\,^{\circ}$C) consists of trigonal bipyramidal molecules, on the basis of IR, Raman and electron diffraction studies as discussed for PCl_nF_{5-n} molecules in Section 2.3. The ^{19}F spectrum should therefore be of A_2B_3X type, i.e. we have two axial F atoms and three equatorial F atoms, whose resonances will be affected by the central ^{31}P nucleus. But at room temperature, and indeed down to about $-100\,^{\circ}$C, the ^{19}F spectrum shows a doublet, consistent with all the fluorines being equivalent but split by spin–spin coupling to ^{31}P. Likewise in the case of PCl_2F_3, with one equatorial and two axial fluorines, the ^{19}F spectrum fails to distinguish the non-equivalent F atoms. This is attributed to *pseudo-rotation*, a common type of fluxional behaviour. Axial and equatorial fluorines undergo rapid exchange, via a square pyramidal intermediate, as shown below:

The difference in chemical shift between two different environments is measured in frequency units (Hz, i.e. s^{-1}). If the rate at which nuclei exchange positions is much faster than this, we see only a single, average chemical shift. In the case of PF_5, it is futile to ask an ^{19}F nucleus the question: 'Are you axial or equatorial?', because in the time it takes to ask the question, an individual nucleus will have switched many times from one environment to the other. In other words, the timescale of the NMR experiment is relatively long, and the chemical environment of a nucleus as determined by its chemical shift is time-averaged. In contrast, the electron diffraction pattern produced by a gaseous sample is processed to give a radial distribution function which reflects the superimposition of many 'snapshots' of different molecules in different configurations, as they execute the various modes of vibration and undergo pseudorotation. The overall picture is somewhat blurred – the peaks in the radial distribution function are relatively broad – but the equilibrium geometry of the PF_5 molecule clearly emerges as a trigonal bipyramid. The interconversion between the trigonal bipyramid and the square pyramidal intermediate can be accomplished via a vibrational mode common to both geometries.

By lowering the temperature, it is often possible to slow down the rate of intramolecular processes of this kind, and hence to obtain NMR spectra consistent with expectation. It is then possible to measure the rate constant for the process, and hence the activation barrier. In the case of PF_5, this barrier amounts to $18 \, kJ \, mol^{-1}$. This kind of experiment gives much useful information about molecular flexibility, and about the relative energies of isomers.

Evidently, we should be wary of drawing conclusions from what may be a deceptively-simple NMR spectrum measured at room temperature; wherever fluxionality seems plausible, it must be suspected. Organometallic compounds involving the cyclopentadienide ligand

$C_5H_5^-$ provide many examples. Ferrocene $Fe(C_5H_5)_2$ was reported in 1951, and was originally assigned the structure (a):

(a)   (b)

An X-ray analysis of ferrocene was not published (a little surprisingly) until 1956; the 'sandwich' structure (b) which it revealed had, however, been suspected by the growing band of organometallic chemists for some time. The IR spectrum suggested a highly-symmetric aromatic ring system, and the proton NMR spectrum showed only a single peak, consistent with structure (b) with all H atoms equivalent. Later, however, it was observed that $Hg(C_5H_5)_2$ also gave only a single NMR line. Mercury forms many molecular compounds HgR_2 with the linear C—Hg—C skeleton; and while MO treatments of ferrocene assured organometallic chemists that they could feel comfortable with its sandwich structure, they definitely ruled out such a structure for $Hg(C_5H_5)_2$. This awkwardness was resolved when it was found that on cooling the NMR spectrum of the mercury compound split up to give resonances consistent with structure (a). It was postulated that the molecule is fluxional; in each ring, the C atom bonded to Hg is rapidly changing, so that on the NMR timescale all protons have the same time-averaged environment:

Many other examples of 'ring whizzers' have been established, and careful NMR studies over a range of temperatures yield much information about the activation barriers and detailed mechanisms.

Limitations of NMR in inorganic chemistry

Two limitations of NMR spectroscopy affect the inorganic chemist far more than the organic chemist. First, all the discussion so far relates to measurements on liquids, solutions and gases, where we are dealing with discrete molecules (or, in solution, perhaps polyatomic ions). NMR of solid samples is much less familiar and subject to experimental restrictions. Some of the earliest NMR studies were made on solids, but the lines are so broad that only limited information can be obtained. With modern techniques, however, rapid progress is being made in solid-state NMR. This opens up the important area of insoluble, non-molecular solids. Thus ^{29}Si NMR is proving useful in examining the structures of silicates; these important materials contain Si–O–Si bridges (non-linear), and an empirical correlation is observed between ^{29}Si chemical shifts and the bridging angles. Since many silicate materials are not amenable to X-ray methods of structure determination, solid-state NMR fills an important gap in the chemist's arsenal.

The second important limitation of NMR is encountered when we are dealing with paramagnetic materials, i.e. where we have unpaired electrons. Interaction between a nuclear spin and an unpaired electron spin usually leads to line-broadening to the extent that no NMR spectrum can be observed. This rarely troubles the organic chemist, or the organometallic chemist, but it restricts the use of NMR in the study of d block coordination chemistry. However, these interactions are far from being a universal nuisance. In some circumstances, the effect of a paramagnetic centre in the molecule under scrutiny is to alter (often dramatically) the chemical shifts of nuclei which interact strongly with the unpaired spin(s), without undue broadening. This is known as a *contact shift*. In the case of a complex organic molecule coordinated to a paramagnetic ion, the proton NMR spectrum may be so spread out that all the lines are completely resolved, so that its interpretation is easier than that of the free molecule. The contact shifts can sometimes be quantitatively interpreted in terms of the delocalisation of unpaired spin density over the molecule, i.e. the nature of the MO in which the unpaired electron resides can be ascertained in some detail. Measurements of this kind provided convincing evidence for the need to modify the simple CF theory of the 1950s.

2.5   Rotational spectroscopy

Rotational quanta are much smaller than vibrational quanta, and correspond to electromagnetic radiation in the microwave region, typically in the range 10^3–10^5 MHz ($1\,\text{cm}^{-1} \equiv 3.3 \times 10^4\,\text{MHz}$). Rotational transitions can be excited directly, in microwave absorption spectroscopy (pure rotational spectroscopy), but can also be observed in the rotational fine structure in high-resolution vibrational or electronic spectra.

In a linear molecule, the rotational states have energies $BJ(J+1)$, where J is an integer (the rotational quantum number) and B is the rotational constant, proportional to the moment of inertia I. Given I and the atomic masses, the internuclear distances can be calculated. The linear molecule has a unique axis with respect to which a moment of inertia can be defined. Non-linear molecules have three principal moments of inertia and hence three rotational constants. Depending on the molecular symmetry, two or even all three rotational constants may be equal.

Pure rotational spectroscopy in the microwave or far IR regions joins electron diffraction as one of the two principal methods for the accurate determination of structural parameters of molecules in the gas phase. The relative merits of the two techniques should therefore be summarised. Microwave spectroscopy usually requires sample partial pressures some two orders of magnitude greater than those needed for electron diffraction, which limits its applicability where substances of low volatility are under scrutiny. Compared with electron diffraction, microwave spectra yield fewer experimental parameters; more parameters can be obtained by resort to isotopic substitution, because the replacement of, say, ^{16}O by ^{18}O will affect the rotational constants (unless the O atom is at the centre of the molecule, where the rotational axes coincide) without significantly changing the structural parameters. The microwave spectrum of a very complex molecule of low symmetry may defy complete analysis. But the microwave lines are much sharper than the peaks in the radial distribution function obtained by electron diffraction, so that for a fairly simple molecule whose structure can be determined completely, microwave spectroscopy yields more accurate parameters. Thus internuclear distances can often be measured with uncertainties of the order of $0.001\,pm$, compared with (at best) $0.1\,pm$ with electron diffraction. If the sample is a mixture of gaseous species (perhaps two or more isomers in equilibrium), it may be possible to unravel the lines due to the different components in the microwave spectrum, but such resolution is more difficult to accomplish with electron diffraction.

Absorption of microwave radiation to excite molecular rotation is allowed only if the molecule has a permanent dipole moment. This restriction is less severe than it may sound, however, because centrifugal distortion can disturb the molecular symmetry enough to allow weak absorption, especially in transitions between the higher rotational states which may appear in the far IR ($c.\ 100\,cm^{-1}$). Microwave spectroscopy can provide a wealth of other molecular data, mostly of interest to physical chemists rather than inorganic chemists. Because of the ways in which molecular rotation is affected by vibration, it is possible to obtain vibrational frequencies from pure rotational spectra, often more accurately than is possible by direct vibrational spectroscopy.

Under high resolution, rotational fine structure can be observed in the IR and Raman spectra of gaseous substances, i.e. a change in the vibrational quantum number for a given normal mode may be accompanied by a change in a rotational quantum number, subject as always to selection rules. The band contours arising from such rotational structure can be of help in the assignment of the vibrational mode. Rotational constants can be obtained in cases where the pure rotational spectrum cannot be observed. For example, the rotational fine structure in the Raman spectrum of gaseous H_2, O_2 etc. (homonuclear diatomics have no dipole moments) yields the internuclear distance with great accuracy.

Rotational fine structure may also be resolved in the electronic spectra (see also Section 2.6) of molecules in the gas phase. This can provide structural information about unstable, short-lived species which cannot be isolated as pure substances. It may also yield structural parameters for molecules in electronic excited states. Such information is relevant to discussion of bonding and electronic structure.

2.6   Electronic spectroscopy

An electronic spectrum measures transitions from the ground state of the absorbing species to excited states which differ from the ground state in the precise distribution of electrons among orbitals. Thus an electronic transition usually involves one or more 'electron jumps' between orbitals; but if the ground state has two or more unpaired electrons, electronic transitions may be associated with changes in total spin, without changes in orbital occupancy. The absorbing species may be a discrete molecule or ion; but in the case of a crystalline substance where there are strong interactions among neighbouring atoms/ions/molecules, the whole crystal may be seen as the absorbing species.

Electronic transitions occur over a wide range of the electromagnetic spectrum; some have been measured well into the IR region, among vibrational transitions, while others occur far into the ultraviolet (UV). However, to most inorganic chemists, electronic spectroscopy covers the range $10\,000$–$50\,000\,cm^{-1}$ (1000–$200\,nm$), accessible using the familiar visible/UV spectrophotometers which are to be found in every chemistry laboratory. Measurements are usually made on solutions, but solid samples can be tackled in a number of ways.

Electronic spectra may be used (as in organic chemistry) as finger-prints, and they are very important in kinetic studies. The change in the electronic spectrum of a reaction mixture as the reaction proceeds is often the best way of following its rate, and quite elaborate methods are available for measuring very fast reaction rates. However, the application which the reader is most likely to encounter in more advanced texts is in the area of coordination compounds of the transition elements, whose electronic spectra may yield information about structure and bonding.

Such measurements, together with magnetic studies (see Section 2.8), provide much of the experimental basis for CF/LF theory. The electronic spectra of complexes are thoroughly treated in most standard inorganic texts – *ad nauseam*, some may think – so only the briefest sketch is attempted here.

Configurations and states

A specified electronic *configuration* is a statement of how many electrons are present in each energy level, described in terms of orbitals. For example, if we say that the Ti^{2+} ion has the d^2 configuration, we are simply saying that there are two electrons in the fivefold-degenerate 3d subshell. The statement does not specify whether these electrons are in the same 3d orbital, or in two different orbitals; it does not specify which orbital(s) is/are occupied, or the relative spins of the electrons if they are in different 3d orbitals.

A spectroscopic *state* is a more precise description than a configuration. It defines the spins and orbital occupancies of the electrons; in quantum-mechanical terminology, it corresponds to a many-electron wave function having a unique energy, whereas a configuration is a one-electron description and neglects interelectron repulsion. A given configuration may give rise to several states, whose relative energies can be expressed in terms of interelectron repulsion parameters. Thus the energy change accompanying an electronic transition is usually expressible as the sum of a change in orbital energy (if there is a change in configuration) and a change in interelectron repulsion.

It is not always possible to relate a given spectroscopic state to a unique configuration. If two or more states, arising from different configurations, have the same symmetry (denoted by group-theoretical symbols such as E_g, T_{2u} etc.), they become mixed up. In quantum-mechanical language, the pure (unmixed) states have wave functions that are not all mutually orthogonal (see Sections 1.4 and 7.1), and we have to take linear combinations to make them acceptable (cf. the construction of MOs and AOs of the same symmetry as discussed in Chapter 7). This is called *configuration interaction*.

Three types of electronic transition can be distinguished among compounds of the d block transition elements; d–d bands, charge transfer (or electron-transfer) bands and intra-ligand bands. Configuration interaction may make the distinctions rather hazy, however.

d–d transitions

The neatness with which these can be interpreted, often quantitatively, by CF/LF theory is one reason for their popularity among inorganic chemists; another, of course, is the fact that they are largely (though by no means wholly) responsible for the fascinating colours of so many compounds in d block chemistry.

d–d transitions are observed in compounds where d block atoms/ions having partly-filled nd subshells are present. They involve the transfer of electrons between nd orbitals and/or changes in spin states. To the extent that the orbitals concerned are pure nd orbitals (as in the pure, electrostatic CF model), such transitions are localised upon the d block atom. CF theory shows how to work out the spectroscopic states which arise when a d^n ion finds itself in a chemical environment of given symmetry. These states are classified by group-theoretical symbols, and can be written as n-electron wave functions based on pure nd orbitals. Their energies are written in terms of orbital splitting parameters and interelectron repulsion parameters; although many attempts have been made to calculate theoretical values for these parameters, coordination chemists are content to find empirical values which best fit the experimental d–d transition energies.

d–d bands are relatively weak, as electronic transitions go. The mechanism for electronic absorption is essentially the same as for IR absorption in vibrational spectroscopy (Section 2.3).

In atomic spectroscopy, transitions between orbitals having the same quantum number l are forbidden (Laporte selection rule). In a compound where the chemical environment of a d block ion lacks a centre of symmetry, this rule is somewhat relaxed. Even where there is a centre of symmetry, d–d bands can acquire significant intensity by vibronic coupling; crudely speaking, this means that vibrational modes may disturb the centre of symmetry. A more elaborate treatment shows that vibronic coupling permits the mixing of d–d states with charge transfer states (see below); a small admixture of charge transfer character in a band which is predominantly d–d in nature can cause a considerable enhancement in intensity, especially if the states being mixed are fairly close in energy.

Another selection rule forbids transitions between states of different total spin. 'Spin-forbidden' bands do appear, however; in the case of octahedral high-spin d^5 compounds (which embraces nearly all Mn(II) and many Fe(III) compounds) all d–d transitions are necessarily spin-forbidden, because the ground state is the only state which can have five unpaired electrons, all having the same spin. All other states arising from the d^5 configuration necessarily have a total spin S of $\frac{1}{2}$ or $\frac{3}{2}$ as opposed to $S = \frac{5}{2}$ for the ground state. The spin selection rule is relaxed by the effects of spin–orbit coupling, i.e. the interaction between spin and orbital angular momenta. This tends to degrade the integrity of spin as a valid quantum number for the classification of states, and hence to diminish the significance of selection rules making explicit reference to spin. Spin–orbit coupling tends to increase with increasing atomic number. Thus it is particularly important towards the right-hand side of each transition series, and is very large for the 4d and 5d atoms/ions, compared with their 3d analogues.

d–d bands are usually rather broad, so that closely-spaced transitions may be difficult to resolve. The broadness arises from complex vibrational structure; recall that vibronic coupling, the disturbance of the chemical environment from high symmetry, is largely responsible for the appearance of d–d transitions where the atom/ion lies at a centre of symmetry, e.g. in octahedral and square planar complexes. However, spin-forbidden d–d bands where the ground state and the excited state arise from the same configuration can be very sharp. For example, the ground state of an octahedral d^3 system, designated $^4A_{2g}$, arises from the configuration $(t_{2g})^3$ with each of d_{xy}, d_{xz} and d_{yz} singly occupied and all spins parallel. A twofold-degenerate doublet state ^2E_g arises from the same configuration. The transition to this excited state from the ground state is observed as weak, sharp line. In emission, the transition $^2E_g \rightarrow {}^4A_{2g}$ is the line produced by the ruby laser.

Routine measurements of d–d spectra are performed on solutions. If a suitable solvent cannot be found for a solid sample, a diffuse reflectance spectrum of a powdered sample can be taken. This is actually an absorption spectrum of the surface layers of the sample and is subject to a number of anomalies and artefacts. It is much better to study microscopic single crystals, preferably at low temperatures. Large crystals (if they can be grown) tend to absorb too strongly around band maxima; small, thin ($c. 0.01$ mm) plates are best. It is usually necessary to condense the incident beam by means of a lens in order to obtain detectable intensities of transmitted radiation. Thus the technique is more difficult and time-consuming than the familiar, routine solution measurement; but it can provide much more information.

At low temperatures, crystal spectra reveal great detail. Ill-resolved bands can be distinguished, and vibrational fine structure can be analysed. The use of plane-polarised light can be of value; if the symmetry of the ligand field is lower than cubic – i.e. if it is lower than tetrahedral or octahedral – the spectrum may depend upon the orientation of the plane of polarisation of the incident beam with respect to the axes which define the nd orbitals as d_{xy}, d_{z^2} etc. The selection rules may tell us that a particular d–d transition is allowed in one such polarisation, but not in another, at right angles to it. Such observations are of great value in the assignment of d–d spectra.

d–d spectra were used for fingerprinting purposes, and for the drawing of structural inferences from empirical correlations, for some time before their detailed analysis became a popular pursuit among inorganic chemists. The quantitative interpretation of a d–d spectrum, using CF/LF theory, yields two types of parameters: orbital splitting parameters, and interelectron repulsion parameters (spin–orbit coupling is often thrown into the calculation as another variable). In the case of a cubic ligand field (octahedral or tetrahedral), there is only one splitting parameter Δ. Its magnitude is of some significance in magnetic and thermodynamic arguments, as well as its intrinsic theoretical interest. In cases of lower symmetry, three or four parameters may be necessary to define the d orbital splitting. If these can be reliably determined, useful information about bonding can be extracted. For example, it is thoroughly established (from single-crystal polarised spectra measured at liquid helium temperature) that the relative energies of the copper 3d orbitals in the square planar ion $CuCl_4^{2-}$ is $d_{z^2} < d_{xz,yz} < d_{xy} < d_{x^2-y^2}$. The quantitative interpretation of this observation has aroused much activity and debate.

For a given d block atom in a given oxidation state, the relative values of the octahedral parameter Δ for a series of different ligands follow a sequence – the *spectrochemical series* – which differs little as the d block atom is changed. The theoretical interpretation of this series is discussed in Section 8.2.

The interelectron repulsion terms which arise from analysis of a d–d spectrum are usually expressed in terms of the Racah parameters B and C, of which the former is more important and more often reported. The Racah parameter thus obtained is practically always smaller than the value for the free, gaseous ion inferred from atomic spectra. This is the *nephelauxetic* (i.e. cloud-expanding) *effect*. It can be interpreted in terms of covalency in the bonding to the notional d block ion which gives rise to the d–d spectrum. Consider, for example, tetrahedral $CoCl_4^{2-}$. Crystal field theory views this as a Co^{2+} ion surrounded by four Cl^- ions. In an MO or VB treatment – in which orbital overlap and covalent bonding are explicitly considered – the bonding involves overlap between filled orbitals on the Cl^- ions and empty orbitals on the Co^{2+}. The resulting electron distribution must lead to a partial positive charge on the Co atom of rather less than two units, since electron density has effectively been transferred from Cl^- to Co^{2+}. Compared with the free Co^{2+} ion, the relative size of the Co atom in the complex ion (as measured by the radial extension of its valence orbitals) will increase; cf. the effect on the radius of a cation of increasing positive charge, discussed in Section 4.2. This means that the 3d electrons will be able to keep further away from each other, and their mutual repulsion will be decreased. Note that the observation of a nephelauxetic effect does not necessarily invalidate the crystal field approach as far as the 3d orbitals are concerned; even if the 3d orbitals play no part whatsoever in bonding, involvement of the Co 4s and 4p orbitals in covalent bonding would lead to a nephelauxetic effect.

A nephelauxetic series – analogous to the spectrochemical series – can be constructed. It is quite different from the spectrochemical series and is consistent with the interpretation of the effect given in the previous paragraph. Thus the largest nephelauxetic effects are observed with ligands whose donor atoms are of relatively low electronegativity, i.e. where an ionic description is least appropriate.

Charge transfer transitions

These are electronic transitions between orbitals that are largely localised on different atoms. In coordination compounds ML_n where M is a d block element and L_n represents the ligands, we can distinguish two types of charge transfer: $L \rightarrow M$ and $M \rightarrow L$. These are depicted schematically in Fig. 2.2, along with d–d transitions and intra-ligand transitions.

Charge transfer transitions are best discussed in the language of MO theory (Section 8.2). An $L \rightarrow M$ transition involves the transfer of an electron from a nonbonding or weakly-bonding MO localised mainly on the ligand moieties L_n, to a vacancy in the partly-filled nd subshell of M. Such a transition is likely to appear in the visible or near UV spectrum if the energy separation between the donor and acceptor orbitals is relatively small. This will be the case if:

(*a*)   M is in a relatively high oxidation state, and is readily reduced; the higher the oxidation state, the lower in energy will be the nd orbitals.

(*b*)   L has donor atoms of relatively low electronegativity, in which case its highest occupied orbitals will be fairly high in energy; or,

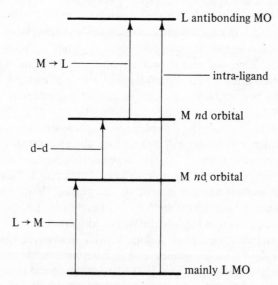

Fig. 2.2. Schematic illustration of the types of electronic transition found in coordination compounds of the transition elements.

to put it another way, the donor atoms have relatively low ionisation potentials.

The other type of charge transfer transition, $M \rightarrow L$, is likely to appear at a relatively low energy – in the visible or near UV – if M is in a relatively low oxidation state (i.e. M is easily oxidised) and if the ligands have low-lying empty MOs. The latter circumstance occurs most often with conjugated organic molecules/ions having extended π systems. The delocalised π bonding MOs are matched by empty π antibonding MOs, labelled π^*, which may be sufficiently low in energy to be effective electron-acceptors; in other words, such ligands are easily reduced, and the combination of an easily-oxidised central atom M and an easily-reduced ligand L is likely to lead to facile $M \rightarrow L$ electron transfer. Thus, for example, we might expect to see $M \rightarrow L$ charge transfer bands in the electronic spectra of Fe(II) complexes with aromatic ligands like 2,2'-bipyridyl or 1,10-phenanthroline, which are respectively (1) and (2) below:

(1)   (2)

On the other hand, with Fe(III) complexes we are more likely to observe $L \rightarrow M$ charge transfer bands. Fe(II) tends to be oxidisable to Fe(III) and Fe(III) is reducible to Fe(II), but Fe(III) is not easily oxidised.

Charge transfer bands are usually very intense, two or three orders of magnitude more so than d–d bands. Thus where such bands occur at relatively low energies, they usually blot out the d–d spectrum. In complexes of the 4d and 5d elements, where ligand field splittings are usually much greater than in analogous compounds of the 3d elements, d–d bands are mostly well into the UV where they are likely to be occluded by the much stronger charge transfer bands. This is one of the reasons why d–d spectra are less useful in studying complexes of the 4d and 5d elements. (Another is the strong spin–orbit coupling, which complicates the d–d spectra.)

In principle, charge transfer spectra should be full of valuable information, because they depend on the properties of two atoms, bonded together. However, they have proved to be far less amenable to detailed analysis than d–d bands, mainly because the changes in inter-electron repulsion which accompany charge transfer are not so easily parameterised. Single crystal measurements, using polarised light at low temperatures, are important and valuable in the study of d–d spectra, but charge transfer bands are often too intense for such measurements; a solution can be diluted to taste, but a crystal cannot be diluted unless a transparent host lattice, isomorphous with the crystal under examination, can be found. For example, the intense purple colour of the MnO_4^- ion is caused by a charge transfer band (L→ M), which can be studied in detail by diluting $KMnO_4$ in the isomorphous and transparent $KClO_4$. Likewise, the intense charge transfer bands in Cs_2CuBr_4 (which contains the (approximately) tetrahedral $CuBr_4^{2-}$ ion) can be measured in a crystal of Cs_2ZnBr_4 which has been deliberately contaminated with some Cu(II). But relatively few such studies have been made, and it is probably true to say that most inorganic chemists regard charge transfer bands as something of a nuisance, although they have some value in fingerprinting, in spectrophotometric analysis, and in following reaction rates via changes in the charge transfer spectrum as a reaction proceeds. Thus the visible spectrum of MnO_4^- is unmistakable, even at concentrations as low as $10^{-5} \, mol \, l^{-1}$. An obvious way of following the rate of an oxidation by permanganate would be to monitor the optical absorbance of the reaction mixture at 530 nm (the absorption maximum) as the reaction proceeds. Permanganate is unusual in that vibrational structure is resolvable in the charge transfer transition even in solution, at room temperature. The aqueous spectrum shows seven peaks, with a constant separation of $744 \, cm^{-1}$. This corresponds to the Mn–O stretching frequency in the excited state. The Raman spectrum of MnO_4^- shows the symmetric stretch at $850 \, cm^{-1}$, and we may infer that the bonding is somewhat weaker in the charge transfer excited state. An MO treatment shows that the highest occupied MO in the ground state is a nonbonding orbital composed of oxygen 2p orbitals; in the course of the transition, an electron jumps from this into the lowest empty orbital. This is a somewhat antibonding MO consisting mainly of the Mn d_{z^2} and $d_{x^2-y^2}$ orbitals, the degenerate e level of crystal field theory. Information derived from vibrational structure is usually obtained from low-temperature studies on single crystals. Analogous information for gas phase molecules can be obtained from photoelectron spectroscopy (Section 2.7).

Intra-ligand transitions

If the free ligand has low-lying electronic excited states giving rise to transitions in the visible/near UV region, we may expect to see these transitions – somewhat modified in position and intensity – in the complex. Such ligands are mostly those of the kind mentioned above in connection with M→L charge transfer transitions: organic molecules with extensive delocalised π bonding. The perturbation of intra-ligand absorption gives a useful indication that complex formation in solution has in fact occurred, and quantitative measurements of such shifts are used to study both the thermodynamics and kinetics.

One of the most familiar examples of intra-ligand absorption is found in haemoglobin, the red iron complex in mammalian blood (see also Section 9.8). The colour arises from strong $\pi \rightarrow \pi^*$ transitions of the porphyrin ligand. The slight difference in colour between oxygenated and deoxygenated blood arises from the effect of O_2 coordination to Fe atoms. On exposure to the atmosphere, the iron in haemoglobin is oxidised from the II to the III state. There are now charge transfer states (one of them in the near IR, at $c.$ 10 000 cm^{-1} or 1000 nm) which mix with the $\pi \rightarrow \pi^*$ states to give a rather different spectrum. Similarly, myoglobin – the red pigment in mammalian muscle which stores O_2 in the form of an iron(II) complex – is oxidised on exposure to air to the dark-brown Fe(III) form. The bright red colour of Fe(II) myoglobin can be restored by treatment with sodium nitrite. This does not reduce the iron to the II state; the coordination of NO_2^- to Fe(III) leads to shifts in the positions of the charge transfer bands leading to a colour similar to that of the Fe(II) form.

2.7   Photoelectron spectroscopy

Two types of photoelectron spectroscopy can be distinguished:
(a) Ultraviolet photoelectron spectroscopy (UPS), nearly always performed on gases.
(b) X-ray photoelectron spectroscopy (XPS), most often performed on solids.

Both involve the irradiation of a sample with a monochromatic source, leading to the expulsion of electrons whose kinetic energies are measured by deflection in an electromagnetic field.

Ultraviolet photoelectron spectroscopy (UPS)

Here a gaseous sample is assaulted with a monochromatic beam of photons, usually the emission of a helium lamp at 58.4 nm, or 171 180 cm^{-1}. More commonly, the photon energy is expressed as 21.22 eV, in conformity with the units used for atomic ionisation potentials (see Section 4.3). This is called He(I) radiation, and corresponds to transitions from an excited state of the He atom arising from the 1s2p configuration, back to the ground state. We can also obtain He(II) emission from a helium lamp; this involves the 2p → 1s transition of the He$^+$ ion, and gives a line at 40.8 eV. A 21.22 eV photon has more than enough energy to expel the least tightly-bound electrons from any molecule. An electron that has been forcibly evicted by such a photon – a photoelectron – will have a kinetic energy E_K which is given by:

$$E_K = (21.22 - E_B)\,eV$$

where E_B is the binding energy of the electron in the target molecule. Thus E_B is the ionisation potential associated with the MO occupied by the electron prior to its ejection. On a conventional MO energy level diagram, the energy of the MO is equal to $-E_B$, i.e. an energy E_B is needed to raise the energy of the electron to zero, at which point it can escape; any surplus energy takes the form of kinetic energy. Any electron for which E_B is less than 21.22 eV is liable to be expelled by He(I) radiation; under suitable conditions, the He(II) spectrum can also be obtained, in which we observe the expulsion of electrons having binding energies up to 40.8 eV.

A photoelectron spectrometer analyses the beam of photoelectrons according to kinetic energy, i.e. it measures the E_K values. A photoelectron spectrum is usually recorded as a plot of intensity of the analysed photoelectron beam versus the binding energy E_B. It may also be tabulated as observed E_B values. Thus we obtain, in effect, the MO energy level diagram for the molecule. Although UPS can be used for fingerprinting and to obtain structural information, it has few advantages in this respect over other well-established spectroscopic methods and its main application is in providing information about bonding and electronic structure.

UPS and MO theory are inextricably linked. It is impossible to interpret UPS measurements without at least a qualitative MO treatment, and it is now commonplace to perform quantitative calculations for comparison with the UPS data. The development of UPS techniques since about 1960 helped to popularise MO theory, and semi-empirical MO methods are often calibrated by appeal to UPS data. Thus an approximation which greatly simplifies an MO calculation is held to be justifiable if, over a fair range of molecules, the calculated orbital energies are in good agreement with UPS binding energies.

According to some writers, UPS 'proves' that VB theory is 'wrong'. For example, UPS of H_2O reveals four ionisations below 40.8 eV (three of them are below 21.2 eV). This is in agreement with simple MO theory (see Section 7.3), which leads to four filled MOs constructed from the hydrogen 1s orbitals and the oxygen 2s and 2p orbitals. It is asserted that VB theory predicts only two ionisations, because it places the eight valence electrons in two equivalent bond orbitals and two equivalent nonbonding orbitals (or lone pairs). This is not quite fair. Perhaps the best way of looking at photoelectron spectroscopy is to see it as an extension of electronic absorption spectroscopy; in UPS, we are looking at transitions from the ground state of the molecule X to electronic states of the cation X^+. The lowest ionisation potential is a transition from the ground state of X to the ground state of X^+. The subsequent ionisations (not to be confused with successive ionisation potentials I_n of atoms – see Section 4.3) observed in the spectrum are from the ground state of X to excited states of X^+. If we shift our reference point to the ground state of X^+, the photoelectron spectrum of X is in fact an electronic absorption spectrum of X^+. It may then be argued that the experiment is telling us more about the electronic structure of X^+ than about X (see Fig. 2.3). If X is a closed-shell species (with all electrons paired), X^+ is necessarily an open-shell system and VB theory – as explained in Section 1.4 – is not really applicable. In other words, VB theory does not fail in the sense that its predictions could lead us astray; it makes no prediction at all, because of inbuilt limitations.

The remarks in the previous paragraph are particularly relevant to UPS studies of coordination compounds where the central atom has a partly-filled nd subshell. This area is restricted by the low volatility and/or thermal stability characteristic of most coordination compounds; UPS is

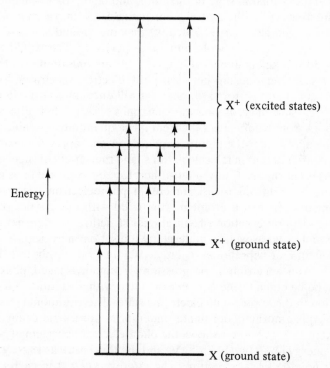

Fig. 2.3. Schematic illustration of the transition energies measured by photoelectron spectroscopy of the molecular substance X; the dashed lines indicate electronic transition energies of X^+.

limited to gaseous samples. However, considerable progress has been made and the results have aroused much comment.

You might expect that a molecule ML_n, where M is a d block atom and L are ligands, should have as its highest occupied orbitals the partly-filled nd subshell. Thus the first ionisation observed in the spectrum, corresponding to the removal of an electron of lowest binding energy E_B, should be from the nd subshell. In fact, it often happens that the first ionisation is apparently from an MO mainly localised on the ligands. This would imply that a filled ligand orbital lies higher in energy than the partly-filled subshell, which seems absurd; electrons should surely fall from higher to lower levels spontaneously, leading to reduction of M and oxidation of L. What such an observation really means is that in the cation ML_n^+, the ground state has a 'hole' in a predominantly ligand MO, and the highest occupied ligand orbitals are higher in energy than the nd subshell. In such circumstances, ML_n^+ is likely to be unstable with respect to an internal redox reaction, with spontaneous electron transfer from the ligands to the central atom. But this implies nothing about the electronic structure of ML_n. It is well known that d orbitals are very sensitive in energy to the oxidation state of the atom, and fall markedly in energy as the positive charge on the central atom increases. A simple example is the observation (via atomic spectroscopy) that the ground states of the V atom and the V^+ ion are $3d^34s^2$ and $3d^4$ respectively. This would suggest that the 3d subshell is stabilised relative to 4s on ionisation.

The precise relationship between photoelectron spectra and orbital energies is not straightforward, and you will encounter in the literature much discussion of 'reorganisation energies', 'correlation effects' and 'deviations from Koopmans's theorem'. The quantitative significance of 'orbital energy' in MO theory depends on the approximations and assumptions inherent in the calculations, and the variety of such approximations is enormous. Caution must therefore be exercised in using MO calculations as an aid to assignment in photoelectron spectroscopy. Assignment is indeed a problem, compared with other types of spectroscopy. The observation of vibrational structure is often helpful. A sample molecule, as well as undergoing ionisation, may acquire one or more quanta of vibrational energy. We may then observe in the photoelectron spectrum a progression of equally-spaced peaks, the spacing being equal to the energy associated with a vibrational mode of the cation in the appropriate electronic state. The vibrational mode will be one whose normal coordinate undergoes appreciable change upon ionisation. In Fig. 2.4(a) is shown the photoelectron spectrum of N_2. The lowest-energy ionisation at 15.5 eV, and the third peak at 18.8 eV, show little or no vibrational structure. The inference is that in each case an electron is being expelled from a nonbonding MO so that there is little difference in electronic structure, and in the equilibrium internuclear

(a)

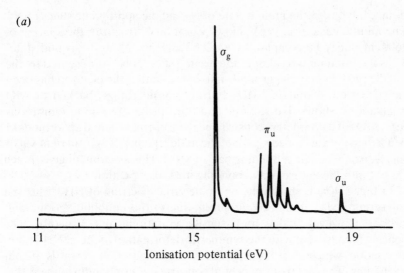

(b)

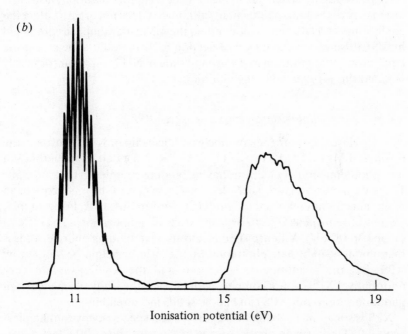

Fig. 2.4. Photoelectron spectra of (a) N_2 and (b) NH_3; reproduced with permission from Ebsworth, E. A. B., Rankin, D. W. R. and Cradock, S. (1987). *Structural Methods in Inorganic Chemistry*. Oxford: Blackwell.

distance, between the ground state of N_2 and the appropriate states of N_2^+. The middle peak at c. $17\,eV$ shows vibrational structure; the spacing of $1600\,cm^{-1}$ may be compared with $2087\,cm^{-1}$ for N_2 in its ground state. Thus the bonding is weaker in the excited state of N_2^+ represented by the middle peak than in the ground state of N_2, so that the electron has been expelled from a bonding MO. Further consideration, backed by MO calculations, shows that the first and third peaks are due to ionisations from MOs of σ symmetry, constructed from nitrogen 2s and $2p_z$ orbitals (z being the internuclear axis), while the middle peak arises from removal of an electron from the π bonding pair of MOs. This assignment gives much insight into the nature of the bonding in N_2 (see Section 7.2).

In Fig. 2.4(b) is shown the photoelectron spectrum of NH_3. The first ionisation peak, at c. $11\,eV$, is surely due to the nonbonding lone-pair orbital. However, it has an extensive vibrational fine structure. The spacing is consistent with the symmetric deformation mode of NH_3^+, i.e. the mode which takes the pyramidal configuration towards planar triangular. The fact that the orbital concerned is nonbonding means that there is essentially no change in the N–H bond upon ionisation; however, occupancy of the lone-pair orbital – although it does not greatly affect the equilibrium N–H distance – does affect the molecular shape, as predicted by VSEPR and VB theories (see Section 6.1). It would appear that the equilibrium configuration in the ground state of NH_3^+ is planar (or nearly so), and this is reflected in the vibrational structure.

X-ray photoelectron spectroscopy (XPS)

Here the source is a monochromatic beam of X-rays, usually the K_α lines of Mg ($1254\,eV$) or Al ($1487\,eV$). With the detection methods in use, photoemission of electrons having binding energies in the range 50–$1200\,eV$ can be observed. Such electrons form part of the inner core of an atom, and are not involved in bonding to any great extent. For example, we may observe peaks due to oxygen 1s ($532\,eV$), phosphorus 2p ($135\,eV$) or zinc 3p ($87\,eV$). Valence shell electrons have binding energies of less than about $40\,eV$; their photoionisation can in principle be studied by XPS, but the resolution is very poor and the results are not very informative. XPS cannot compete with UPS for the study of valence shell photoionisation; but XPS can handle solids more readily.

XPS is sometimes labelled ESCA (electron spectroscopy for chemical analysis). This term emphasises some of the versatility of the technique. The core electron binding energies of a particular atom in the accessible range provide a distinctive fingerprint of the element, and the composition of a sample – even if it is a complex mixture of different substances – can often be determined at a glance. For example, a peak corresponding to a binding energy of $270\,eV$ can be unmistakably attributed to ionisation from the 2s orbital of chlorine, even if the origin of the sample offers no hint that chlorine might be present.

The measured binding energy of an electron in a given core orbital of a given atom is not constant, and can vary from one substance to another by several eV. This is often called a 'chemical shift', by analogy with the more familiar NMR shifts. The greater the fractional positive charge on the atom, the greater will be the binding energies of its electrons, whether they be in the valence shell or in the core. Thus the binding energy of the carbon 1s orbital will be greater if the atom is bonded to atoms of high electronegativity (e.g. O, F) than if it is bonded to less electronegative atoms such as H or Si. In complexes of the d block elements, the binding energy will increase as the oxidation state of the central atom increases. For example, it is possible to determine the oxidation state of osmium by measuring the binding energy of its 4f level (50–55 eV). A number of products which were believed (on the basis of elemental analysis) to contain Os(V) have been shown to be mixtures of Os(IV) and Os(VI) compounds by this method.

Over a range of compounds, it is possible – relatively, if not absolutely – to assign fractional charges to atoms. In coordination/organometallic chemistry, these may be used to infer the relative electron-donating or electron-withdrawing characteristics of different ligands.

XPS is more useful than UPS for the purpose of characterisation, although in this role it rarely provides information that cannot be obtained from cheaper and more readily-accessible techniques. It is subject to considerable experimental difficulties and sources of error (e.g. surface effects) which limit its popularity. However, the growing body of XPS data is a most valuable contribution to inorganic chemistry. Compared with UPS, XPS gives poor resolution and vibrational structure is rarely observed. On the other hand, as previously noted, XPS is amenable to the study of solids while UPS is largely restricted to gases.

2.8   Magnetic susceptibility and electron spin resonance (ESR)

These complementary techniques are applied by inorganic chemists to the study of materials containing unpaired electrons. In such materials, we can identify ions (usually of the d or f block elements) as paramagnetic centres. The unpaired electrons may be appreciably delocalised over other atoms, however, and there may be considerable interaction between electrons on neighbouring centres, often to the extent that we have to look at a whole crystal and not at individual ions/molecules in describing the magnetic properties.

Consider a substance where the unpaired electrons are deemed to reside in d orbitals, localised – as in CF theory – upon ions of a transition element, n on each centre. The total spin S of each ion is equal to $n/2$. If a magnetic field is now applied to the sample, the ground state of the ion is split into $2S + 1$ (i.e. $n + 1$) magnetic states, designated by the quantum numbers M_s. Thus, for example, where $n = 2$, we obtain states with M_s values of 1, 0 and -1. For $n = 1$, we have the simpler situation shown below:

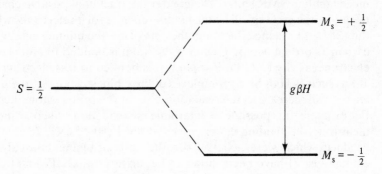

This shows the *Zeeman effect*. The splitting of the two magnetic states is equal to $g\beta H$, where β is a universal constant, H is the field strength and g is the magnetogyric factor – more commonly called the g-factor, or g-value – which should take the value 2.0023 if the ground state has no orbital angular momentum ($L = 0$). For a magnetic field of 1 T, this leads to a splitting of about $1\,\mathrm{cm}^{-1}$, or two orders of magnitude greater than the nuclear spin splitting in a typical NMR experiment. At room temperature, the higher state will have an appreciable thermal population; about 49.8% of the ions will have $M_s = +\frac{1}{2}$, while 50.2% have $M_s = -\frac{1}{2}$, given a Zeeman splitting of $1\,\mathrm{cm}^{-1}$. Magnetic susceptibility measurements on a bulk sample analyse the populations of the magnetic states and the consequences thereof. Electron spin resonance (ESR: also known as electron paramagnetic resonance (EPR)) induces spectroscopic transitions between the magnetic states. The two give different but complementary kinds of information about magnetic states, unpaired electrons and their associated orbitals.

Magnetic susceptibility

This is a classical measurement, dating back to the Faraday era of the mid-nineteenth century. It can be performed on solids or solutions by several techniques, of which the Gouy method is the most familiar. In the example given above of a substance consisting of $S = \frac{1}{2}$ ions in a magnetic field of 1 T, there was a small preponderance of ions in the lower energy magnetic state at room temperature. This means that the sample is slightly more stable when the field is switched on than when it is switched off. The sample is therefore attracted by the applied magnetic field. In the

Gouy method, the force acting on the sample when placed in a magnetic field is measured by weighing the sample in and out of the field. At absolute zero temperature, all the ions should be in the lower state ($M_s = -\frac{1}{2}$); at a very high temperature, the populations of the two states become effectively equal, and there is no stabilisation of the sample in a magnetic field. Another way of putting it is to imagine each unpaired electron as generating a magnetic dipole, which is permitted by quantum theory to take up an alignment either in the same direction as the applied field, or opposed to the field; the former corresponds to the lower state $M_s = -\frac{1}{2}$ and the latter to $M_s = +\frac{1}{2}$. At absolute zero, there is perfect alignment of the dipoles with the field. At an infinitely high temperature, thermal energy causes complete destruction of the alignment and the dipoles are distributed equally and randomly between the two allowed alignments. The magnetic susceptibility depends on the extent of alignment per unit of field strength, and on the magnitude of each magnetic dipole. A material whose magnetic behaviour can be adequately described in terms of a Boltzmann distribution among the Zeeman states, with negligible interaction between unpaired electrons on different atoms, is said to be *paramagnetic*: the term is more loosely applied to all materials which are relatively weakly attracted into an applied magnetic field, including substances where such behaviour is not associated with electron spin (see below). A positive magnetic susceptibility indicates paramagnetism. Substances which have no unpaired electrons are usually *diamagnetic*, and have negative susceptibilities, i.e. they are repelled by a magnetic field. This is a property of electron pairs, and is also present in all paramagnetic materials except atomic hydrogen. In most such cases, the diamagnetic effect is much smaller than the paramagnetism, although it must be corrected for if we wish to measure the paramagnetism quantitatively.

It should be apparent that the magnetic susceptibility χ of a paramagnetic sample (corrected for diamagnetism) should be infinite at absolute zero (with perfect alignment of the magnetic dipoles, or 100% population of the lowest Zeeman state) and zero at infinite temperature (random alignment of dipoles, equal population of Zeeman states). An obvious expression which embodies these conditions is:

$$\chi = C/T$$

where T is the absolute temperature.

This is the Curie law (Pierre, not Marie), C being a constant for a given sample. Better agreement with experimental data is obtained with the modified Curie–Weiss expression:

$$\chi = C/(T + \theta)$$

The constant θ is usually positive but it may be negative; its sign and magnitude are determined by various effects which are not allowed for in the simple treatment leading to the Curie expression.

The magnetic susceptibility is thus a temperature-dependent property of a bulk substance. It is meaningless to specify the susceptibility of a single molecule or complex ion. For convenience, inorganic chemists prefer to answer the question: 'How paramagnetic is this substance?' in terms of the *magnetic moment* μ_{eff} defined by the relation:

$$\mu_{eff} = 2.84(\chi_M T)^{\frac{1}{2}} \text{ B.M.}$$

where χ_M is the susceptibility per mole of substance (corrected for diamagnetism) and B.M. stands for Bohr magnetons, the unit of magnetic moment. Provided that there are no significant interactions between different paramagnetic centres (i.e. atoms or discrete ions on which unpaired electrons reside), this quantity can be interpreted in terms of the electronic structure of the isolated paramagnetic centre. If the Curie law is obeyed – if the constant θ in the Curie–Weiss equation is close to zero – μ_{eff} is independent of temperature, and can be expressed as:

$$\mu_{eff} = g[J(J + 1)] \text{ B.M.}$$

where J is the total angular momentum quantum number of the ground state of the paramagnetic centre. If (as is often the case) the ground state has no orbital angular momentum, J is the same as S, and is therefore equal to $n/2$ where n is the number of unpaired electrons on each paramagnetic centre. In these circumstances, g takes the 'free electron' value of 2.0023; approximating this as 2, we obtain the famous 'spin-only' formula:

$$\mu_{eff} = [n(n + 2)]^{\frac{1}{2}} \text{B.M.}$$

The determination of n from measurement of μ_{eff} is the most familiar application of magnetic susceptibility measurements to inorganic chemists. To the extent that the spin-only formula is valid, it is possible to obtain the oxidation state of the central atom in a complex. Thus an iron complex with a μ_{eff} of 5.9 B.M. certainly contains Fe(III) (high-spin d^5) and not Fe(II). The diamagnetism of AgO rules out its formulation as silver(II) oxide, because Ag^{2+} has an odd number of electrons (d^9) and should be paramagnetic; it contains Ag(I) and Ag(III), in equal amounts. There are, however, a number of pitfalls, especially if reliance is placed on a single measurement at room temperature. The Curie law is rarely obeyed within the limits of experimental error. This means that the measured μ_{eff} is somewhat temperature-dependent. A number of factors can be responsible for deviations from ideal Curie (or even Curie–Weiss) behaviour, and/or from the spin-only formula.

(1) The second-order Zeeman effect leads to the equal stabilisation of all the magnetic states arising from the Zeeman splitting of the ground state. This leads to a contribution to the susceptibility that is independent of temperature, but is proportional to the field H; hence the term TIP (temperature-independent paramagnetism). The contribution made to μ_{eff} will obviously be temperature-dependent. TIP is found in a number of compounds which have no unpaired electrons and hence no first-order Zeeman effect; $KMnO_4$ is a well-known example. Rarely, if ever, is the TIP in such a case large enough to mislead us into attributing the wrong number of unpaired electrons to a paramagnetic centre.

(2) There may be a significant orbital contribution to the magnetic moment; the spin-only formula, as the label implies, does not take this into account. The quantitative treatment of orbital contributions – which effectively alters the g-value from the free-electron value of 2.0023 – was one of the earliest and most important tasks of CF theory. The following generalisations can be made on the basis of the experimental results and the theoretical treatment; for simplicity, we consider only octahedral and tetrahedral fields.

(i) If the octahedral t_{2g} (or tetrahedral t_2) orbitals are unevenly occupied in the ground state – i.e. if they have a total of one, two, four or five electrons – the ground state has an intrinsic orbital angular momentum. This will be coupled to the spin angular momentum arising from the unpaired electron(s) (spin–orbit coupling). The magnetic moment is strongly temperature-dependent, and is sensitive to small distortions (which are usually present) from ideal octahedral or tetrahedral geometry about the central ion. If the temperature is relatively high, and the spin–orbit coupling weak, the magnetic moment is somewhat lower than the spin-only value if the threefold-degenerate level is less than half-filled (e.g. for octahedral Ti^{3+}, V^{3+}); the moment is greater than the spin-only value if the t_{2g} level (t_2 for tetrahedral complexes) is more than half-filled (e.g. octahedral low-spin Fe^{3+}, tetrahedral Ni^{2+}). At very low temperatures, and where the spin–orbit coupling is strong, the magnetic moment should approach zero in some cases; the orbital and spin angular momenta cancel.

(ii) If the t_{2g} (or t_2) level houses zero, three or six electrons in the ground state – i.e. if it is empty, half-filled or filled – the ground state has no intrinsic orbital angular momentum. However, with the exception of the ground state with five unpaired spins arising from the configuration $(t_{2g})^3(e_g)^2$ or $(e)^2(t_2)^3$, some orbital angular momentum will be 'borrowed' from excited states through the agency of spin–orbit coupling. This has the effect of modifying the g-value for the Zeeman splitting. If the d subshell is less than half-filled, g is reduced from its free-electron value of 2.0023; if the subshell is more than half-filled, g is increased. In either case, the modification to g is inversely proportional to the orbital splitting Δ, and proportional to the spin–orbit coupling constant which increases as we go from left to right across each transition series. Thus for octahedral Cr(III) complexes (d^3) we find magnetic moments up to about 0.2 B.M. lower than the spin-only value of 3.87 B.M.; with octahedral Ni(II) systems, moments up to about 0.4 B.M. greater than the spin-only value of 2.84 B.M. may be observed. In the absence of other effects which lead to departures from Curie law behaviour, the magnetic moments of such systems remain independent of temperature.

The orbital contribution may be large enough to invalidate the use of the spin-only formula as a reliable means of determining the number of unpaired electrons per paramagnetic centre; a single measurement at room temperature is of little value for complexes in category (i) above, especially where spin–orbit coupling is strong, as is the case among the 4d and 5d series elements. Thus the IrF_6^- ion, which is a low-spin d^4 system, ought to have two unpaired electrons. But the magnetic moment of $Cs[IrF_6]$ at 300 K is 1.26 B.M., less than that expected for one unpaired electron. In the range 80–300 K, its magnetic moment is proportional to $T^{\frac{1}{2}}$. A detailed analysis could yield valuable information about the ground state and the ligand field, but the spin-only formula is useless. The μ_{eff} for a tetrahedral Ni(II) complex $[NiX_2L_2]$ where X is an anion such as Cl^- or NCS^- and L is a neutral ligand such as PX_3 may be as high as 4.0 B.M., which according to the spin-only formula is consistent with three and not two unpaired electrons. However, we should not be deceived by this since a large and positive orbital contribution is expected for a tetrahedral d^8 system. Magnetic susceptibility measurements are in fact of great value in the characterisation of such Ni(II) complexes; they may be either tetrahedral or square planar, and the latter geometry leads to diamagnetism.

(3) So far we have considered only 'magnetically-dilute' systems, where the unpaired spins on neighbouring paramagnetic centres do not interact with each other. This is usually the case if the centres are insulated from each other by surrounding ligands. If, however, we are dealing with binary compounds such as oxides or fluorides – where the anions are small and the cations fairly close together – or a binuclear complex in which two cations are bridged by suitable ligands and 'see' each other clearly, such interactions may be significant. Examples of the latter case include copper(II) complexes of the type $Cu_2X_2L_{2n}$, where X is a bridging ligand (see Section 8.1) such as Cl^- or OH^-, i.e.

$$L_n Cu \underset{X}{\overset{X}{\diagdown\diagup}} CuL_n$$

In a magnetically-dilute system, the distribution of paramagnetic centres over the available Zeeman states is determined by Boltzmann statistics, and the magnetic state adopted at any instant by one centre is in no way influenced by its neighbours. Deviations from such behaviour may be classified as *ferromagnetic* or *antiferromagnetic*.

In a ferromagnetic material, the interaction between unpaired electrons on different centres tends to encourage the spins on one atom to align themselves with respect to the applied field in conformity with their neighbours. The result is a great enhancement of the magnetic susceptibility, which is now dependent on the field strength. All ferromagnetic materials exhibit essentially normal paramagnetic behaviour above a certain temperature, the Curie temperature T_C, where the ferromagnetic coupling becomes comparable in importance to thermal energy. Above the Curie temperature, the thermal energy largely destroys the alignment due to ferromagnetism. For metallic iron, T_C is 768 °C; many other ferromagnetic substances have T_C values close to absolute zero and behave as normal paramagnets around room temperature.

Antiferromagnetism is perhaps of more interest to inorganic chemists. Here, coupling between unpaired spins on neighbouring atoms leads to opposite alignments with respect to the applied field. The magnetic susceptibility is thus lower than would be expected. As with ferromagnetism, there is a critical temperature – the Néel temperature T_N – above which the substance is paramagnetic because thermal energy is able to disrupt the antiferromagnetic coupling. Antiferromagnetism is associated with overlap between the orbitals which accommodate the relevant electrons. Suppose we have on each paramagnetic centre one unpaired electron, and that the centres are paired-off such that we can envisage overlap between the orbitals housing the unpaired spins for each pair. The result will be the formation of a bonding and an antibonding combination. The ground state of each dimer – for we are now considering the substance in terms of structural units, each of which contains two paramagnetic centres – may be a singlet ($S = 0$) or a triplet ($S = 1$), as depicted below:

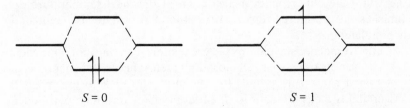

$S = 0$ $\qquad\qquad\qquad\qquad$ $S = 1$

If the energy separation δ between the bonding and the antibonding MOs is large, the ground state will be a singlet with the triplet as an excited state lying much higher in energy. If δ is small, the ground state may be the triplet if the lower repulsion between the electrons outweighs the higher orbital energy associated with occupancy of the antibonding level. Four situations can be envisaged, as shown below:

$$S = 0 \text{———} \quad \text{———} S = 1$$

$$S = 0 \text{———} \qquad\qquad\qquad \text{———} S = 1$$
$$S = 1 \text{———} \qquad\qquad\qquad \text{———} S = 0$$

$$S = 1 \text{———} \quad \text{———} S = 0$$

    (i)   (ii)   (iii)   (iv)

In (i) and (iv), the energy separation between the states is sufficiently small that the upper state is appreciably populated at room temperature. In (ii) and (iii), the splitting is so large that there is no significant population of the higher state at room temperature. Situation (ii) corresponds to strong ferromagnetic interaction, and the substance behaves like an $S = 1$ paramagnet; the bulk substance may not be ferromagnetic unless there is further interaction among neighbouring dimers. In situation (iii), the substance is diamagnetic. Situation (i) corresponds to a weak ferromagnetic interaction; at high temperatures the magnetic susceptibility is consistent with isolated, magnetically-dilute $S = \frac{1}{2}$ centres, as is also the case for (iv). The classic antiferromagnetic interaction is represented by (iv). At high temperatures we observe 'normal' paramagnetic behaviour as if there were no antiferromagnetic coupling; as the temperature is lowered, the susceptibility increases according to the Curie law but reaches a maximum at the Néel temperature and vanishes at absolute zero, where the $S = 1$ state is completely depopulated.

Antiferromagnetism has aroused great interest among inorganic chemists for a variety of reasons. Of most direct relevance to this chapter is the fact that it can create pitfalls for chemists who seek to use magnetic susceptibility measurements for the routine characterisation of products. But there has been an increasing appreciation that antiferromagnetic coupling is a particularly subtle and delicate manifestation of interactions involving electrons – and these, ultimately, are what chemistry is all about. Situation (iii) shown above is an extreme case of antiferromagnetic coupling and represents a typical covalent bond. In other words, the interaction between (say) two hydrogen atoms to form the H_2 molecule could be regarded as a form of strong antiferromagnetic coupling. There is therefore much to be learned about electrons and their mutual interactions from studies of antiferromagnetism.

Electron spin resonance (ESR)

As already noted, the separation between the Zeeman states of an ion having unpaired spin(s) with a typical magnet (1 tesla) is about $1 \, cm^{-1}$, which corresponds to microwave radiation. With an experimental arrangement similar to that used in NMR (but with much higher-frequency radiation sources) we can observe transitions between the magnetic states. The g-value can thus be measured directly; in bulk susceptibility measurements made over a range of temperatures, g is one of a number of parameters which are varied until the best fit to the experimental data is obtained and may be subject to some uncertainty. If the ligand field about each paramagnetic centre has octahedral or tetrahedral symmetry, the g-value (strictly speaking, the g-tensor) is isotropic, i.e. it is independent of the orientation of the paramagnetic molecule/ion with respect to the applied field because the three Cartesian axes which define the coordination geometry are equivalent. If the symmetry is lower, there may be two or three g-values, depending on whether the direction of the applied field is parallel or perpendicular to the molecular plane. The different g-values can all be obtained from ESR measurements on oriented single crystals of known structure. This kind of experiment gives information about the symmetry of the ligand environment about the paramagnetic centre. The anisotropy of g is sometimes amenable to further theoretical analysis, yielding parameters relevant to the bonding of the ligands.

Two types of fine structure can be seen in favourable cases. Hyperfine structure is caused by interaction between the unpaired electron spin(s) and the nuclear spin on the paramagnetic centre. For example, ^{55}Mn (natural abundance 100%) has a nuclear spin of $\frac{5}{2}$; a high-spin complex of Mn(II) has a ground state with $S = \frac{5}{2}$, and the coupling between these gives rise to an ESR signal having 30 lines arising from the allowed transitions among the 36 (i.e. 6×6) states Apart from the fact that this spectrum is an unmistakable fingerprint for high-spin Mn(II), the magnitude of the hyperfine splitting – expressed as a coupling constant – can provide valuable information. For example, the coupling is much smaller for tetrahedral Mn(II) complexes compared with octahedral. Compounds having the composition $(cat)_2MnCl_4$ (where cat is a singly-charged cation) may contain discrete, tetrahedral $MnCl_4^{2-}$ ions, or may have a polymeric structure based on $MnCl_6$ octahedra, depending on the cation. The ESR spectrum of a powdered sample provides a quick and reliable determination of which structure is present. Further information relevant to the electronic structure and bonding at the paramagnetic centre can be extracted from hyperfine coupling constants.

Another type of hyperfine interaction leading to the observation of fine structure in the spectrum is sometimes called superhyperfine coupling. Here we see the effects of interaction between the unpaired spin(s) and the nuclear spins on the ligands. Such observations show that the unpaired electrons are in MOs having at least some ligand character. It is possible – although the theory involves some serious approximations – to estimate the coefficients of the constituent atomic orbitals in the relevant MOs from such ESR data. There are cases where the unpaired electrons belong wholly to the ligands; here, the g-value is close to the free-electron value of 2.0023.

ESR is subject to some serious limitations. For reasons which need not concern us, resonance may be difficult or impossible to observe if the paramagnetic centre has an even number of unpaired electrons. This means, for example, that ESR can make little contribution to the extensive chemistry of nickel(II). Another problem arises from interactions between paramagnetic centres in magnetically non-dilute samples; the resonance may be very broad and uninformative. ESR is best performed on magnetically-dilute samples, which means that we cannot obtain the kind of information furnished by bulk susceptibility measurements about ferromagnetic and antiferromagnetic interactions.

On the credit side, however, ESR can be an extremely sensitive technique. It is widely applied to the study of short-lived free radicals, which can often be trapped in very low concentrations in glasses etc. ESR is extremely useful in the burgeoning field of bio-inorganic chemistry. Many proteins, enzymes etc. contain paramagnetic cations such as Fe^{3+} or Cu^{2+}. ESR can provide useful information about the binding sites of these in the biological material, even where only very small quantities of the sample are available or where there is only one paramagnetic ion in a molecule of molecular weight 10^5–10^6. Thus we may be able to say that a certain protein contains Cu(II) in a distorted tetrahedral environment, at least one of its ligands being an S-donor. If the protein contains cations such as Mg^{2+} or Zn^{2+} which have no ESR spectra, it may be possible to replace these with paramagnetic ions, e.g. Co^{2+}, and hence obtain information about the binding sites; assuming, of course, that the paramagnetic ions adopt the same coordination geometries with the same ligands as the diamagnetic ions.

2.9 Miscellaneous methods

Here we mention briefly some methods whose applicability in the study of inorganic substances is more limited than those already covered; this may be because they can be applied only to a very limited number of samples, or because the information – valuable though it may be – is less varied.

Mass spectrometry

Many would disagree with the consignment of mass spectrometry to this section and would argue that it deserves a section to itself. Until fairly recently, this technique was restricted to volatile, molecular substances and was therefore obviously of much less interest to inorganic chemists than to their organic colleagues.

Mass spectrometry requires a means of generating from the sample a beam of cations, which can be analysed according to their masses by deflection through an electromagnetic field. The various ways of producing the ions need not concern us. The resulting mass spectrum gives us the masses of the molecular ion – the cation X^+ where the sample consists of molecules X – and of other species. The latter are mostly fragmentation products arising from the rupture of bonds in the X molecule. The mass of the molecular ion (not always observed) gives us, of course, the molecular weight of a molecular substance, obviously of value in its characterisation. This can be obtained with great precision, better than 0.001 a.m.u. For example, CO and N_2 have the same mass number (28) but their molecular masses are significantly different: 27.9949 for $^{12}C^{16}O$ and 28.0061 for $^{14}N^{14}N$. Starting from a carbonyl complex $ML_m(CO)_n$, where M is one of the d block elements, L is a ligand and m, n are integers, an inorganic chemist might attempt the preparation of a complex in which one or more CO ligands are replaced by N_2. Mass spectrometry would be useful in characterising the products of such attempts.

Identification of the several fragments from their masses is also useful in establishing the constitution of the sample. For example, a gaseous substance having the empirical formula $Si_2H_3F_3$ might be one of two isomers, $H_3Si—SiF_3$ or $H_2FSi—SiHF_2$. The appearance in the mass spectrum of mass numbers 85 (SiF_3^+) and 31 (SiH_3^+) would indicate that the former is present. Sometimes, however, rearrangements occur leading to the observation of fragments which are not present in the neutral molecule; sometimes, too, there may be additional peaks due to the products of ion–molecule reactions, so that the mass spectrum of X—Y might reveal mass numbers corresponding to X_2Y^+. Mass spectrometry has been very useful in the important field of carbonyl complexes. A host of substances contain molecules which can be represented as $M_xL_y(CO)_z$, where M is a d block atom, L is a ligand other than CO and x, y and z are integers. The CO groups are lost, one at a time, in the mass spectrometer, so that we usually observe a series of peaks separated by 28 mass units corresponding to $M_xL_y(CO)_n^+$ ($n = 0$ to z) and perhaps $M_xL_{y-1}(CO)_n^+$, etc. This is the most convenient way of counting the number of CO ligands in the complex molecule.

Extended X-ray absorption fine structure (EXAFS)

It was noted in Section 2.7 that X-rays of appropriate wavelength can cause the ejection of core electrons from a target sample, and measurement of the kinetic energies of the photoelectrons produced by bombardment with monochromatic X-radiation is the technique of X-ray photoelectron spectroscopy (XPS). In EXAFS, we have a source of continuous X-radiation and we measure an absorption spectrum in the region where photoionisation of a particular atom occurs. For example, the X-ray spectrum of an iron-containing substance shows a peak at about 7400 eV, corresponding to the ejection of an iron 1s electron. The low-energy side of the absorption peak is very abrupt and steep – an absorption edge – but on the high-energy side the absorption falls off slowly with increasing energy and shows a characteristic 'squiggle' just beyond the maximum. This 'squiggle' arises from interference of the photoelectron wave by the atoms surrounding the atom from which the electron has been ejected. Analysis of the 'squiggle' leads to a radial distribution function rather similar to that obtained from electron diffraction; each peak corresponds to a vector between the central atom (the subject of the X-ray absorption experiment) and a neighbouring atom. Hence we obtain information about the chemical environment of the target atom.

EXAFS is available in only a score or so of laboratories around the world. A synchrotron is the only suitable source of continuous X-radiation; this is a very large and expensive facility, requiring the great resources of a particle physics research establishment. The other drawback is that it gives bond lengths, but not interbond angles. Thus it may tell us that the Fe atom in a complex is bonded to two S atoms and two N atoms, but other methods are needed to tell us the geometrical arrangement, which may be tetrahedral, square planar or something in between. On the other hand, it can be applied in situations where other structural techniques are inappropriate.

For example, samples of animal tissue can be investigated without the need to isolate any specific substance. This has been used to study the bonding of platinum atoms in animal organs, following the administration of platinum complexes as anti-tumour agents.

Nuclear quadrupole resonance (NQR)

This is a difficult technique, unfamiliar to most inorganic chemists and of limited applicability. However, in the hands of a small band of enthusiasts, it has produced results of real value.

We noted in Section 2.4 that a nucleus whose spin is greater than $\frac{1}{2}$ has a quadrupole moment. If the chemical environment of the nucleus has less than cubic symmetry (i.e. if the arrangement of atoms around the nucleus under scrutiny is not cubic, octahedral or tetrahedral, and the three Cartesian axes needed to describe the environment are not all equivalent) there is a non-zero electric field gradient at the nucleus. This interacts with the nuclear spin to give a number of nuclear energy levels, separated by energies equivalent to radiofrequency radiation, in the range 10^2–10^6 kHz. The most commonly-studied nuclei have a spin of $\frac{3}{2}$, and this gives rise to two energy levels. One transition can therefore be observed with a suitable radiation source.

NQR is restricted to solids, and fairly large samples are required (of the order of several grams). The spectrum should immediately tell us the number of chemically-distinct sites occupied by the nuclei under examination. For example, the ^{35}Cl spectrum of $[Co(NH_3)_6][CuCl_5]$ shows two peaks, whose intensities are in the ratio $3:2$, consistent with a trigonal bipyramidal $CuCl_5^{3-}$ ion. Solids often undergo phase transformations on heating or cooling from room temperature, and NQR has proved very useful in studying these.

The NQR resonance frequency can be correlated with the nature of the bonding of the nucleus in question. For example, a genuine Cl^- ion has spherical symmetry and should not exhibit NQR. In the Cl_2 molecule, the ^{35}Cl quadrupole coupling constant is 109.7 MHz. If these are taken to represent the extreme cases of 100% ionic and 100% covalent character in the bonding of a Cl atom, we should be able to estimate the percentage of each in molecules or polyatomic ions containing chlorine from the ^{35}Cl NQR spectrum. However, the quantitative theories most often employed involve some rather questionable assumptions and statements like: 'the bonding in $PtCl_6^{2-}$ is 44% ionic', should not be taken too literally.

Mössbauer spectroscopy

The status of Mössbauer spectroscopy is akin to that of NQR. It is restricted to solids, and it yields information about the electronic and chemical environment about certain nuclei. Where diffraction methods are inappropriate, Mössbauer may be the best technique for structure determination.

Nuclei can exist in excited energy states which are accessible by absorption of gamma radiation of appropriate wavelength, and the precise energy required to excite such a transition is dependent on the chemical environment of the nucleus. A nucleus may be formed as a product of radioactive decay in such an excited state, and will drop to the ground state with emission of a gamma photon. Thus ^{57}Co decays to ^{57}Fe (half-life 270 days); the latter isotope is formed in an excited state, and some of it may drop to the ground state via another excited state which lies 14.41 keV above the ground state and has a relatively long lifetime of 10^{-7} s. A sharp resonance absorption of the emitted gamma photon (14.41 keV) by a strategically-placed sample containing ^{57}Fe will occur if the chemical environments of the iron in source and sample are identical. In practice, the sample is oscillated back and forth relative to the source until the Doppler shift associated with its velocity makes its resonance frequency the same as that of the photons emanating from the source. The maximum absorption as a function of the source velocity gives the *isomer shift*, usually expressed in mm s^{-1} relative to some standard (e.g. stainless steel in the case of ^{57}Fe). If the nucleus has a quadrupole moment and a non-zero electric field gradient (as discussed above for NQR), the Mössbauer lines will be split, and we obtain a further parameter, the *quadrupole coupling constant*.

Relatively few nuclei are amenable to Mössbauer spectroscopy. Apart from ^{57}Fe described above, the most important ones are ^{119}Sn, ^{121}Sb, ^{129}Te, ^{129}I, ^{129}Xe and ^{197}Au. Detailed theoretical treatments of the isomer shift and quadrupole coupling constant have been expounded, and the experimental data can be interpreted quantitatively in terms of electron density at the nucleus etc. More often, inorganic chemists make use of empirical correlations to obtain structural information. If for a given nucleus we plot the isomer shift against the quadrupole coupling constant for a number of compounds of known structure, we can identify on the resulting two-dimensional map regions characteristic of definite structural types, e.g. high-spin octahedral Fe(II), or linear, two-coordinate Au(I). Mössbauer spectra thus provide a very useful means of characterising some classes of inorganic compounds in situations where other methods fail. For example, many iron complexes are prepared as insoluble powders, useless for X-ray crystallography. The biological chemistry of gold has aroused much interest recently, and many compounds containing Au(I) or Au(III) bound to complex proteins etc. have been prepared. ^{197}Au Mössbauer studies are important in determining the nature of the binding sites.

2.10   Further reading

An excellent, up-to-date account of nearly all the methods discussed in this chapter is given by Ebsworth, E. A. B., Rankin, D. W. H. and Cradock, S. (1987). *Structural Methods in Inorganic*

Chemistry. Oxford: Blackwell. This book does not cover bulk magnetic measurements. Good elementary accounts of these are found in most inorganic texts. For modern, advanced treatments see Carlin, R. (1986). *Magnetochemistry*. Berlin: Springer; and Gerloch, M. (1983). *Magnetism and Ligand Field Analysis*. Cambridge University Press.

3

Nomenclature, notation and classification of inorganic substances

3.1   Elements, atoms, molecules and substances

'Like all other scientific concepts, that of an element has changed its meaning many times and in many ways during the development of science.' Thus wrote Wilhelm Ostwald (the father of physical chemistry and a positivist philosopher) in the 1911 edition of the *Encyclopaedia Britannica*. This was a time of dramatic developments in physics and chemistry; within a few years, even the most entrenched positivists were beginning to believe in the real existence of atoms and subatomic particles.

Unfortunately, both classical and modern meanings of the term 'element' are still found in the literature, without proper discrimination. Most contemporary dictionaries define an element as a *substance* – a material having measurable bulk properties such as density, melting/boiling points, electrical conductivity, specific heat etc. – which cannot be further divided by chemical means. This stems from the classification by Lavoisier and Dalton (*c*. 1800) of substances into elements and compounds. But in almost any modern text, you will find usage of the term 'element' which may seem confusing and contradictory. The word has retained its classical meaning to imply a substance, but has assumed a much broader significance.

You may, for example, see it written that 'aluminium is one of the most abundant elements in the Earth's crust'. This, of course, does not mean that macroscopic particles of the light, silvery metallic substance from which jumbo jets and saucepans are largely fabricated are to be found in nature. 'Element' has become a collective term, and encompasses all the atoms having a particular atomic number, regardless of their state of chemical combination. We must therefore be careful to refer to an *elemental substance* if that is what we mean. Thus when we say that 'lead occurs in sulphide minerals', we are referring to an element; the statement that 'lead reacts only slowly with dilute hydrochloric acid' obviously refers to the elemental substance. One of our difficulties is that nomenclature and notation have not kept pace with developments; the word 'lead' may imply a substance, an element or a single atom, depending on the context.

In most texts you will find a tabulation headed 'Ionisation potentials (or energies) of the elements'. This turns out to be a collection of data for isolated atoms (if we are talking about ionisation potentials) or for atomic substances (in the case of ionisation energies – see also Section 4.3). Likewise you may find tables of 'Electronic configurations of the elements', which strictly apply only to isolated atoms. Perhaps more defensible is the heading 'Electronegativities of the elements'. Electronegativity (see Section 4.4) is defined as a property of an *atom* in a *molecule*, but the molecule is not specified, and if the electronegativity of an atom is deemed to be independent of its state of chemical combination (this is open to dispute; see Section 4.4) the use of the term 'element' may be justified. It is perfectly correct to say that 'the alkaline earth elements always exhibit the II oxidation state in their compounds', it being understood that the oxidation number of an atom in an *elemental substance* is always zero.

The distinction between elements and compounds as stressed by Lavoisier and Dalton has become less important. For the purpose of classification, it is useful to place elemental substances in a separate category inasmuch as all the atoms therein have the same atomic number. However, apart from the noble gases, all elemental substances at room temperature contain chemical bonds; the principles which govern their structures and properties are no different from those which apply to compounds. Unfortunately, some terms which strictly apply to substances have found their way into the chemical vocabulary in other contexts. For example, the term 'metal' refers to a class of substances (by no means all elemental) and implies certain characteristic properties which arise from the nature of the bonding among the atoms which constitute the substance. It therefore seems absurd to speak of 'the metal atom' in a coordination compound, or of 'metal–metal bonds' in, for example $(CO)_5Mn—Mn(CO)_5$. Such terminology is derived from the classical (Lavoisier/Dalton) definition of 'element'. Thus manganese is a metallic element, i.e. the elemental substance is a metal, one of the 'transition metals', and chemists find it difficult to eschew the use of the word 'metal' out of its proper context. It is far too late to discourage such usage, but it will be avoided as far as possible in this book.

As noted in Section 1.3, the terms 'molecule' and 'substance' tend to be used interchangeably by organic chemists, who are usually dealing with substances which consist of discrete molecules. There is a tendency among students to do likewise in inorganic chemistry. This harks back to the late nineteenth century, when inorganic chemists were accustomed to writing down molecular structures, satisfying the usual valency numbers, for inorganic substances generally. Care must be taken to distinguish between microscopic entities and bulk substances.

3.2   Names and formulae

Systematic names and stoichiometric formulae are intended to convey as much information as possible in the smallest space concerning the composition and nature of a substance. The internationally-agreed (though not universally-obeyed) rules for the naming of inorganic compounds are set out briefly in most inorganic texts, and in *The Handbook of Chemistry and Physics*; they need not be repeated here.

The student may well feel that the need to name inorganic compounds is a relic of the past, when stoichiometric formulae were often uncertain, and that a formula will convey a more concise specification of a substance. This attitude is not without merit, especially where we are dealing with rather complex substances whose systematic names are hopelessly clumsy and lengthy. For example, the important substance rendered as $IrCl(CO)(PPh_3)_2$ on the printed page has the systematic name *trans*-chlorocarbonylbis(triphenylphosphine)iridium(I). Although hundreds of papers have been published on the reactions of this compound and tens of thousands of students have heard of it, the systematic name is rarely used. In speech, and often in print, it is given the trivial name 'Vaska's compound'. In any research laboratory, you may hear talk of 'Judy's compound', or 'Ralph's anion'.

Yet systematic names, however clumsy, have their uses. In the case of Vaska's compound, the name informs the initiated that it is a molecular substance having the square planar structure:

$$Cl - Ir \begin{array}{c} P(C_6H_5)_3 \\ | \\ \\ | \\ P(C_6H_5)_3 \end{array} - CO$$

The prefix *trans* tells us that the triphenylphosphine groups occupy opposite corners of a square rather than adjacent corners (*cis*). The use of 'chloro' in the name (instead of 'chloride') implies the absence of discrete chloride ions, as in $[Ir(CO)(PPh_3)_2]^+Cl^-$. Names, as opposed to formulae, are easier to index or to store in a data bank for the rapid retrieval of information on compounds having specific structural features. If a chemist wished to collect the available thermodynamic data for inorganic nitrites, a computer search of formulae containing NO_2 or some multiple thereof would produce a vast number of organic nitro-compounds of no interest. More productive would be a search for articles whose titles contained the word 'nitrite'.

A further advantage of names as opposed to formulae is that the latter may refer to either a molecule or a substance (which may or may not consist of discrete molecules), while a name can specifically imply a substance. For example, the formula $BeCl_2$ could apply to the gas-phase triatomic molecule or to the crystalline solid (which does not contain discrete molecules – see Section 3.3). However, if we write beryllium(II) chloride, a substance (which may be a solid, liquid or gas, depending upon temperature and pressure) is implied. The distinction between molecules and substances is important when we are discussing the instability or 'nonexistence' of a substance. Molecules such as CuF or AlO can be observed in gaseous mixtures at high temperatures, or trapped in inert matrices at low temperatures, but no pure substances having these stoichiometries can be obtained. We may say that AlO exists, in the sense that its molecular properties – bond length, bond energy, vibrational spectrum etc. – have been measured, but aluminium(II) oxide does not exist. It would be too dogmatic to insist upon names rather than formulae where we are dealing with substances as opposed to molecules; but where a formula is written, it is important that the context should make clear the distinction.

Where a substance is known to consist of molecules which are oligomeric – i.e. the molecules can be written as $(X)_n$ where n is a small integer, as in B_2Cl_4, Nb_4F_{20} etc. – some inconsistencies are to be found in the literature. Thus the preferable notation B_2Cl_4 and name diboron tetrachloride are generally used, but reference will be found to NbF_5 to denote niobium(V) fluoride. The origins of such inconsistencies are to be found in the habit of writing down molecular structures consistent with classical valency numbers. In the absence of structural information, a substance whose empirical formula was found to be BCl_2 would naturally be written as $Cl_2B—BCl_2$. However, a substance having the composition NbF_5 requires no oligomeric formulation to be consistent with the usual valency numbers of Nb and F. Formulae which first attained general currency in the literature before the structures of the substances were determined have a tendency to persist. In older books, and even in some quite recent reference works, you may see copper(I) chloride rendered as Cu_2Cl_2. This formulation arose from an early study of copper(I) chloride in the vapour phase; the vapour density indicated a molecular weight consistent with a dimeric molecule, and it was inferred that such molecules were also present in the solid. We now know that crystalline copper(I) chloride does not contain discrete molecules; it has a polymeric structure (see Section 3.3). At the boiling point, the vapour is a mixture of roughly equal amounts of monomeric CuCl and cyclic trimeric Cu_3Cl_3 molecules. For similar reasons, you may still encounter such quaint formulations as V_2O_2, V_2O_4 and Mg_2Cl_4; these can be observed as molecules in the vapour phase, but their use with reference to the crystalline substances vanadium(II) oxide, vanadium(IV) oxide and magnesium(II) chloride is indefensible.

Thus substances which are known to contain $(X)_n$ molecules ($n = 2, 3, 4, \ldots$) and where X offends the simple rules of valency should be named and formulated accordingly, e.g.:

B_2Cl_4 diboron tetrachloride (not BCl_2 boron(II) chloride)
O_2F_2 dioxygen difluoride (not OF oxygen monofluoride)
S_4N_4 tetrasulphur tetranitride (not SN)

But where the substance contains $(X)_n$ molecules under ordinary conditions and X is plausible as a stable monomeric species, names and formulae such as:

$GaCl_3$ gallium trichloride or gallium(III) chloride
ICl_3 iodine trichloride or iodine(III) chloride
SO_3 sulphur trioxide or sulphur(VI) oxide

are acceptable for the substances which actually contain Ga_2Cl_6, I_2Cl_6 and (in one crystalline form) S_3O_9 molecules.

Compounds in which an element exhibits more than one oxidation state can cause problems. Such 'mixed-valence' compounds should be given names which clearly convey their nature, and formulae which avoid any confusion. Thus the compound whose empirical formula is $GaCl_2$ actually consists of an array of Ga^+ and $GaCl_4^-$ ions, and should be named gallium(I) tetrachlorogallate(III). It is best represented by the formula $Ga[GaCl_4]$, rather than $GaCl_2$ or Ga_2Cl_4; the latter would imply a dimeric structure $Cl_2Ga\!-\!GaCl_2$, while the former implies a gallium(II) compound of unspecified structure. This is a fairly straightforward case since there is a discrete complex ion in the structure. Other mixed-valence compounds are more awkward. For example, SbO_2 is not antimony(IV) oxide; it is a complex substance containing Sb(III) and Sb(V) in equal proportions. The common formulation Sb_2O_4 should be taken to imply $Sb(III)Sb(V)O_4$, rather than a molecular dimeric structure.

3.3   Classification of inorganic substances

So far we have encountered substances which consist of discrete molecules, of discrete ions, and polymeric substances. It is now time to set out a systematic classification scheme.

Most inorganic substances can be classified as belonging to one (sometimes more than one) of the following categories:

(a)   Atomic substances.

(b)   Ionic substances.
(c)   Metallic substances.
(d)   Molecular substances.
(e)   Polymeric covalent substances.

Atomic substances

The noble gases are the only substances which consist of free atoms under ordinary conditions (i.e. at room temperature and atmospheric pressure). They remain as such in the liquid and solid phases, whose structures are maintained by weak London (or dispersive) forces of attraction. These can be described as 'induced dipole–induced dipole' interactions. An isolated atom has spherical symmetry with respect to its electronic charge distribution. When two atoms – whether or not any bond is formed – come into close proximity, the outer electrons on one atom induce a dipole on the other, by deformation of the spherical electron cloud. These atomic dipoles oscillate back and forth in a correlated manner such that there is a time-averaged net attraction between neighbouring atomic dipoles. The London attraction in the solids ranges from $0.1 \, kJ \, mol^{-1}$ for helium (which can be solidified only under pressure) to $15 \, kJ \, mol^{-1}$ for xenon. The boiling points (ranging from 4 K for He to 164 K for Xe) are much lower than for molecular substances of comparable mass; molecules are much more polarisable than atoms, other things being equal, and this enhances the London attraction. The London attraction increases with increasing atomic number and increasing atomic size.

All elemental substances are atomic at a sufficiently high temperature; the high entropy of an atomic gas will ultimately lead to a $T\Delta S$ term for atomisation which outweighs the unfavourable enthalpy term. Elemental mercury boils under atmospheric pressure at 357 °C to give a monoatomic gas. Most other elemental substances are molecular in the vapour phase at the boiling point (e.g. Na_2, P_4, S_8) but these molecules break down to smaller fragments and eventually to atoms on heating.

Ionic substances

Ionic substances are invariably solids at room temperature and atmospheric pressure. They have three-dimensional structures in which no finite molecules can be discerned. For example, in crystalline NaCl each Na has as its nearest neighbours six Cl atoms, arranged octahedrally; each Cl is surrounded by six Na atoms, similarly arranged. Since a given Na atom is not uniquely associated with one particular Cl atom, we cannot identify NaCl molecules in the lattice.

The properties of such substances are consistent with (but do not definitely prove) their formulation as arrays of cations and anions. The structures adopted can be rationalised in terms of the best compromise between the need to maximise the attraction between ions of opposite charge while minimising the repulsions between ions of like charge, in order to optimise the net attraction. The structures of ionic solids cannot be easily described in terms of localised electron-pair bonds (although this applies to many covalent substances as well).

Ionic solids may contain simple ions (e.g. Na^+, Ca^{2+}, O^{2-}, F^-) or complex ions (e.g. PCl_4^+, $Pt(NH_3)_4^{2+}$, SO_4^{2-}, $AlCl_4^-$). The bonds within the complex ions are essentially covalent; but the structure of an ionic substance containing complex ions is primarily determined by the packing of cations and anions. For example, NH_4PF_6 consists of NH_4^+ and PF_6^- ions arranged as in the NaCl structure. The most common structures for ionic solids are summarised in Table 3.1. Diagrams depicting these structures are to be found in many books, and their reproduction here seems superfluous (see the references at the end of this chapter, and in the Appendix). The structure adopted by an ionic solid of given stoichiometry can be rationalised to some extent by the relative sizes of the cation and anion. A large cation can accommodate a large number of anions around it without the latter 'touching'. (Since cations are nearly always smaller than the associated anions, the packing of anions around cations is more restrictive than the reverse.) Ionic radii are discussed in more detail in Section 4.2. Attempts have been made to establish

Table 3.1 *Principal three-dimensional structures adopted by ionic substances AB_n*

	Description	Coordination
AB	Zinc blende ⎫	4:4
	Wurtzite ⎬ ZnS	4:4
	Sodium chloride NaCl	6:6
	Nickel arsenide NiAs	6:6
	Caesium chloride CsCl	8:8
AB_2	Quartz (and variants) SiO_2	4:2
	Rutile TiO_2	6:3
	Fluorite CaF_2	8:4
AB_3	Aluminium fluoride AlF_3	6:2
AB_4	Zirconium fluoride ZrF_4	8:2

Note: More complex three-dimensional 'ionic' structures are known where the A and/or B atoms are not all in equivalent sites having the same coordination numbers/geometries.

quantitative 'radius ratio rules' for ionic structures. But because (as we shall see in Section 4.2) *absolute* ionic radii are subject to some uncertainties, and because these radius ratio rules are subject to many violations, they are of dubious value.

The ordered structure of an ionic solid obviously leads to a low entropy, and the disarray of the liquid and gaseous phases should lead to highly favourable entropy changes for melting and vaporisation. However, these processes must overcome the large number of strong electrostatic attractions between ions of opposite charge which are lost upon the breakdown of the ordered crystal lattice. Thus ionic solids have characteristically high melting/boiling points which correlate (subject to a number of anomalies) with the lattice energies (see Section 5.1). In the molten state, typical ionic substances exhibit the high electrical conductivities which would be expected of liquids consisting of mobile charged particles. However, many ionic solids sublime or decompose without melting.

Ionic solids tend to dissolve (to varying degrees) in polar solvents such as water. Such solutions contain the separated ions, surrounded by solvent molecules. Ion–dipole attraction (in some cases with appreciable covalency as well) provides the energic compensation for the loss of electrostatic attraction which must accompany the dissolution of an ionic solid. Consider, as an example, the following thermochemical data:

	$\Delta H^0 \,(\text{kJ mol}^{-1})$
$NaCl(s) \rightarrow Na^+(g) + Cl^-(g)$	787
$Na^+(g) + Cl^-(g) \rightarrow Na^+(aq) + Cl^-(aq)$	-784
$NaCl(s) \rightarrow Na^+(aq) + Cl^-(aq)$	$+3$

The energy required to decompose crystalline NaCl into the gaseous ions is fortuitously balanced by the attraction between ions and solvent molecules. The enthalpy of solution of NaCl is slightly endothermic, but there is a small positive entropy change which renders the solid soluble to the extent of $6.14 \,\text{mol}\,\text{l}^{-1}$ in water at $25\,^\circ\text{C}$. It should be apparent that the balance between the exothermic and endothermic terms is delicate, and entropy considerations are important. A useful generalisation is as follows. An ionic solid consisting of large cations and large anions, or of small cations and small anions, tends to have a relatively low solubility in water. Salts where the cation is small and the anion large (or vice versa) tend to be freely soluble. Thus the solubilities of the Group 2 sulphates decrease as we go down to the group from Be to Ba, while the solubilities of Group 1 fluorides increase from LiF to CsF. Where the anion is of intermediate size – Cl^-, NO_3^- and CO_3^{2-} come into this category – there is little correlation between solubility and cation size. Thus the great majority of chlorides and nitrates are freely soluble in water, while carbonates – apart from those of Group 1 – are generally insoluble.

One might expect the dissolution of an ionic solid in water to be accompanied by a highly favourable entropy change, since the highly-ordered three-dimensional lattice is destroyed and the ions are free to move in the solvent. However, this effect is opposed by the ordering of the solvent molecules which are attracted around the ions. The higher the charges on the ions, and the smaller their size, the more important is this term. Consider the thermodynamic data for the following dissolution processes (all in kJ mol^{-1} at 25 °C):

	ΔH^0	ΔG^0
$KClO_4(s) \rightarrow K^+(aq) + ClO_4^-(aq)$	+51	+11
$CaSO_4(s) \rightarrow Ca^{2+}(aq) + SO_4^{2-}(aq)$	−18	+24
$KNO_3(s) \rightarrow K^+(aq) + NO_3^-(aq)$	−5	0
$CaCO_3(s) \rightarrow Ca^{2+}(aq) + CO_3^{2-}(aq)$	−13	+47

Comparing the 'isoelectronic' species $KClO_4$ and $CaSO_4$, we see that entropy assists the dissolution of $KClO_4$ but hinders that of $CaSO_4$. With the large, singly-charged cations K^+ and ClO_4^-, the solvent-ordering effect is relatively small and the entropy change is positive overall. But with the doubly-charged ions Ca^{2+} and SO_4^{2-}, the solvent-ordering term dominates ΔS^0. The nitrate and carbonate ions are rather smaller than perchlorate and sulphate, so that solvent-ordering is stronger and the ΔS^0 term less favourable to solubility.

Ionic solids tend to be less soluble in polar organic solvents (such as ethyl alcohol, acetone) than in water. However, LiCl, LiBr and LiI are very soluble in a wide range of such solvents. This is sometimes cited as evidence of considerable covalent character in their bonding. A better explanation is to be found in the fact that the small Li^+ ion forms quite strong complexes with molecules containing O and other donor atoms. Complex formation will generally favour solubility; even such an insoluble salt as $BaSO_4$ will dissolve in strongly-complexing organic media.

The properties of ionic solids are consistent with their description as such, but do not offer conclusive proof that the bonding is purely ionic. For example, the fact that molten sodium chloride is a good conductor of electricity does not prove that ions are present in the solid state. A fairly simple electrostatic model can account almost quantitatively for the thermochemistry of many ionic solids; this seems impressive at first glance, but could well be fortuitous. A similar electrostatic calculation, based on the supposition of ionic bonding, gives equally good results for a number of molecular species, where the bonding is surely covalent. The most direct evidence for the nature of the bonding in crystalline compounds such as NaCl comes from electron density maps, which can be obtained from accurate X-ray diffraction measurements. These reveal, even in the most typically ionic solids, appreciable electron density in the internuclear region, suggesting overlap between the charge clouds of the formal cation and anion (see also the discussion of ionic radii in Section 4.2). The term 'ionic solid' should therefore not be taken too literally; it is properly applied to a class of substances having certain characteristic properties and structures which can be conveniently described in terms of ionic bonding but there is little doubt that the 'pure' ionic model is an oversimplification.

The thermochemistry of ionic solids is discussed in detail in Chapter 5.

Metallic substances

The most familiar metals are elemental substances such as iron, tin, aluminium etc. However, many compounds are metallic. As well as 'intermetallic' compounds such as AgCd and NaTl, and a huge number of non-stoichiometric alloys, many oxides, sulphides, halides etc. have metallic properties. For details of structure and bonding in metallic substances, see Section 7.5.

Metallic substances vary in hardness from, e.g., sodium and gallium which can easily be cut with a knife, to very hard materials such as tungsten, which finds applications in making drills etc. Their melting and boiling points are equally variable, and correlate fairly well with their hardness. Thus caesium and gallium melt just above room temperature, while tungsten melts at 3380 °C. The most characteristic properties of metals are:

(*a*) Malleability and ductility.
(*b*) Opaqueness and metallic lustre (high reflectivity).
(*c*) High thermal conductivity.
(*d*) High electrical conductivity, which decreases with increasing temperature.

The last of these is perhaps the best criterion for metallic characteristics. Most other types of inorganic substance are effectively insulators, or semiconductors whose conductivity increases with temperature. Semiconductors constitute a borderline area; some lie between metals and three-dimensional polymers (e.g. GaAs, isoelectronic and isostructural with elemental germanium), while others have structures which could be described as ionic (e.g. PbS, which has the NaCl structure).

Typical metallic substances have structures which cannot readily be described either in terms of directional covalent bonds or as arrays of cations and anions. Metallic elemental substances exhibit high coordination numbers, and can – to a first approximation – be viewed as arrays of cations embedded in a 'sea' or 'glue' of electrons completely delocalised over the crystal. This model helps to explain the characteristic mechanical, thermal and electrical properties of metals. It is also consistent with the preference for metallic structures of elemental substances whose atoms have low ionisation potentials. Intermetallic compounds and alloys have structures similar to those of metallic elemental substances, with varying degrees of disorder in the distribution of atoms over the lattice sites. In metallic binary compounds M_aM_b (M being of low electronegativity, while X is O, S, Cl etc.), the structures often exhibit short M–M distances, not greatly different from those in the elemental substance M. In some cases, the structure can be identified as essentially a metallic structure into which foreign atoms have been accommodated, with varying degrees of distortion.

Molecular substances

Molecular substances contain discrete, finite molecules under ordinary conditions. They may be solids, liquids or gases at room temperature and atmospheric pressure, depending upon the magnitudes of the Van der Waals interactions between molecules. These can be divided into:

(a) London (dispersive) attraction, i.e. induced dipole–induced dipole interaction as discussed above for atomic substances. These are rather weak, but increase rapidly with the atomic numbers of the atoms involved.

(b) Dipole–dipole attraction between polar molecules. If the centres of gravity of positive and negative charge in a molecule do not coincide, the molecule can be described as a dipole, having a positive end and a negative end. In a crystal, such molecules will tend to pack in such a way that the electrostatic attraction between these dipoles is maximised. These interactions are often much stronger than the London attraction, so that substances which consist of polar molecules in the solid state tend to have higher melting and boiling points than would be expected if London forces were wholly responsible for the Van der Waals attraction. Note, however, that intermolecular attraction can be strong in a molecular substance where individual bonds have a high polarity, even if the molecule as a whole has no dipole moment. Thus linear CO_2 has no dipole moment, while bent O_3 has a small dipole moment. But ozone is rather more volatile than carbon dioxide; evidently the relatively high polarity of the C=O bonds is responsible.

The hydrogen bond can be regarded as a particularly strong dipole–dipole interaction. The situation depicted as X—H....Y commonly occurs where X is an atom or group of higher electronegativity than H (usually N, O, F, Cl) and where Y has a lone pair and is therefore a source of high electron density. This suggests an electrostatic description of the H....Y bond, i.e.:

$$\delta-\quad\delta+\quad\delta-$$
$$X—H....Y$$

However, there is a growing body of evidence – both theoretical and experimental – for significant covalency in the H....Y bond, especially in the strongest and shortest hydrogen bonds. That a hydrogen atom can form bonds to two other atoms has become apparent from exhaustive studies of multicentre bonding in 'electron-deficient' hydrides (Section 7.4). The strengths of hydrogen bonds are mostly in the range $10–60\,kJ\,mol^{-1}$.

Hydrogen bonding between molecules in the solid and liquid phases naturally tends to enhance the melting/boiling points of molecular substances, and plays an important part in determining the three-dimensional structures of molecules and crystals.

Molecular substances tend to dissolve in non polar solvents; attraction between the molecules and the solvent can easily balance the small positive ΔH^0 for the breakup of the lattice, and entropy will favour dissolution. Molecular substances that consist of polar molecules, or of molecules having polar bonds, may also dissolve in polar solvents.

Polymeric covalent substances

These are solids whose structures, like those of ionic substances, reveal no finite molecules. They are viewed as containing infinite molecules (subject, of course, to the boundaries of the crystal). Such structures can be classified as:

(*a*) One-dimensional (chain) polymers. These consist of molecules which extend infinitely in one direction; neighbouring chains are held together in the crystal by London and (where appropriate) electrostatic forces of attraction. Examples include:

Elemental tellurium

Silicon disulphide

The arrangement about the Si atoms is tetrahedral. Beryllium(II) chloride has a similar structure.

Mercury (II) oxide

The same structure is exhibited by gold(I) chloride.

(*b*) Two-dimensional (layer) polymers. Infinite molecules extend in two directions, forming layers or sheets, which are held together by Van der Waals forces. Some examples:

Mercury (II) iodide (red form)

Each atom is tetrahedrally bonded to four I atoms, while each I atom is bonded to two Hg atoms (non-linearly).

Cadmium (II) chloride

Each Cd atom is octahedrally bonded to six Cl atoms, while each Cl atom is pyramidally bonded to three Cd atoms. This (or the rather similar CdI_2

structure) is adopted by most dichlorides, dibromides and diiodides of the 3d series elements.

Chromium (III) chloride

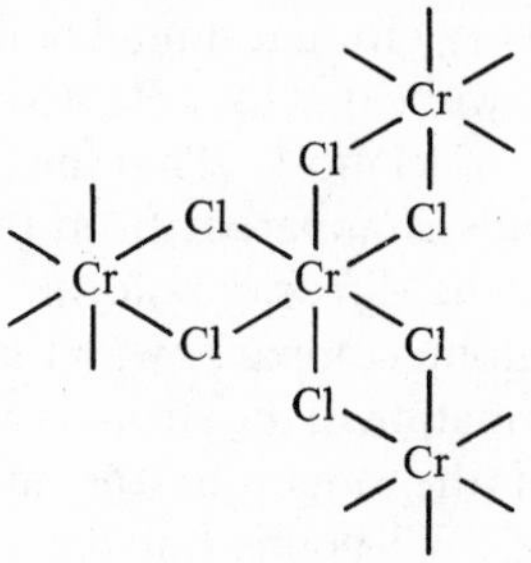

Each Cr atom is octahedrally bonded to six Cl atoms, while each Cl atom is non-linearly bonded to two Cr atoms. This structure is adopted by many trichlorides, tribromides and triiodides.

(c) Three-dimensional polymers (network solids). The structure extends infinitely in three dimensions, and the entire crystal may be seen as a 'giant molecule'. The classic example is silicon dioxide, whose crystal chemistry is described in detail in most texts. All forms (except a high-pressure modification, which has the rutile structure) contain SiO_4 tetrahedra connected by sharing corners. Each O atom is bonded to two Si atoms; the O–Si–O angle is between 144° and 155° depending on the crystalline form.

Many compounds of the type AB have structures in which each A is tetrahedrally bonded to four B, and vice versa. The zinc blende and wurtzite structures are of this type, and differ in details which need not concern us. Examples include ZnS, CuI, BeO, BN and AlP. The diamond lattice is of the zinc blende type.

Polymeric solids tend to have higher melting and boiling points than comparable molecular solids, since covalent bonds are broken in the course of these processes. As an example, consider $BeCl_2$, a chain polymer having the same structure as that illustrated above for SiS_2. It melts at 405 °C to give a non-conducting liquid which contains Be_nCl_{2n} molecules, where n is a small integer. At the boiling point of 547 °C, the vapour consists mainly of Be_2Cl_4 molecules; dissociation to monomeric $BeCl_2$ molecules is complete at about 900 °C.

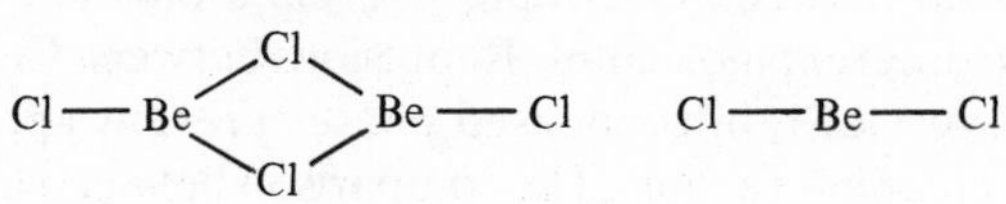

Relevant thermochemical data are as follows:

$$BeCl_2(s) \rightarrow \tfrac{1}{2}Be_2Cl_4(g) \qquad \Delta H^0 = 73 \, kJ \, mol^{-1}$$
$$\tfrac{1}{2}Be_2Cl_4(g) \rightarrow BeCl_2(g) \qquad \Delta H^0 = 64 \, kJ \, mol^{-1}$$

It can be deduced that the Be–Cl bond energy for a terminal bond (as in the $BeCl_2$ molecule) is about 463 kJ mol^{-1}, while that for a Be–Cl bridging bond (as in the crystalline solid) is about 263 kJ mol^{-1}. That the bridging bonds are weaker than the terminal bonds is apparent from the bond lengths: 202 and 177 pm respectively. In the dimeric molecule Be_2Cl_4, there are only three Be–Cl bonds per Be atom, compared with four in the solid; this is partly compensated by the formation of the stronger terminal bonds. Likewise, the decomposition of the dimer to the monomer involves a reduction in the number of Be–Cl bonds but the enthalpy change is relatively small compared with a covalent bond energy. The point being made here is that the breakup of a polymeric structure into discrete molecules is not as endothermic as might be expected, given the need to break covalent bonds. Even so, polymeric substances are less volatile than would be the case if only Van der Waals forces were responsible for the cohesive energy of the crystal. In the case of crystalline $BeCl_2$, these contribute only about 10–20 kJ mol^{-1}.

Melting and boiling points tend to increase as we go from one- to two- and three-dimensional polymers. Copper(I) chloride, zinc(II) chloride, gallium(III) chloride and germanium(IV) chloride provide a useful comparison. In the solid state, all have structures based on ECl_4 tetrahedra (E = Cu, Zn, Ga, Ge). CuCl is a three-dimensional polymer, $ZnCl_2$ has a layer structure (two-dimensional), $GaCl_3$ consists of discrete Ga_2Cl_6 molecules (see also Section 9.2), while $GeCl_4$ is monomeric in all phases:

	CuCl	$ZnCl_2$	$GaCl_3$	$GeCl_4$
m.p. (°C)	430	313	78	−50
b.p. (°C)	1690	732	201	84

The simplest rationale is that in $GaCl_3$ (i.e. Ga_2Cl_6) and $GeCl_4$, only weak Van der Waals forces need to be overcome in order that the higher entropies associated with the liquid and gaseous phases dominate the thermodynamics. In CuCl and $ZnCl_2$, covalent bonds need to be broken, and the disruption of a three-dimensional structure (CuCl) is more costly than that of a two-dimensional structure ($ZnCl_2$). In the case of Ga_2Cl_6, the molecules can be stacked in the crystal in such a way as to optimise the electrostatic attraction between electropositive Ga atoms and electronegative Cl atoms in different molecules. Repulsions between Cl atoms in different molecules of $GeCl_4$ in condensed phases prevent appreciable Ge . . . Cl intermolecular interaction. The comparison between CuCl and

$ZnCl_2$ is complicated by the formation of trimeric Cu_3Cl_3 molecules in the liquid and gaseous phases; gaseous zinc dichloride is apparently monomeric, and the structure of the liquid phase is uncertain.

Many three-dimensional polymeric substances are particularly refractory, insoluble and unreactive. One- and two-dimensional polymers tend to be more soluble. For example, dichlorides and trichlorides of the 3d elements are generally quite soluble in weakly-polar organic solvents such as alcohols, ethers and ketones. The driving force here is the formation of complexes with the solvent molecules. These compounds are also soluble in water, with some degree of hydrolysis. Aluminium(III) chloride (which has a layer structure similar to that of $CrCl_3$) dissolves in some non polar organic solvents, such as benzene, in which it forms Al_2Cl_6 dimers.

There is an ill-defined boundary between molecular and polymeric covalent substances. It is often possible to recognise discrete molecules in a solid-state structure, but closer scrutiny may reveal intermolecular attractions which are rather stronger than would be consistent with Van der Waals interactions. For example, in crystalline iodine each I atom has as its nearest neighbour another I atom at a distance of 272 pm, a little longer than the I–I distance in the gas-phase molecule (267 pm). However, each I atom has two next-nearest neighbours at 350 and 397 pm. The Van der Waals radius of the I atom is about 215 pm; at 430 pm, the optimum balance is struck between the London attraction between two I atoms and their mutual repulsion, in the absence of any other source of bonding. There is therefore some reason to believe that the intermolecular interaction amounts to a degree of polymerisation, and the structure can be viewed as a two-dimensional layer lattice. The shortest I–I distance between layers is 427 pm, consistent with the Van der Waals radius. Elemental iodine behaves in most respects – in its volatility and solubility, for example – as a molecular solid, but it does exhibit incipient metallic properties.

Another troublesome borderline area is that between ionic solids and three-dimensional polymers. The distinction cannot be made from the structure alone. Electrical conductivity in the molten state does not, as already mentioned, necessarily demonstrate the presence of ions in the solid state; and such a test is inapplicable where, as often happens, the substance sublimes or decomposes before melting. There can rarely be any objective means of assigning a compound to one category or the other. We are often persuaded towards one description on aesthetic grounds. For example, the structure of sodium chloride cannot easily be rendered in terms of localised, electron-pair bonds (but this is true also of many unequivocally covalent compounds). Its structure is eminently plausible for an array of cations and anions, however.

Polymeric structures can be described in terms of electron-pair bonds, for example:

CuCl

$$
\begin{array}{c}
\text{Cl} \\
\downarrow \\
\text{Cu} \\
\text{Cl}\nearrow \;\; \nwarrow\text{Cl} \\
\text{Cl}
\end{array}
\qquad
\begin{array}{c}
\text{Cu} \\
\uparrow \\
\text{Cl} \\
\text{Cu}\nearrow \;\; \nwarrow\text{Cu} \\
\text{Cu}
\end{array}
$$

ZnS

$$
\begin{array}{c}
\text{S} \\
\downarrow \\
\text{Zn} \\
\text{S}\nearrow \;\; \nwarrow\text{S} \\
\text{S}
\end{array}
\qquad
\begin{array}{c}
\text{Zn} \\
\uparrow \\
\text{S} \\
\text{Zn}\nearrow \;\; \nwarrow\text{Zn} \\
\text{Zn}
\end{array}
$$

$CdCl_2$

$$
\begin{array}{c}
\text{Cl} \\
\text{Cl}\searrow \; \downarrow \; \swarrow\text{Cl} \\
\text{Cd} \\
\text{Cl}\nearrow \; \uparrow \; \nwarrow\text{Cl} \\
\text{Cl}
\end{array}
\qquad
\begin{array}{c}
\text{Cl} \\
\text{Cd}\nearrow \;\; \nwarrow\text{Cd} \\
\text{Cd}
\end{array}
$$

SiO_2

$$
\begin{array}{c}
\text{O} \\
\mid \\
\text{Si} \\
\text{O}\nearrow \;\; \nwarrow\text{O} \\
\text{O}
\end{array}
\qquad
\begin{array}{c}
\text{O} \\
\text{Si}\nearrow \;\; \nwarrow\text{Si}
\end{array}
$$

The bonding of each atom as rendered above is consistent with its behaviour in discrete molecules or ions. For example, the pyramidal three-coordination exhibited by Cl in $CdCl_2$ is found in simple species such as $HClO_3$. The tetrahedral coordination of Si in SiO_2 is found in many molecules like SiH_4, $SiCl_4$ etc. and bent, two-coordinate O is common.

There is reason to believe that silver(I) chloride is not to be treated as an ionic solid, although it has the NaCl structure. It can be described as a polymeric solid, with two-centre electron-pair bonds, in terms of VB theory. The octahedral environment of each Ag atom demands d^2sp^3 hybridisation (see Chapter 6). If we place six electrons in the three nonbonding 4d orbitals, five electrons have to be accommodated in the six hybrids; five of these will be singly occupied, and the sixth empty. The Cl atom requires the same hybridisation; with seven valence electrons, we will have five singly-occupied hybrid orbitals with the sixth filled. The required single bonds can then be formed by overlap of the singly-occupied Ag and Cl orbitals, plus the formation of one dative bond. Resonance will account for the equivalence of the six bonds formed by each atom. The situation can be depicted as for the polymeric structures described above:

$$
\begin{array}{cc}
\text{Cl} & \text{Ag} \\
\text{Cl}\diagdown\uparrow\diagup\text{Cl} & \text{Ag}\diagdown\uparrow\diagup\text{Ag} \\
\text{Ag} & \text{Cl} \\
\text{Cl}\diagup\ |\ \diagdown\text{Cl} & \text{Ag}\diagup\ |\ \diagdown\text{Ag} \\
\text{Cl} & \text{Ag}
\end{array}
$$

The absurdity of describing the Ag atom in silver chloride as 'univalent' should be apparent. Octahedral six-coordination for Ag(I) is admittedly unusual but such coordination is known for Cl in the ClF_6^+ cation and is common for I in, e.g., $IO(OH)_5$ and IO_6^{5-}.

One might describe the bonding in CaF_2 – which has $8:4$ coordination – in terms of covalent bonding in which the F atoms each form four bonds (including three dative bonds):

$$
\begin{array}{c}
\text{Ca} \\
\uparrow \\
\text{F} \\
\text{Ca}\diagup\ \diagdown\text{Ca} \\
\text{Ca}
\end{array}
$$

but no such situation is encountered for fluorine in any simple species; for example, FO_4^- is unknown, unlike ClO_4^-. Thus one feels inclined to describe the bonding in CaF_2 as ionic.

One of the advantages of the ionic model is the relative ease of performing quantitative calculations. For example, a simple electrostatic approach to the bonding in NaCl yields a calculated enthalpy of formation which is within a few kJ mol^{-1} of the experimental value (see Chapter 5). But for solids such as CuCl, ZnS and $CdCl_2$ the ionic model fares rather poorly, and is unable to account for the strength of the bonding in these crystals. This suggests (but does not prove) that a description based on covalent bonding is more appropriate.

The above considerations do not provide hard-and-fast criteria for distinguishing between ionic and polymeric solids. There is little doubt that covalency makes some contribution to the bonding even in crystals such as NaCl, while there is likely to be some ionic component in solids such as $CdCl_2$ or SiO_2. We may imagine that for any such crystalline compound, there is a wave function which can be expressed as a combination of an ionic and a covalent wave function, i.e.:

$$a\psi \text{ (covalent)} + b\psi \text{ (ionic)}$$

The relative magnitudes of a and b will determine whether the bonding is better described as ionic or covalent. If the energy E_c associated with ψ (covalent) is lower than E_i, the energy associated with ψ (ionic), a will be greater than b, and vice versa; ionic–covalent resonance ensures that the true energy E of the system will be lower than either E_c or E_i. However, it seems that in many cases E_c and E_i are not very different and either a covalent or an ionic description affords a reasonable approximation.

Another hazy boundary separates polymeric and metallic substances. We have already noted the case of iodine, which can be described as a molecular solid but which might also be viewed as a two-dimensional polymer having incipient metallic properties. Elemental tellurium, whose chain structure was described earlier in this section, has pronounced metallic properties. Each Te atom is bonded to two others at a distance of 284 pm, and this connectivity leads to a helical chain. However, each Te atom is bonded to four more in other chains, at a distance of 350 pm. These longer Te–Te contacts are apparently responsible for the metallic properties.

3.4   Coordination and organometallic compounds

Most coordination or organometallic compounds fall into the categories listed in Section 3.3. It seems proper, however, to regard these as distinct types of inorganic compounds; coordination and organometallic chemists constitute two of the largest tribes in chemistry, and there has been much intercourse between them.

In Section 1.3, we saw how coordination compounds were recognised as apparently violating the rules of valency insofar as these were understood in the late nineteenth century. In the 1920s and 1930s, the nature of the bonding in coordination compounds and complex ions was clarified by the postulate of the coordinate or dative bond. Present-day definitions of coordination compounds are sometimes so broad that practically any molecule or polyatomic ion could qualify as 'complex'. It is perhaps most useful to combine the original definition – a compound which cannot be given a molecular structure in terms of classical valency numbers – with the more modern view that at least some of the ligands are required to form coordinate bonds to the central atom, as discussed further in Section 8.1.

Many writers make a fetish of enclosing discrete complex species in square brackets. This is often useful and sometimes essential. To give an example from Section 3.2, the formulation $Ga[GaCl_4]$ instead of $GaCl_2$ or Ga_2Cl_4 conveys more clearly the nature of this substance. To write $Cs_2[CoCl_4]$ is to assert that the compound contains discrete $CoCl_4^{2-}$ ions; but Na_2MnCl_4 should be rendered without square brackets because the structure reveals no finite $MnCl_4^{2-}$; the anion is polymeric, with octahedral six-coordination about each Mn atom. However, the reader will find many inconsistencies in the literature. In writing equations for the reactions of complex ions in solution, species such as $Fe(CN)_6^{3-}$ or $Co(NH_3)_6^{3+}$ may or may not have their formulae enclosed in square brackets, according to the whim of the author or editor. Square brackets are, of course, also used to indicate the concentrations of species in equilibria. Another possible source of confusion can arise with formulations such as $Ni[P(OMe)_3]_4$. This does not contain discrete ions; it consists of tetrahedral molecules, with four-coordinate nickel(0). In this book, square brackets are used only where necessary to indicate a discrete complex molecule or ion, and to specify the atoms or groups which constitute its coordination sphere. Some examples will illustrate the usage of square brackets.

(a) $[Ni(NH_3)_4Br_2]$ and $[Pd(NH_3)_4]Br_2$

The former contains octahedral complex molecules with six-coordinate Ni(II). The latter contains square planar $Pd(NH_3)_4^{2+}$ cations; the bromide ions complete an ionic lattice but are not part of the coordination sphere of Pd(II).

(b) $[Cr(NH_3)_4F_2]Cl$ and $[Cr(NH_3)_4FCl]F$

These are *ionisation isomers*, consisting respectively of $[Cr(NH_3)_4F_2]^+$ and Cl^- ions, and of $[Cr(NH_3)_4FCl]^+$ ions. They differ in the distribution of halide ions inside and outside the coordination sphere of Cr(III).

(c) $[PCl_4][VCl_5]$ and $[PCl_4]CrCl_4$

Both contain discrete, tetrahedral PCl_4^+ cations. $[PCl_4][VCl_5]$ contains also distorted trigonal bipyramidal VCl_5^- ions but – contrary to statements you may find in other books – $[PCl_4]CrCl_4$ does not contain discrete $[CrCl_4]^-$ ions; the visible spectrum clearly indicates octahedral six-coordination about Cr(III), so that the anion must be polymeric.

There are some cases of ambiguity. For example, the placement of square brackets in the formulation $[Ni(NH_3)_4(NO_2)_2]$ properly indicates a discrete molecule in which we have octahedral six-coordination about Ni(II). But in the corresponding copper(II) complex the octahedron is appreciably distorted: the NO_2 groups are further away from the Cu atom, the Cu–N distance being 265 pm compared with 207 pm for the nickel complex (see Fig. 8.3). Such distortion is usual for d^9 complexes, and the distant NO_2^- ligands are sometimes described as 'semi-coordinated', the coordination number of Cu(II) being $(4 + 2)$. It is therefore a moot point whether we should write $[Cu(NH_3)_4(NO_2)_2]$ or $[Cu(NH_3)_4](NO_2)_2$, and it may be best to delete the square brackets altogether in this case.

The best rule for the student is to take care *not* to use square brackets where they are inappropriate, or where they may cause confusion; if in doubt, leave them out.

Organometallic compounds are most strictly defined as containing direct bonds between carbon atoms of organic groups and atoms of elements which, as elemental substances, are metallic or semi-metallic. This definition is often interpreted with considerable latitude. One of the leading journals in this field will consider for publication papers dealing with boron hydrides and their organic derivatives, complexes of the transition elements with carbon monoxide (but not with isoelectronic cyanide), and hydrides of the more electropositive elements; but carbonates and carbides are excluded, and although organophosphorus compounds are admitted to the club organosulphur compounds are discouraged. A more practical definition would embrace compounds containing bonds between carbon atoms of organic groups and atoms of lower electronegativity than carbon. The term 'organoelement chemistry' is favoured by some chemists, especially in the Soviet Union.

Organometallic chemistry does have its own distinctive flavour. Organometallic compounds are often (but not invariably) air- and moisture-sensitive and their handling requires much care and skill. In the interplay between kinetic and thermodynamic considerations, organometallic chemistry falls somewhere between inorganic and organic chemistry. Most organometallic compounds of the transition elements can also be classified as coordination compounds, and are dealt with in more detail in Chapter 8. (See also Section 10.5.)

Finally in this chapter, we consider the use of arrows (as opposed to straight lines) to denote coordinate (or dative) bonds in depicting the structures of coordination/organometallic compounds. The complex $Be(NH_3)_2Cl_2$ could be represented by any of (i), (ii) or (iii):

(i)   (ii)   (iii)

Structure (i) could be misleading; it is doubtful whether the Be^{2+} ion has any independent existence under ordinary conditions and the bonds between Be and Cl can be rendered simply as 'ordinary' rather than coordinate bonds. Structure (iii) is objectionable because the lack of arrows could imply that the bonds are all 'ordinary' covalent bonds to which each atom contributes equally, leading to untidy electronic book-keeping; it is not immediately apparent that the Be and N atoms obey the octet rule which is almost invariably followed by atoms in that Period (see Section 6.1). Structure (ii) is to be preferred.

But in complex compounds of the d block elements, the use of arrows to denote coordinate bonds is falling into disuse and many inconsistencies will be found in the literature. For example, the anion $Cr(CO)_5H^-$ could be rendered in a number of ways, including:

(i)   (ii)

Structure (i) is less likely to be encountered in the modern literature than (ii). As we shall see in Chapter 8, the bonds between Cr and CO cannot be viewed simply as single coordinate bonds. Lacking any simple represen-tation such bonds are usually written as M—CO. M—H bonds are always written as such, although for the purpose of assessing the oxidation state of M we regard H as being in the form of H^-. There is no problem with the book-keeping in resorting to what may seem to be sloppy notation; if the electron-counting rules set out in Section 8.6 are followed, there is no need to specify the exact type of bonds formed between the Cr atom and its ligands. (Note in structure (ii) above the use of the truncated square bracket in the pictorial representation of a large, complex ion.)

3.5   Further reading

Huheey (1983) (see Section A.3 of the Appendix) gives in Appendix J a good, concise summary of the rules for inorganic nomen-clature. For the classification of substances according to structural type, see the books listed in Section A.6. Wells (1984) is specially recommended.

4

Periodicity and atomic properties

4.1   The Periodic Table

In 1922, the English-born (and New Zealand-educated) chemist J. W. Mellor published the first of 16 volumes which constituted his *Comprehensive Treatise on Inorganic and Theoretical Chemistry* – perhaps the greatest single-handed effort in the whole of the chemical literature. In explaining the arrangement of his material, he wrote: 'We now flatter ourselves that the Periodic Law has given inorganic chemistry a scheme of classification which enables the facts to be arranged and grouped in a scientific manner. The appearance of order imparted by that guide is superficial and illusory. Allowing for certain lacunae in the knowledge of the scarcer elements, prior to the appearance of that Law, the arrangements employed by the earlier chemists were just as satisfactory and in some cases, indeed, more satisfactory than those based on the Periodic Law. The arrangement of the subject matter of inorganic chemistry according to the periodic scheme is justified solely by expediency and convention. It has a tendency to make teachers over-emphasise unimportant and remote analogies, and to underestimate important and crucial differences.'

These views were certainly worthy of respect in 1922; Mellor, who was already well known as an author, probably knew as much factual inorganic chemistry as anyone alive. Should we take heed of Mellor today?

Before going any further, two important points should be noted; in 1922, the theoretical basis for the Periodic Law was only just beginning to emerge, and the version of the Periodic Table to which Mellor was referring was not quite the same as the Table in common use today. The *short form* of the Table, derived directly from Mendeleev's revised Table of 1871, is shown in Fig. 4.1. The familiar *long form* – which actually dates back to the 1890s – gained ground after the quantum revolution of the 1920s had established the concept of orbitals. It was then clear that elements in the same Group had similar ground state electron configura-

Group 0	I A	I B	II A	II B	III A	III B	IV A	IV B	V A	V B	VI A	VI B	VII A	VII B	VIII
	H														
He	Li		Be		B		C		N		O		F		
Ne	Na		Mg		Al		Si		P		S		Cl		
Ar	K		Ca		Sc		Ti		V		Cr		Mn		Fe, Co, Ni
		Cu		Zn		Ga		Ge		As		Se		Br	
Kr	Rb		Sr		Y		Zr		Nb		Mo		Tc		Ru, Rh, Pd
		Ag		Cd		In		Sn		Sb		Te		I	
Xe	Cs		Ba		La*		Hf		Ta		W		Re		Os, Ir, Pt
		Au		Hg		Tl		Pb		Bi		Po		At	
Rn	Fr		Ra		Ac		Th		Pa		U		Np		Pu, Am, Cm

*La includes the lanthanides

Fig. 4.1. 'Short' form of Periodic Table, based upon Mendeleev's table of 1871 but including elements discovered since.

tions in the atomic state, and that these configurations tended to determine chemical behaviour. The long form of the Periodic Table is shown in Fig. 4.2. The division into s, p, d and f blocks emphasises the grouping of elements according to atomic electron configurations. This is very convenient for the student as an aid to writing down these configurations; it does, however, have some disadvantages from the viewpoint of grouping the elements as a basis for the orderly study of inorganic chemistry. A multitude of alternative forms of the Table have been devised but none has superseded the conventional long form.

Some of the evils referred to by Mellor arose from the short form of the Table in vogue at the time. For example, chemists sought to find similarities between Cu, Ag and Au on the one hand, and Li, Na, K etc. on the other. There were attempts to seek analogies between manganese and the halogens, both appearing in Group VII. Apart from some similarities between perchlorates and permanganates, these are few and far between. Some chemists insisted, however, that Mn and Re (Tc was unknown until the 1940s, although its existence was, of course, predictable) ought to form anions Mn^- and Re^-, as do the halogens. This suggestion appeared to be vindicated by the preparation of a compound formulated as $KRe.4H_2O$. This, however, turns out to be a hydrido-complex $KReH_8$, in which Re is in the VII oxidation state and not $-I$.

There are a number of 'anomalies' in the long form, whose exposition has caused the spilling of much ink. One of Mendeleev's greatest triumphs was his prediction of a new element ('eka-silicon') between Si and Sn in his Table. Mendeleev had the audacity to predict some chemical properties of this new element, and his prophecies were substantially fulfilled a few years later by the isolation of germanium and a preliminary exploration of its chemistry. These predictions were made simply by interpolation between Si and Sn. Chemists, and chemistry students, have come to expect that the chemical properties within a Group follow monotonic trends; properties can be predicted by interpolation and extrapolation. Experimental observations which do not fit such simple trends lead to the identification of 'anomalies'. At one extreme, there may be a tendency to sweep such anomalies discreetly under the carpet, or even to question the validity of the data; at the other extreme, strenuous efforts are made to account for anomalies by means of elaborate and sometimes fanciful theorising.

Paradoxically, germanium is a case in point. In recent years, there has been much discussion of the 'middle element anomaly' in the p block; Ga, Ge, As, Se and Br do not quite 'fit in' with the other members of their respective Groups, although the anomaly does fade away towards the right-hand side of the Table. Anomalies have also been noted in the heavier p block elements Tl, Pb etc.

It must be remembered that close chemical relationships are to be

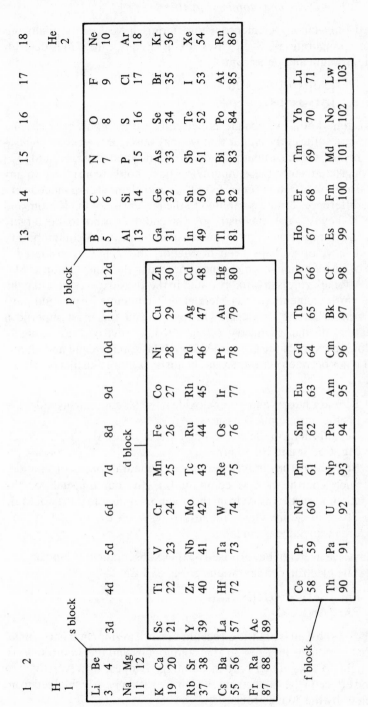

Fig. 4.2. 'Long' form of Periodic Table, with new numbering scheme for Groups.

expected only among elements whose atoms have exactly analogous electron configurations. Such cases are relatively rare. Consider, for example, sodium and potassium:

Na: $[Ne]3s = [He]2s^2 2p^6 3s$
K: $[Ar]4s = [Ne]3s^2 3p^6 4s$

The two atoms both have a single ns electron outside a noble gas core, and the cores are sufficiently similar that we may describe the two configurations as being exactly comparable; the difference between [He] and [Ne] can be neglected, since these inner cores need not be considered in any chemical argument. The chemistries of the two elements are indeed very similar; differences can be attributed to the larger size of the K atom and K^+ ion. Likewise, the chemistry of Ca could be said to be largely predictable from a knowledge of magnesium chemistry. A contemporary of Mendeleev, or perhaps even of Mellor, might have considered a scheme of classification in which zinc appeared in the same Group as Mg, Ca etc. After all, zinc is invariably found in the II oxidation state, like the alkaline earth elements (or, as Mendeleev's contemporary would have put it, zinc is always divalent). However, even the most superficial examination of the chemistry of zinc would reveal it as a highly anomalous member of the Group. Although in atomic weight and atomic number it lies between Ca and Sr, its chemistry is more like that of Be, for example:

(i)   Be and Zn form amphoteric oxides; the alkaline earth oxides are (as the term implies) basic.

(ii) Be and Zn form the interesting basic acetates $M_4(CH_3COO)_6O$; Mg, Ca, Sr and Ba do not.

(iii) Be is almost invariably found in tetrahedral four-coordination. Such coordination is common for Zn, but unusual for the alkaline earths (aside from the organometallic chemistry of Mg).

(iv) There are some close similarities between Be and Zn in their organometallic chemistries.

Nowadays, we need not be surprised by these observations. Compare the ground state electron configurations of Mg and Zn:

Mg: $[Ne]3s^2 = [He]2s^2 2p^6 3s^2$
Zn: $[Ar]3d^{10}4s^2$

The filled 3d subshell is of little direct importance in zinc chemistry, in the sense that it remains intact in all zinc compounds and the 3d orbitals are not thought to take much part in bonding. It does, however, have a profound effect on the 4s electrons, which are rather poorly shielded from the nucleus by the 3d electrons.

No 'anomaly' is represented by zinc because modern versions of the

Periodic Table place it in the d block, rather than the s block. This is not entirely satisfactory, because the triad Zn, Cd and Hg have more in common with their immediate neighbours in the p block than they do with their neighbours on the other side, in the d block. Textbook writers have always found difficulty in dealing with these elements. In this book, they will be placed with either the transition elements or the Main Group elements, as the occasion demands.

Zinc's immediate neighbour on the right, gallium, is placed in the p block and appears in the same Group as B and Al. We need not be surprised, however, to discover that Ga is an anomalous member of the Group; its filled 3d subshell is still capable of exerting an effect on the 4s and 4p orbitals. Just as we noted some similarities between Zn and Be, so we find that in some respects Ga resembles boron more than aluminium. The 'anomaly' is still apparent with germanium, which often behaves more like carbon than silicon or tin. For example, GeH_4 (like CH_4) does not undergo hydrolysis by dilute alkalies. The tetrachlorides of Si and Sn react violently with water to give the hydrated dioxide and copious fumes of HCl. CCl_4, however, does not react with water and $GeCl_4$ undergoes only partial, reversible hydrolysis. The anomaly is less evident by the time we reach bromine. Even so, some features of bromine chemistry have aroused surprise. The perbromate ion was elusive until recently, for example.

Many other anomalies – some glaring, others more subtle – can be rationalised (if not fully understood) by recognition that we should compare 'like with like' if we are to expect orderly periodic behaviour. The filling of the 4f subshell produces anomalies which persist for quite a long way beyond the lanthanides. Even radium, 17 elements on from the last of the lanthanides, is affected.

Notation and nomenclature in the Periodic Table

With the modern form of the Periodic Table, some of the inconsistencies and sources of confusion associated with the old form have been largely removed. Until recently, the Groups were numbered in essentially the same way as in the old short form. Thus the elements of the s block – i.e. the elements whose atoms in their ground states have just one or two ns electrons outside a noble gas configuration – were classified as Group IA (Li, Na etc.) and Group IIA (Be, Mg etc.). The triads of the d block were numbered Groups IIIA (Sc, Y, La), IVA (Ti, Zr, Hf) etc. but the triads headed by Fe, Co and Ni constituted together Group VIII. The copper triad (Cu, Ag, Au) then became Group IB, the zinc triad Group IIB, and so on until the halogens were labelled Group VIIB and the noble gases Group O. This has obviously unsatisfactory features – three triads in Group VIII, for example – and some writers insisted on using an alternative A and B subgroup notation so that, e.g., Sc, Y and La were Group IIIB and B, Al etc. Group IIIA. This difficulty can be removed by refraining from giving numerical designations to the d and f block elements and using Group I, II etc. for the s and p blocks, the Main Group elements. However, most international bodies have now approved a new system of notation which should resolve all previous difficulties – the student will need to be familiar with the older schemes too, however, for some time to come.

Groups are numbered from 1 to 18. Group 1 comprises H, Li, Na etc. while Group 2 is Be and the alkaline earths. The d block elements are labelled from Group 3d (Sc, Y, La, Ac) to 12d (Zn, Cd, Hg), while the f block elements are assigned to Groups 3f (Ce, Th) to 16f (Lu, Lr). Thus Fe, Ru and Os constitute Group 8d while Eu and Am are in Group 8f. The boron Group – B, Al, Ga, In, Tl – become Group 13, and the remaining Main Groups continue up to Group 18, the noble gases (including He). Thus the numbering of the Groups in the p block differs from the old scheme by adding 10 and using Arabic instead of Roman numerals.

The main reason for the change was to eliminate the confusion which was being caused by inconsistencies in the use of A and B designations. This can equally well be achieved by reserving Group numbers for the Main Groups – in which case A and B terminology becomes redundant – and referring to, e.g., the iron Group for triads in the d block. The extension of the numbering scheme to the f block leads to a serious anomaly. In Fig. 4.2, care was taken to list the f block elements in such a way that they appear in the same columns as their d block analogues having the same number of valence electrons. Thus the relationship of Ce and Th to Ti, Zr and Hf is clear. In many ways, thorium behaves like a 6d element rather than a 5f element and is to be regarded as a member of the same family as Ti, Zr and Hf. But the new scheme places it in Group 3f; La and Ac are in Group 3d, while Ti, Zr and Hf constitute Group 4d. Group designations are unlikely to become either useful or popular for the f block elements.

A number of trivial, yet useful and descriptive, designations continue to be used, e.g. alkali metals, alkaline earths (Be does not qualify, strictly speaking, as an alkaline earth because its amphoteric oxide is very weakly alkaline), the coinage metals (Cu, Ag, Au) and the noble (or inert, or rare) gases.

In accounts of descriptive inorganic chemistry – especially in more elementary texts – it is common practice to classify elements as 'metals' or 'nonmetals', with 'semi-metals' or metalloids as a borderline case, according to the nature of the elemental substance. The chemistry of an element is, to some extent, broadly predictable from this classification. Metallic elements tend to form ionic oxides and halides; they form cationic species in aqueous solution; and their atoms are of relatively low electronegativity. Nonmetallic elements form covalent halides and acidic oxides; they form anionic species in solution, and their atoms have relatively high electronegativities. These generalisations are not without value; but there are many inconsistencies and exceptions. For example, beryllium forms almost exclusively covalent compounds but the elemental substance is definitely metallic. Molybdenum and tungsten form practically no cationic species in aqueous solution, yet these are also metals in the elemental state. Gold and mercury are also typical – in some respects quintessential – metals in the elemental state but their chemistry has few 'metallic' characteristics; the Au atom is fairly high on the scale of electronegativity, it has a high electron affinity and there is some evidence for the formation of the Au^- ion. In this book, the adjective 'metallic' and the noun 'metal' are reserved for *substances* having the physical properties associated with metals (see also Section 3.1).

To summarise this section: we should not be too dogmatic in our classification of the elements according to the Periodic Table. We have already mentioned the case of the zinc Group, which strictly belong to the d block but have more in common with their p block neighbours. We might mention also the scandium Group, at the other end of the d block, whose chemistry could be usefully discussed alongside that of aluminium. Lanthanum falls in the d block, but is better classed as one of the lanthanides. Scandium and yttrium are sometimes described along with the lanthanides in inorganic textbooks. Hydrogen and helium are not easily accommodated in either the s or p blocks. These problems and uncertainties of classification need not cost the reader any sleep. In the study of descriptive inorganic chemistry, any classification which makes the subject palatable and which leads to an intellectually-stimulating organisation of the facts will serve the purpose.

4.2   Radii: the sizes of atoms and ions

It is, of course, impossible to measure the absolute size of an isolated atom; its electron cloud extends to infinity. It is possible to calculate the radius within which (say) 95% of its total electron cloud is confined; but most measures of atomic/ionic size are based upon experimental measurements of internuclear distances in molecules and crystals. This means that the measurement is dependent on the nature of the bonding in the species concerned, and is a property of the atom or ion under scrutiny in a particular substance or group of substances. This must always be borne in mind in making use of tabulated radii of atoms or ions. The most important dictum to remember is that radii are significant only insofar as they reproduce experimental internuclear distances when added together. The absolute significance of a radius is highly suspect, although a collection of such radii may meaningfully convey the relative sizes of the atoms/ions.

Atomic radii

The *atomic radius* of the atom X is defined as half the length of an X–X single bond. This can be obtained experimentally from the structures of elemental substances containing molecules X_n where the X–X bond order is believed to be unity, e.g. Cl_2, P_4, S_8. It may also be obtained from the X–X distances found in molecules such as HO—OH, H_2N—NH_2 etc. for atoms which form multiple bonds in the elemental substance. Such atomic radii may be termed *covalent radii*. For atoms which form metallic elemental substances, *metallic radii* are obtained. These are usually standardised for 12-coordination of each atom, which is the most common situation in metals. Corrections can be made in the cases of metals which adopt other structures.

Some writers feel it important to distinguish carefully between covalent and metallic radii. Others suggest that a 'self-consistent' set of atomic radii – some covalent, others metallic – can be devised. Such a collection is presented in Table 4.1. In cases where both a covalent and a metallic radius can be obtained, the agreement is variable. For example, the metallic radii of atoms of the Group 1 elements are 20–30 pm greater than the corresponding covalent radii, taken from the M–M distances in $M_2(g)$. The Mn–Mn distance in $(CO)_5Mn$—$Mn(CO)_5$ is 293 pm, which compares with 274 pm calculated from the metallic radius of Mn. The electronic environment of the Mn atom in the carbonyl complex is, of course, very different from that in the elemental substance.

The value of atomic radii is circumscribed by the fact, noted in the previous sentence, that a given radius may be affected by some peculiarity of the bonding in the elemental substance, which may not necessarily apply in compounds. For example, the internuclear distance in the F_2 molecule is found experimentally to be 142 pm, giving a covalent radius of 71 pm for the F atom. However, even after making allowance for various bond-shortening effects in fluorides EF_n, this seems too large and a value of 60 pm is often quoted. There is some reason to believe that the F–F bond is weakened by the repulsions between lone pairs and that the internuclear distance in F_2 is not a fair reflection of the size of the F atom in other situations. In the case of the H atom, the covalent radius based on the internuclear distance in the H_2 molecule is 37 pm, but a value of about 30 pm works better in practice. The metallic radii of the lanthanide atoms Eu and Yb seem to be decidedly out of step with those of their neighbours. As discussed further in Section 7.5, this arises from a rather different type of metallic bonding and structure in the elemental substances. No covalent radii can be established for these atoms, but the

Table 4.1 *Atomic (covalent/metallic) radii (pm)*

1	2	3	4	5	6	7	8	9	10	11	12	13	14	15	16	17
H 37																
Li 157	Be 112											B 80	C 77	N 74	O 74	F 71
Na 191	Mg 160											Al 143	Si 118	P 110	S 103	Cl 99
K 235	Ca 197	Sc 164	Ti 147	V 135	Cr 129	Mn 137	Fe 126	Co 125	Ni 125	Cu 128	Zn 137	Ga 153	Ge 139	As 120	Se 116	Br 114
Rb 250	Sr 215	Y 182	Zr 160	Nb 147	Mo 140	Tc 135	Ru 134	Rh 134	Pd 137	Ag 144	Cd 152	In 167	Sn 158	Sb 161	Te 143	I 133
Cs 272	Ba 224	La 188	Hf 159	Ta 147	W 141	Re 137	Os 135	Ir 136	Pt 139	Au 144	Hg 155	Tl 171	Pb 175	Bi 182	Po	At

Ce 183	Pr 182	Nd 181	Pm 181	Sm 180	Eu 199	Gd 179	Tb 176	Dy 175	Ho 174	Er 173	Tm 173	Yb 194	Lu 172
Th 180	Pa 161	U 139	Np 140	Pu 151	Am 140								

ionic radii of Eu^{n+} and Yb^{n+} ($n = 2, 3$) are not anomalous compared with other lanthanide ions.

Bearing in mind these points, let us look at periodicity in atomic (covalent or metallic) radii. In going down a Group, we might expect an increase in atomic size. The size of an atom – as measured by its covalent or metallic radius – is determined by the radial extension of its outermost occupied orbitals. Going down any Group, the effective nuclear charge experienced by these outer electrons increases, since the shielding afforded by other electrons is less than perfect. At the same time, however, the kinetic energy of an electron increases as the principal quantum number n of its orbital increases. The overall effect is an increase in size down the Group. In traversing any Period, there is also an increase in effective nuclear charge, but no overriding increase in the kinetic energy, so that we tend to observe a decrease in atomic size from left to right across a Period.

There are inevitably some anomalies. In the p block, the increases in going from the second member to the third, and from the fourth to the fifth, are relatively small. This, of course, reflects discontinuities in the electron configurations of the atoms. Between the second and the third members, the 3d subshell is filled. Since 3d electrons are relatively inefficient in shielding the 4s and 4p electrons from the nucleus, there is a large increase in effective nuclear charge in going from, e.g., Al to Ga which is only just compensated by the higher kinetic energies of the outer electrons. Between the fourth and fifth members of a p block Group, e.g. In and Tl, the 4f subshell has been filled; 4f electrons are even less efficient than 3d electrons in shielding. The contraction in size along the lanthanides themselves – allowing for the anomalies of Eu and Yb already mentioned – is not particularly spectacular, nor are its chemical consequences within the lanthanides. The 'lanthanide contraction', however, has profound effects on the chemical behaviour of the elements which follow the completion of the 4f series, and its manifestations are even apparent in the following Period, after Rn.

It will be noted in Table 4.1 that the metallic radii along the 5d series are very close to the values for the corresponding atoms in the 4d series. This observation is related to the remarkable chemical similarities between Zr and Hf. The effect of the lanthanide contraction on metallic radii persists to the end of the 5d series, but its chemical consequences become less marked as we pass from left to right. It is important not to make too much of similarities in metallic radii in chemical arguments, however. For example, the triad Mn, Tc and Re all have practically equal metallic radii, but their chemical behaviour shows dramatic differences. The near-equivalence of their metallic radii is less marked when we look at bond lengths in molecules. For example, the M–M distances in $(CO)_5M—M(CO)_5$ are respectively 293, 304 and 302 pm for M = Mn, Tc and Re.

Towards the end of each of the three series which constitute the d block, we observe a marked increase in radius. For reasons discussed in more detail in Section 7.5, the filled nd subshell tends to weaken the bonding in metallic elemental substances, leading to longer internuclear distances.

We should also mention the fact that an atomic radius is sensitive to the type of hybridisation of the atom's orbitals. It is well established that the covalent radius of a carbon atom increases with increasing p character in its hybrid orbitals, viz. $sp < sp^2 < sp^3$.

Ionic radii

Ionic radii are ultimately obtained from experimentally-determined internuclear distances in crystals that are deemed to be ionic; the sum of the radii of cation and anion should be close to the observed shortest distance in the crystal. The problem in obtaining the radii of individual ions from the mass of structural data for crystals is how to partition the internuclear distance between cation and anion. The classical methods (dating back to the 1920s) of achieving this need not be described in detail. Pauling's scale, still reproduced in modern books and quoted in the contemporary literature, is based on the assumption that an ionic radius is inversely proportional to the effective nuclear charge experienced by the outermost electrons; some adjustments have to be made for multiply-charged ions. Pauling radii work reasonably well, although published tables still include calculated values – having no empirical basis – for a number of ions such as H^-, Cu^+ and Ag^+; these give very poor agreement with experimental data if used to calculate internuclear distances in crystals, being much too large.

Other 'traditional' scales, such as that of Goldschmidt, are more empirically based and rely upon experimental internuclear distances found in oxides and fluorides (which are expected to be particularly ionic); assumed values for the radii of F^- and O^{2-} are needed. A very complete tabulation, based on a large amount of modern data for oxides and fluorides, both binary and complex, has been published by Shannon and Prewitt using values of 133 pm for F^- and 140 pm for O^{2-}. One of the merits of this collection is that it recognises the dependence of ionic radii on the coordination number. For example, in the case of Cd^{2+} radii are quoted for coordination numbers 4, 5, 6, 7, 8 and 12 as shown below:

Coordination number	$r(Cd^{2+}, pm)$
4	78
5	87
6	95
7	103
8	110
12	131

It will be apparent that the ionic radius increases with the coordination number; this effect is much more marked for cations than for anions. Since cations are mostly smaller than the counter-anions in the ionic solid (excluding, of course, solids containing complex ions), the cation–anion distance is usually determined by the need to pack anions around cations without the former bumping into each other. As the coordination number is increased, the cation–anion distance tends to increase in order to make room for the larger number of anions around the cation.

The Shannon–Prewitt tabulation also distinguishes between different spin states for ions of the transition elements. For example, the radius of Fe^{2+} in octahedral six-coordination is 17 pm smaller for the low-spin state as opposed to high-spin. If you study bio-inorganic chemistry in a more advanced text, you will find that this fact is of great importance in understanding the mechanics of the haemoglobin molecule (see Section 9.8).

The Shannon–Prewitt tables probably provide the most useful and most comprehensive collection of ionic radii. Values are included for improbable species such as Br^{7+}; the sum of its radius and that of O^{2-} gives the observed internuclear distance in BrO_4^-. Shannon–Prewitt radii for the more plausible ions in octahedral six-coordination are given in Table 4.2. These are useful for most practical purposes (see Chapter 5) except where octahedral six-coordination is uncommon for the ion in question.

If we were to take a pair of nearest-neighbour ions in a lattice and plot the electron density along the internuclear axis, we would expect this to become effectively zero at a point marking the boundary between cation and anion, if the 'hard sphere' ionic description is perfectly valid. A number of accurate experimental measurements (based on X-ray data) have been made; it is found that the electron density does not fall to zero at any point between the nuclei, which may be taken to indicate appreciable overlap between the charge clouds of cation and anion, i.e. there is some degree of covalency. Such a plot is shown schematically in Fig. 4.3. The minimum may be taken as the point where the cation ends and the anion begins, so that absolute ionic radii can be determined (subject to some reservations concerning the real existence of ions in 'ionic' crystals, which the method must arouse). Only a limited number of crystals have been dealt with in this way, but the results suggest that cations are consistently about 14 pm larger, and anions 14 pm smaller in radius than the 'traditional' values. The radii of Shannon and Prewitt can be modified accordingly, and many contemporary writers prefer these to the older values.

The various scales of ionic radii are in fact about equally good, within the criterion that, when added together, they should reproduce observed internuclear distances in crystals. The 'realistic' radii derived from electron density plots probably convey better the relative sizes of cations and anions. This is important, since 'radius ratio rules' have been widely invoked in crystal chemistry. The ratio $r(\text{cation})/r(\text{anion})$ will obviously vary significantly between 'traditional' and 'realistic' values.

For the moment, we are interested in periodic trends in the relative sizes of ions. The most important may be summarised as follows:

(1) The radius of an ion decreases with increasing positive charge. Compare, for example, the following isoelectronic series (all radii in pm):

Te^{2-}	I^-	Cs^+	Ba^{2+}	La^{3+}	Ce^{4+}
221	220	167	135	103	87

It will be noted, however, that the radii of Te^{2-} and I^- are practically equal. The same is true for the other pairs of isoelectronic Group 16 $(2-)$ and Group 17 $(1-)$ ions.

Where an element may exist in two or more oxidation states, the ionic radius decreases with increasing (positive) oxidation state; compare, for example, Fe^{2+} and Fe^{3+}, or Pb^{2+} and Pb^{4+} in Table 4.2.

(2) As with atomic radii, ionic radii increase as we go down any Group, for essentially the same reasons.

(3) Radii for ions having the same charge do not necessarily decrease in going across a Period. Along the 3d series, the variation for M^{2+} and M^{3+} ions is rather irregular. It would appear that crystal field effects are at work here. Remembering that the radii in Table 4.2 are those appropriate for octahedral six-coordination, crystal field stabilisation energies (CFSE) reach maxima for d^3 and d^8 configurations and are zero for d^0, d^5 (high-spin) and d^{10}. Comparing Ca^{2+} (d^0), Mn^{2+} (high-spin d^5) and Zn^{2+} (d^{10}), we observe a steady contraction; the same is true for Sc^{3+}, Fe^{3+} (high-spin) and Ga^{3+}. If we turn to the 4f series, there is now a monotonic decrease in ionic radii for M^{2+} and M^{3+} across the series. This may be attributed to the lack of any significant crystal field effects for the lanthanides; 4f orbital splittings are two orders of magnitude smaller than 3d splittings.

It may also be noted that 'inert pair' cations such as Tl^+ and Pb^{2+} are relatively large; Pb^{2+} is larger than Hg^{2+}, and not much smaller than Ba^{2+} despite the filling of the 4f and 5d subshells.

Table 4.2 *Ionic radii for six-coordination (pm)*

(*a*) Main Group ions									
Li^+					O^{2-}	F^-			
74					140	133			
Na^+	Mg^{2+}	Al^{3+}			S^{2-}	Cl^-			
102	72	53			184	184			
K^+	Ca^{2+}	Ga^{3+}			Se^{2-}	Br^-			
138	100	62			198	196			
Rb^+	Sr^{2+}	In^{3+}	Sn^{4+}		Te^{2-}	I^-			
149	116	80	69		221	220			
Cs^+	Ba^{2+}	Tl^{3+}	Pb^{4+}						
170	136	88	78						
		Tl^+	Pb^{2+}	Bi^{3+}					
		150	118	102					
(*b*) d block ions									
	Ti^{2+}	V^{2+}	Cr^{2+}	Mn^{2+}	Fe^{2+}	Co^{2+}	Ni^{2+}	Cu^{2+}	Zn^{2+}
	86	79	73(LS)	67(LS)	61(LS)	65(LS)	70	73	75
			82(HS)	82(HS)	78(HS)	74(HS)			
Sc^{3+}	Ti^{3+}	V^{3+}	Cr^{3+}	Mn^{3+}	Fe^{3+}	Co^{3+}			
75	67	64	62	58(LS)	55(LS)	55(LS)			
				65(HS)	65(HS)	61(HS)			
	Ti^{4+}	V^{4+}	Cr^{4+}	Mn^{4+}					
	61	59	55	54					

Y³⁺ 89	Zr⁴⁺ 72	Nb⁵⁺ 64	Mo⁴⁺ 65	Tc⁴⁺ 64	Ru⁴⁺ 62	Rh³⁺ 67	Pd²⁺ 86	Ag⁺ 115	Cd²⁺ 95
La³⁺ 105	Hf⁴⁺ 71	Ta⁵⁺ 64	W⁴⁺ 66	Re⁴⁺ 63	Os⁴⁺ 63	Ir³⁺ 68	Pt²⁺ 80	Au⁺ 137	Hg²⁺ 102
							Pt⁴⁺ 63	Au³⁺ 85	

LS = low-spin
HS = high-spin

(c) f block ions

Ce³⁺ 101	Pr³⁺ 101	Nd³⁺ 98	Pm³⁺ 98	Sm³⁺ 96	Eu³⁺ 95	Gd³⁺ 94	Tb³⁺ 92	Dy³⁺ 91	Ho³⁺ 90	Er³⁺ 89	Tm³⁺ 88	Yb³⁺ 86	Lu³⁺ 85
Ce⁴⁺ 80	Pr⁴⁺ 78			Sm²⁺ 117	Eu²⁺ 117		Tb⁴⁺ 76	Dy⁴⁺ 107			Tm²⁺ 103	Yb²⁺ 102	

Th⁴⁺ 94	Pa⁴⁺ 90	U⁴⁺ 89	Np⁴⁺ 87	Pu⁴⁺ 86	Am⁴⁺ 85	Cm⁴⁺ 85	Bk⁴⁺ 83	Cf⁴⁺ 82
	Pa⁵⁺ 78	U⁶⁺ 73	Np⁵⁺ 75	Pu³⁺ 100	Am³⁺ 98	Cm³⁺ 97	Bk³⁺ 96	Cf³⁺ 95

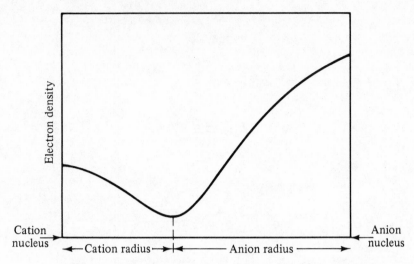

Fig. 4.3. Schematic plot of electron density in a typical 'ionic' crystal between cation and anion nuclei.

4.3   Ionisation potentials and electron affinities (or ionisation energies and electron-attachment energies)

The term *ionisation potential* (IP) refers to an intensive property of a single atom or ion in the gas phase. Measured spectroscopically, it is equal to the energy difference between E^{n+} and $E^{(n+1)+}$ ($n = 0, 1, 2, \ldots$), both species being in their respective ground states. Loosely, but perhaps more usefully, the ionisation potential of an atom or ion is equal to the energy required to remove its outermost electron. The ionisation potential is usually measured in electron volts (eV), although it is sometimes tabulated in the spectroscopic unit of reciprocal centimetres (cm^{-1}); $1\,\mathrm{eV} \equiv 8066\,\mathrm{cm}^{-1}$.

The term *ionisation energy* (IE) refers to a gaseous *substance* E^{n+}(g) ($n = 0, 1, 2, \ldots$), and is therefore an extensive property whose magnitude depends on the amount of substance present. The ionisation energy is therefore measured in kJ (or MJ) mol^{-1} ($1\,\mathrm{eV} \equiv 96.5\,\mathrm{kJ\,mol^{-1}}$). Thermodynamically, the ionisation energy of E^{n+}(g) is equal to ΔU^0 at absolute zero for the process:

$$E^{n+}(g) \rightarrow E^{(n+1)+}(g) + e$$

where both species are in their ground states.

Ionisation energies (as opposed to potentials) have become more fashionable in recent years, mainly because the electron volt is not recognised as an SI unit of energy. However, ionisation potentials are perfectly legitimate, and it is wrong to state, for example, that the ionisation energy of the sodium atom is $496\,kJ\,mol^{-1}$, or that it is $5.14\,eV$. It is correct to state that the ionisation potential of the Na atom is $5.14\,eV$, or that the ionisation energy of sodium (in the gas phase) is $496\,kJ\,mol^{-1}$.

One advantage of ionisation energy as opposed to ionisation potential is its applicability in thermochemical arguments, where we may be performing an energetic analysis of some reaction. Here, we are dealing with substances and the appropriate unit is $kJ\,mol^{-1}$. Usually, however, we want enthalpy changes ΔH^0 at 25 °C, rather than ΔU^0 at absolute zero. For the process:

$$E(g) \rightarrow E^{n+}(g) + ne(g)$$

we have the relation:

$$\Delta H^0 = \Delta U^0 + (5/2)nRT$$

The RT term amounts to $6.2n\,kJ\,mol^{-1}$ at 25 °C. This is small in relation to ionisation energies/enthalpies but is often far from trivial in thermodynamic arguments. However, it is rarely necessary to convert ionisation energies into enthalpies because, as we shall presently see, the RT terms ultimately cancel out. A number of tabulations of thermochemical data give enthalpies (or heats) of formation for ions in the gas phase; these include the RT correction.

The *electron affinity* of an atom or ion is the counterpart of the ionisation potential. It is an intensive property, defined as the energy *released* when the atom in its ground state accepts an electron, i.e. the difference in energy between the ground state of E and that of E^- with the sign convention that exothermic electron affinities are positive. Electron affinities, like ionisation potentials, are expressed in eV.

The *electron attachment energy* is an extensive property of a gaseous substance E(g) or $E^{n\pm}(g)$. It is defined in the case of E(g) as ΔU^0 at absolute zero for the process:

$$E(g) + e \rightarrow E^-(g)$$

It has the usual energy units of $kJ\,mol^{-1}$, and differs in sign from the electron affinity; for example, the electron affinity of the H atom is $0.75\,eV$, but the electron attachment energy of H(g) is $-73\,kJ\,mol^{-1}$. In order to convert electron attachment energies from ΔU^0 quantities at 0 K to ΔH^0 quantities at 298 K, it is necessary to make a correction:

$$\Delta H^0 = \Delta U^0 - (5/2)nRT$$

The sign of the RT term is opposite to that in the corresponding equation which converts ionisation energies into enthalpies. In a thermochemical analysis involving ionisation enthalpies, electron attachment enthalpies are sure to occur also, unless electrons are to appear in the overall stoichiometric equation under analysis. For example, if we are looking at the factors which determine ΔH^0 for the reaction:

$$Zn(s) + 2H^+(aq) \rightarrow Zn^{2+}(aq) + H_2(g)$$

among the steps in the thermochemical analysis will be:

$$Zn(g) \rightarrow Zn^{2+}(g) + 2e$$

and

$$2H^+(g) + 2e \rightarrow 2H(g)$$

The first of these involves the removal of two electrons from zinc atoms in the gas phase, and ΔH^0 for the process will be equal to the sum of the first and second ionisation enthalpies of $Zn(g)$, designated I_1 and I_2 respectively. For the second process, ΔH^0 will be equal to twice the electron attachment enthalpy of $H^+(g)$ (which is the same as $-2I$, where I is the ionisation enthalpy of atomic hydrogen). Obviously, we could use ionisation and electron attachment energies (instead of enthalpies), without correction for the RT terms which cancel out when we sum up the enthalpy changes for all the steps in the analysis. Thus for the process:

$$Zn(g) + 2H^+(g) \rightarrow Zn^{2+}(g) + 2H(g)$$

ΔH^0 is equal to the sum of the appropriate ionisation and electron attachment *energies* and the corresponding enthalpies need not be explicitly considered. Throughout this book, ionisation and electron attachment energies may, where appropriate, masquerade as enthalpies on the understanding that the RT corrections which ought to have been made ultimately cancel.

Periodic trends in ionisation energies

The reader should be familiar with the broad trends in the ionisation potentials of atoms as viewed within the periodic classification. The first ionisation energies I_1 for the atomic Main Group (s and p block) elemental substances are depicted in Fig. 4.4. Ionisation energies are expected to fall as we go down any Group, for much the same reasons as atomic radii increase; the increasing positive kinetic energy of the least strongly-bound electron tends to offset the effect of the increasing effective nuclear charge. In the p block, there are the expected discontinuities between the second and third members of each Group, as a consequence of the filling of the 3d subshell. This is most marked in Group 13, where gallium and aluminium have almost equal ionisation energies, but becomes less evident in succeeding Groups. There is a greater discontinuity between the fourth and fifth members, a consequence of filling the 4f subshell. Thus Tl and Pb have higher values of I_1

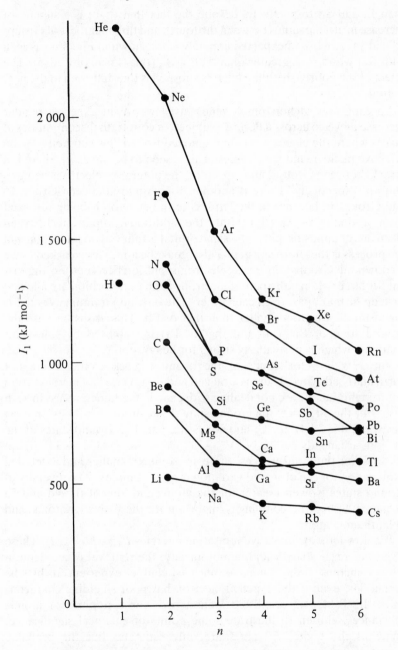

Fig. 4.4. First ionisation energies I_1 of Main Group atomic substances, plotted against n, the horizontal Period.

than In and Sn respectively, despite the fact that there is a significant increase in atomic radius between the fourth and fifth members of Groups 13 and 14. A more unexpected anomaly – not shown in Fig. 4.4 – is seen with Ra, whose I_1 is greater than that of Ba (this is true of I_2 also). The effect of the poorly-shielding filled 4f subshell is thus felt even in the next Period.

Looking now at horizontal trends in I_1, we might expect a regular increase as we go across a Period – subject, of course, to discontinuities at places where the electron is being removed from a new subshell – as the effective nuclear charge increases. This is seen to be the case. There is a 'break' between Group 2 and Group 13, the outermost electrons being ns and np respectively. There is another discontinuity between Group 15 and Group 16, because in the latter case the electron is being removed from a doubly-occupied orbital; the additional repulsion between electrons of opposite spin in the same orbital is relieved on ionisation and the process is therefore assisted. This 'break' becomes less marked as we go down the Groups, since interelectron repulsion becomes less important for larger atoms; the pair of electrons in an outer orbital are allowed to keep further apart from each other. The only major anomaly is the I_1 for bismuth, which is smaller than that of Pb. This arises because the ground state of Bi$^+$ (but not that of Bi) is stabilised by spin–orbit coupling, which is particularly strong for heavy atoms.

The I_1 values for atoms of the d and f block elements are not particularly edifying. There is a ragged and relatively slow increase from left to right, but we are not dealing with exactly the same process in each case. For the 3d series, the ground state of the atom M is $3d^n4s^2$ in most cases, but it is $3d^54s^1$ for Cr and $3d^{10}4s^1$ for Cu. The ground states of the M$^+$ ions are mostly $3d^n4s^1$, but V$^+$ is $3d^4$, Cr$^+$ is $3d^5$ and Cu$^+$ is $3d^{10}$. Even allowing for these differences, there are some anomalies and a detailed explanation for the observed variation is quite complex. The diversity of ground states is even greater for the atoms and ions of the 4d and 5d elements. Spin–orbit coupling is important for these heavier atoms, and in lanthanide atoms.

We now look at successive ionisation energies $I_1, I_2, I_3, \ldots, I_n$. These always increase sharply and monotonically; the removal of an electron always increases the effective nuclear charge experienced by the remainder, even if the expelled electron has poor shielding characteristics. In Table 4.3 are listed values of I_n for a number of atomic substances, chosen to illustrate some points of chemical significance. Some of the higher I_n values are only estimates but the probable uncertainty will not affect the general conclusions. The following points are noteworthy:

(1)   The atomic substances chosen all have similar values of I_1, but they differ considerably when we compare successive I_n values.

Table 4.3 *Successive ionisation energies* I_n *for selected atomic substances kJ mol*$^{-1}$. Values in parentheses are only approximate

	I_1	I_2	I_3	I_4	I_5	I_6	I_7	I_8
Sn	709	1412	2943	3930	6973			
Pb	716	1450	3081	4083	6638			
Mn	717	1509	3248	4940	6990	(9200)	11 508	18 956
Fe	759	1561	2957	5290	7240	(9600)	(12 100)	14 575
Ru	711	1617	2747	(4400)	(6500)	(8200)	(9 600)	(11 000)

(2) In the case of Sn, note the large jumps between I_2 and I_3, and between I_4 and I_5. The first two ionisations involve the removal of 5p electrons, the next two 5s electrons, and thereafter the 4d subshell is violated. As we shall discuss further in Chapter 5, a stable oxidation number n is often associated with a large increase in ionisation energy between I_n and I_{n+1}. The first four IEs of Pb are all greater than those of Sn, notwithstanding the greater size of the Pb atom. This, of course, is a consequence of the filling of the 4f subshell, whose poor shielding characteristics cause a large increase in effective nuclear charge between Sn and Pb. The effect is more marked on the 6s electrons than on the 6p, the former being more penetrating and hence more sensitive to the actual nuclear charge.

(3) For Mn and Fe, it is more difficult to discern any discontinuities of the sort noted for Sn and Pb, apart from the large jump between I_7 and I_8 for Mn which accompanies the violation of the argon core. In each case, the first two IEs involve the removal of 4s electrons, while succeeding IEs remove 3d electrons. The most noteworthy feature is the relatively low value of I_3 for Fe, the only I_n which is lower for Fe than for Mn (apart from I_8). (This effect is illustrated in a different way in Fig. 5.2 in Section 5.2.) The 'break' at the half-filled shell has the same origins as the corresponding discontinuity in the Main Group atoms, e.g. from N to O. The I_3 for Fe involves the removal of an electron from a doubly-occupied orbital ($d^6 \rightarrow d^5$) while that for Mn ($d^5 \rightarrow d^4$) involves the removal of an electron from a singly-occupied orbital. The higher I_3 of Mn is obviously connected with the instability of Mn(III) species compared with Fe(III).

(4) The successive IEs for Ru increase less sharply than those of Fe. The effect seems to be even more marked in the 5d series, although accurate higher IEs are not available. Thinking in terms of the ionic model, you may feel that higher oxidation states should be more acces-

sible for Ru than for Fe. This is certainly the case, although the ionic model is unlikely to be appropriate for such high oxidation states.

Further discussion of ionisation energies and the periodicity thereof will appear in Chapter 5; the case of Cu, Ag and Au is particularly instructive (Section 5.2).

Periodic trends in electron attachment energies

Nearly all the ionisation energies cited in this book have been quite accurately measured by experiment; those which have not been measured can usually be estimated by interpolation or extrapolation and are thought to be reliable enough for most purposes. Electron attachment energies, on the other hand, have been obtained by a number of methods, some of them of dubious accuracy. The 'best' available data for the electron attachment energies E_1 of the Main Group atomic substances are given in Table 4.4. The values for members of Groups 1, 16 and 17, and for a few others (B, C, Si, P, Cu, Ag, Au), have been directly determined by means of electron impact experiments on the gaseous atomic substances with a high degree of accuracy (the estimated uncertainty is less than $1 \, kJ \, mol^{-1}$). In the case of H, a quantum-mechanical calculation of the ground state energy of H^- can be performed (and is probably as accurate as any experimental determination could be). In most other cases, the electron attachment energies are obtained by extrapolation from ionisation energy data.

If we plot the ionisation energies of A^-, A, A^+, A^{2+} etc. against charge, a smooth curve is obtained and can be fitted to a polynomial or other simple function. Where no reliable value for E_1 is available, a curve obtained from the ionisation energies I_1, I_2 etc. can be extrapolated to yield the ionisation energy of A^-, i.e. $-E_1$ for A. Uncertainties of 10–$50 \, kJ \, mol^{-1}$ are estimated for this procedure.

A little surprisingly perhaps, most electron attachment energies appear to be negative, i.e. the addition of an electron is exothermic and the electron affinity is positive. In other words, a nucleus having n protons is usually able to hold $(n + 1)$ electrons, unless the additional electron has to go into a higher subshell. The electron attachment energies of the Group 2 and Group 12d elemental substances and the noble gases are often quoted as zero, but are more likely to be positive.

Given the experimental uncertainties and the relatively low magnitudes of electron attachment energies compared with ionisation energies, it is not worthwhile to attempt any extensive rationalisation of their periodicity. We will restrict ourselves to a few points of some chemical significance. The most negative values are found in Group 16 and Group 17. This is not surprising if we note that the effective nuclear

Table 4.4 *Electron attachment energies (in kJ mol^{-1}) for Main Group atomic substances*

H							He
−73							−21
Li	Be	B	C	N	O	F	Ne
−60	−37	−8	−122	−20	−141	−328	−29
Na	Mg	Al	Si	P	S	Cl	Ar
−53	21	−44	−134	−72	−200	−349	36
K	Ca	Ga	Ge	As	Se	Br	Kr
−48	186	−30	−115	−77	−195	−325	41
Rb	Sr	In	Sn	Sb	Te	I	Xe
−47	146	−30	−120	−100	−190	−294	44
Cs	Ba	Tl	Pb	Bi	Po	At	Rn
−46	46	−30	−105	−105	−186	−270	—

charges experienced by outer electrons increase as we go along a Period, and these atoms have vacancies in low-lying orbitals.

The values for S and Cl are more negative than for O and F, which may seem a little surprising. Since the latter have higher ionisation energies, their 2p orbitals clearly lie at lower energy than the 3p orbitals of S and Cl; they should, you may think, be more willing to accept an additional electron. But in all arguments involving electrons, we must consider interelectron repulsion as well as orbital energy. In the small F$^-$ ion, the repulsion among the 2p electrons assists the expulsion of one of their number, notwithstanding the high effective nuclear charge they experience; this effectively reduces the ionisation energy of F$^-$, i.e. it makes the electron attachment energy of F less negative.

Second electron attachment energies E_2 are the values of ΔU^0 at absolute zero for the process:

$$A^-(g) + e \rightarrow A^{2-}(g)$$

As might be expected, these are large and positive; the addition of an electron to a negatively-charged species is always endothermic. No direct experimental measurements have been made, but estimates can be obtained from thermochemical data where 2− ions are formed in crystalline solids. The values of E_2 for oxygen and sulphur are approximately 800 and 600 kJ mol^{-1} respectively. These endothermic values do not necessarily preclude the existence of O^{2-} and S^{2-} as chemical entities, of course; far greater energies are needed to obtain cations bearing more than one positive charge.

Finally in this section, we should note the relatively exothermic electron attachment energies of Cu, Ag and Au: −118, −126 and

$-223 \, \text{kJ} \, \text{mol}^{-1}$ respectively. The value for gold might suggest a tendency to form Au^- in crystalline solids. CsAu does have 'ionic' properties, and might be formulated as Cs^+Au^-.

4.4   Electronegativity

Linus Pauling introduced the intuitive concept of electronegativity, defined as 'a measure of the tendency of an atom, in a molecule, to attract electrons to itself'. Few concepts in chemistry have proved to be as useful as electronegativity; yet quantitative measurements of electronegativity and the establishment of a numerical scale have led to much controversy. If Pauling's definition be accepted, an ideal scale should be based upon a molecular property which can be accurately measured and which can be fairly rigorously interpreted in terms of the electron-attracting powers of the atoms therein. None of the many scales which have been devised – three or four are in regular use today – really fits the bill. What is surprising is the remarkable extent of agreement among the various scales, despite the very different methods used to calculate them.

The Pauling method

The original, and still the most familiar scale of electronegativity to most chemists, is that of Pauling (1932), based on bond energy data. (Bond energies are discussed in more detail in Section 6.2.) For simplicity, we will consider diatomic molecules AB, of bond order 1 (e.g., HCl, IBr). Pauling proposed that the dissociation energy D_{AB} (the energy needed to break the A–B bond) should be given by:

$$D_{AB} = \tfrac{1}{2}(D_{AA} + D_{BB}) + \Delta_{AB}$$

where D_{AA} and D_{BB} are respectively the dissociation energies of the homonuclear species A_2 and B_2 respectively and Δ_{AB} represents the enhancement of the bond energy of AB due to ionic–covalent resonance; if the molecule AB has appreciable polarity, it can be represented by the resonance structures:

$$A—B \leftrightarrow A^+B^- \leftrightarrow A^-B^+$$

The excess bond energy Δ_{AB} was postulated to be proportional to $(x_A - x_B)^2$, where x_A and x_B are the electronegativities of A and B respectively. Thus a large difference in electronegativity between A and B will lead to a large contribution from one of the ionic resonance structures A^+B^- and A^-B^+ (whichever places the positive charge on the less electronegative atom) and hence to a large enhancement of the bond energy. If A and B are of equal electronegativity, the ionic structures will make only small (but equal) contributions. The proportionality constant a in the equation:

$$D_{AB} = \tfrac{1}{2}(D_{AA} + D_{BB}) + a(x_A - x_B)^2$$

takes the value $96.5 \, \text{kJ} \, \text{mol}^{-1}$ (i.e. $1 \, \text{eV}$).

The thermochemical data upon which Pauling set up his original scale were sparse and in many cases inaccurate. The electronegativities of the Group 1 atoms were based on the enthalpies of formation of crystalline solids (such as NaCl), rather upon data for molecular substances, since at the time no reliable bond energies were available for species such as NaCl(g). Such data have become available in the past 50 years, and the agreement with Pauling's equation – using his original electronegativities – is very poor.

A further problem was posed by the Group 1 hydride molecules. Thus it was found that the dissociation energy of NaH(g) was lower than the arithmetic mean of the dissociation energies for $Na_2(g)$ and $H_2(g)$. This awkwardness could be rectified by using the geometric rather than the arithmetic mean; but most published tabulations of Pauling electronegativities appear to rely on the original method.

Many chemists would regard the Pauling scale as a worthy pioneering effort, nowadays of largely historical interest, while acknowledging that electronegativities – however they may be obtained – are of value in rationalising bond energies. Other chemists, having been brought up on Pauling's scale and still finding it valuable, stick to the numbers which they committed to memory as students. The Pauling scale is still the most widely-reproduced collection of electronegativities, especially in elementary texts.

The Mulliken method

Mulliken proposed (1934) that electronegativities could be obtained from ionisation potentials and electron affinities (we use here the terms appropriate for atoms, rather than atomic substances). If the bonding in a diatomic molecule AB can be represented by the resonance structures:

$$A{-}B \leftrightarrow A^+B^- \leftrightarrow A^-B^+$$

the relative weights to be given to the charged structures A^+B^- and A^-B^+ will depend on their relative energies; whichever is lower in energy will make the greater contribution to the overall structure. If A is of lower electronegativity than B, A^+B^- will be lower in energy than A^-B^+, and vice versa. It is easy to show that the difference in energy between the ionic structures is given by:

$$E(A^+B^-) - E(A^-B^+) = (I_B + E_B) - (I_A + E_A)$$

where I_A and I_B are the ionisation potentials of A and B respectively, and E_A and E_B are the respective electron affinities. If A and B are of equal electronegativity, the ionic structures will be of equal energy and the right-hand side of the equation will be equal to zero. This suggests that the quantity $(I_A + E_A)$ may be taken as a measure of the electronegativity of A, which seems intuitively reasonable; an atom which has a strong tendency to accept an electron (a large and positive E_A) and resists strongly the loss of an electron (large I_A) is expected to have a high electronegativity.

There are a number of practical difficulties, however, in setting up a Mulliken scale of electronegativity. The quantity I_A in the above analysis is not always the same as the readily-obtained I_1, the first ionisation potential. Consider as an example the HCl molecule, where we might consider the resonance structures:

$$H—Cl \leftrightarrow H^+Cl^- \leftrightarrow H^-Cl^+$$

An important rule in resonance theory insists that the contributing structures must all have the same total spin. Now HCl, H^+, H^- and Cl^- in their ground states are all diamagnetic, all electrons being paired. But Cl^+ in its ground state (isoelectronic with S) has two unpaired electrons whose spins will be parallel in order to minimise repulsion. The ionisation potential I_1 of the Cl atom is 12.97 eV (corresponding to an ionisation energy of 1251 kJ mol^{-1} for the atomic substance) but this refers to the ionisation process where both species are in their ground states. What we need in order to determine the Mulliken electronegativity of Cl is the difference in energy between the Cl atom in its ground state and Cl^+ in an excited state where its total spin is zero. There are two such spectroscopic states of Cl^+, designated ^1S and ^1D, which lie respectively at 3.46 and 1.44 eV above the ground state. Taking a weighted mean of these energies – allowing for the fivefold degeneracy of ^1D – we have to add 1.78 eV to the ionisation potential of Cl to obtain the required I_A. There are cases where the necessary spectroscopic data for making this kind of correction are not available.

A further complication is introduced if we recognise (in VB language) that an atom in a molecule is usually considered to be in a valence state which lies at higher energy than its ground state. We therefore should take this into account in the determination of I_A and E_A. The necessary correction is straightforward if the hybridisation of the atom is clear-cut and if further spectroscopic transition energies have been measured. Thus, for example, we find for carbon electronegativities in the sequence $sp > sp^2 > sp^3$, which is consistent with much experimental evidence. However, for many atoms the required spectroscopic data are lacking, and the hybridisation is uncertain unless a very elaborate quantum-mechanical calculation is performed.

The Mulliken electronegativities have a good linear relationship with the Pauling values, and the former are usually adjusted for consistency with the latter. Mulliken electronegativities can also be specified for polyatomic groups, such as CH_3, CF_3, CN etc. (Pauling values for these are available as well). Objections to the Mulliken method are that it relies upon properties of the isolated atom, rather than on molecular properties, and that many values are unobtainable. On the other hand, it has a better theoretical basis than other methods – indeed, it could be said that the Mulliken method provides the theoretical basis for the very concept of electronegativity.

The Allred–Rochow method

Allred and Rochow (1956) proposed that the electrostatic force acting on an electron at the periphery of an atom was a measure of the atom's electronegativity. This force is proportional to Z^*/r^2, where Z^* is the effective nuclear charge acting on the outer electron and r is the atomic radius. Values of Z^* can be estimated by means of a very simple empirical method due to Slater. The resulting values of Z^*/r^2 have a good linear relationship with Pauling's values and are accordingly adjusted, using the equation:

$$x_A = 3600 Z^*/r^2 + 0.74$$

where r is in pm.

Like the Mulliken method, the Allred–Rochow method defines electronegativities in terms of atomic rather than molecular properties. Another objection is that it fails in the case of hydrogen; the above equation yields a value which would give H an electronegativity comparable with O. The value of 2.2 for hydrogen is rather arbitrarily chosen. Nevertheless, the Allred–Rochow scale has won widespread acceptance among inorganic chemists who find it consistent with many features of chemical periodicity.

Periodic trends in electronegativity

In Fig. 4.5, we show how the Allred–Rochow electronegativities of the Main Group atoms vary along each Period and down each Group. In the s block – apart from a minor anomaly between Li and Na – the electronegativities decrease monotonically down the Groups. In Group 13, however, there is a pronounced discontinuity, with Ga having an electronegativity closer to that of B than Al. Proceeding to the later Groups, this discontinuity becomes less marked. The reason lies, of course, in the poorly-shielding filled 3d subshell; there is a large increase in the effective nuclear charge from Al to Ga but only a small increase in atomic radius. Many features of the aforementioned 'middle element anomaly' (Section 4.1) can be explained in terms of the high electro-

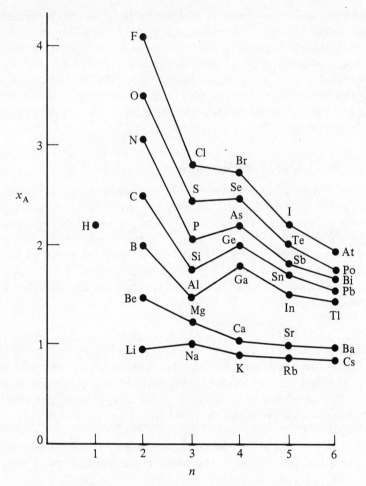

Fig. 4.5. Allred–Rochow electronegativities x_A of Main Group atoms, plotted against Period n.

negativities of their atoms. The small discontinuity between the fourth and fifth members of each Group is a distant effect of the filled 4f subshell.

There is a uniform increase in electronegativity across each Period; the largest increments occur between Group 2 and Group 13 where there is an intervening nd subshell.

Electronegativities are of little value in transition element chemistry and no detailed tabulation need be given. A few useful points are in order, however. Across the 3d series, there is a steady increase, with a maximum of 1.75 at copper, falling slightly to 1.66 at zinc; this arises from the larger radius of Zn, as discussed in Section 4.2, and it would be dangerous to make too much of such variations. The Allred-Rochow values are obviously sensitive to the atomic radius, some values of which may be anomalous because of peculiarities in the elemental substance that have little chemical relevance elsewhere. Along the 4d and 5d series the trend is rather irregular and the Allred–Rochow electronegativities are open to serious doubt. For example, Au and Hg are accorded the relatively low values of 1.42 and 1.44 respectively. Chemical evidence would be consistent with considerably higher values, as obtained by other methods; the Pauling electronegativities for Au and Hg are 2.5 and 2.2 respectively.

The electronegativities of the lanthanides are all around 1.1, or comparable with calcium. Those of the actinides appear to be somewhat larger.

It has been proposed, and it seems intuitively reasonable, that electronegativity increases with the oxidation state. In the Pauling scale, it is sometimes possible to specify different values for, e.g., Sn(II) and Sn(IV). In the Mulliken method, the fact that different oxidation states are usually associated with different valence states makes it possible to incorporate some variability of the electronegativity of an atom from one molecule to another. No such flexibility is possible within the Allred–Rochow method. There is, however, a way of incorporating the concept of electronegativity's dependence on oxidation state without having to quote different values for each state. Many authors have found it useful to consider the idea of electronegativity equalisation. When the atoms come together, charge will flow from less electronegative to more electronegative atoms. As the fractional positive charge on an atom increases, so will its electronegativity; conversely, the electronegativity of an atom will fall as it acquires a fractional negative charge. The equilibrium charge distribution will be reached when all atoms have the same electronegativity. Very roughly, one might take this to be the average of the original electronegativities. Clearly, then, the electronegativity of the S atom in H_2S is less than that in SO_2, which is less than that in SO_3.

4.5   Further reading

See the books listed in Section A.5 of the Appendix. Among the standard inorganic texts (see Section A.3), Huheey (1983) has particularly extensive tabulations of ionic radii and ionisation energies, while Moeller (1982) has the most complete tabulation of atomic/covalent radii. Extensive collections of radii, ionisation energies, electron affinities and electronegativities are given by Ball, M. C. and Norbury, A. H. (1974). *Physical Data for Inorganic Chemists*. London: Longman.

5

The occurrence of simple ions $E^{n\pm}$ in crystals and solutions

5.1 Lattice energies

It is assumed that the reader has some knowledge of the thermodynamics involved in the formation of ionic solids. This topic is adequately treated in all textbooks of inorganic or 'general' chemistry, and detailed accounts are found in the monographs cited at the end of this chapter. The most important exothermic quantity is the lattice energy, or lattice enthalpy, of the ionic crystal. The distinction between lattice energy and lattice enthalpy is somewhat pedantic. The lattice energy is usually defined as ΔU^0 at absolute zero for the formation of the ionic substance from the gaseous ions, while the lattice enthalpy is defined as ΔH^0 at 298 K for the same process. These quantities differ by only a few kJ mol^{-1}; the experimental uncertainty in the enthalpy of formation of the ionic solid is often comparable in magnitude. We therefore use the more familiar term 'lattice energy', although lattice enthalpy would be more appropriate in most of the situations discussed in this chapter.

Given the enthalpy of formation of an ionic solid, an experimental lattice energy can be obtained by thermochemical analysis. For example, the formation of crystalline sodium chloride is broken down as follows:

	ΔH^0
$Na(s) \rightarrow Na(g)$	ΔH^0_{atom}
$Na(g) \rightarrow Na^+(g) + e$	I_1
$\frac{1}{2}Cl_2(g) \rightarrow Cl(g)$	$\frac{1}{2}D$
$Cl(g) + e \rightarrow Cl^-(g)$	E_1
$Na^+(g) + Cl^-(g) \rightarrow NaCl(s)$	U_L
$Na(s) + \frac{1}{2}Cl_2(g) \rightarrow NaCl(s)$	ΔH^0_f

The atomisation enthalpy of elemental sodium ΔH_{atom}^0, the first ionisation energy of atomic sodium I_1, the dissociation enthalpy D of gaseous chlorine, the electron attachment energy E_1 of atomic chlorine and the enthalpy of formation ΔH_f^0 of crystalline sodium chloride can all be taken from standard tabulations of experimental data. An experimental lattice energy U_L is thus given by:

$$U_L = \Delta H_f^0 - \Delta H_{atom}^0 - I_1 - \tfrac{1}{2}D - E_1$$

But if the enthalpy of formation is unknown, the lattice energy has to be calculated. Comparisons between experimental and calculated lattice energies are obviously useful and are often invoked in justification of the ionic model (subject to the reservations set out in Sections 1.4 and 3.3).

Lattice energies are invariably negative in sign; when we speak of one solid having a higher lattice energy than another, it should be understood that 'higher' means 'more negative'.

The conventional treatment of lattice energies, with the non-Coulombic repulsion term rendered as an exponential function $e^{-\varrho r}$ rather than the alternative r^{-n}, leads to the Born–Meyer equation:

$$U_L = -\frac{138800 A z_a z_c}{r}\left(1 - \frac{\varrho}{r}\right)(kJ\,mol^{-1})$$

in which A is the Madelung constant, characteristic of the structure type and obtainable by summing the electrostatic (Coulombic) repulsions and attractions throughout the crystal. Constants for the most common structures adopted by plausible ionic solids are given in Table 5.1. z_a and z_c are the charges on the anion and cation respectively. r is the shortest cation–anion distance in pm. The constant ϱ is usually assigned the value 34.5 pm. Extended expressions include terms dealing with the London attraction (arising from the mutual polarisation of ions) and the zero-point energy.

A simpler approach to lattice energy calculations was suggested by the Russian chemist Kapustinskii. His expression:

$$U_L = -\frac{121400 v z_a z_c}{r_a + r_c}\left(1 - \frac{34.5}{r_a + r_c}\right)(kJ\,mol^{-1})$$

requires no Madelung constant and hence no knowledge of the crystal structure. r_a and r_c are respectively the radii of the anion and cation appropriate for octahedral six-coordination, in pm. v is the total number of ions in the stoichiometric unit which specifies one mole of the

Table 5.1 *Selected Madelung constants*

Structure	Constant
Zinc blende ZnS (4:4)	1.64
Wurtzite ZnS (4:4)	1.64
Sodium chloride NaCl (6:6)	1.75
Caesium chloride CsCl (8:8)	1.76
Quartz SiO_2 (4:2)	2.21
Cadmium chloride $CdCl_2$ (6:3)	2.25
Cadmium iodide CdI_2 (6:3)	2.19
Rutile TiO_2 (6:3)	2.39
Fluorite CaF_2 (8:4)	2.52

substance, e.g. 2 for NaCl, 3 for CaF_2, 5 for Al_2O_3. Any of the various collections of ionic radii will do for r_a and r_c; the sum $(r_a + r_c)$ differs much less among these than the individual (and rather meaningless) ionic radii. For ionic solids AB where 6:6 coordination has been established experimentally, the observed shortest A–B distance may be used for $(r_a + r_c)$.

Kapustinskii's equation is the algebraic sum of two quantities. The attractive (negative) term represents the net effect of the Coulombic, point-charge interactions among all the ions in the lattice. The repulsive term arises from the non-Coulombic interaction between the electron clouds on neighbouring ions. It will be apparent that the latter assumes greater prominence as the radius sum $(r_a + r_c)$ decreases. Kapustinskii's equation works very well for halides and oxides of the s block elements having NaCl, CsCl, rutile or fluorite structures. Results for some representative halides are set out in Table 5.2. Apart from fluorides formally containing highly-charged cations, the 'experimental' or cycle lattice energies are invariably more negative than the Kapustinskii values. The discrepancies are greatest for halides having layer structures. As noted in Chapter 3, these may be better described as polymeric covalent solids rather than as ionic solids; this applies also to the monohalides of Cu and Ag, whose three-dimensional structures can be construed as ionic but whose bonding can also be described in terms of directed, two-centre covalent bonds (see Section 3.3). The discrepancies between calculated and experimental values can be improved – although not completely eliminated in the worst cases – by making corrections for polarisation. The mutual polarisation of ions leads to London-type attraction and enhances the lattice energy. The magnitude of this term depends on the polarisabilities of the ions, and upon their respective

Table 5.2 *'Experimental'* and calculated (Kapustinskii equation) lattice energies of selected crystalline halides. All energies are in kJ mol^{-1}

Substance	Structure	$-U_L$(expt.)	$-U_L$(calc.)	Difference
LiF	NaCl	1034	1001	+33
NaCl	NaCl	786	757	+29
KBr	NaCl	682	660	+22
RbI	NaCl	630	600	+30
CsI	CsCl	604	567	+37
CuCl	ZnS	996	767	+229
AgF	NaCl	967	849	+118
AgCl	NaCl	915	767	+148
MgF_2	Rutile	2957	2955	+2
$MgCl_2$	Layer	2526	2486	+40
$MgBr_2$	Layer	2440	2368	+72
MgI_2	Layer	2327	2200	+127
BaF_2	Fluorite	2352	2361	−9
$BaCl_2$	$PbCl_2$[a]	2056	2048	+8
$BaBr_2$	$PbCl_2$	1985	1966	+19
BaI_2	$PbCl_2$	1877	1848	+29
ZnF_2	Rutile	3032	2921	+111
$ZnCl_2$	Layer	2734	2462	+272
$ZnBr_2$	Layer	2678	2346	+332
ZnI_2	Layer	2605	2180	+425
AlF_3	Ionic	6215	6380	−165
$AlCl_3$	Layer	5492	5308	+184
LaF_3	Ionic	4857	5243	−386
CeF_3	Ionic	8391	9553	−1162

[a] $PbCl_2$ has a three-dimensional structure in which the Pb atom has a coordination number of 7, plus two more at a greater distance, often rendered as (7 + 2).

polarising powers. Polarisability increases with increasing size and charge, for both cations and anions; cations which do not have noble gas configurations (e.g. Ag^+, Pb^{2+}) are much more polarisable than those which do. Since cations tend to be less polarisable than anions, the polarisation of the anion by the cation is usually dominant. The polarising power of a cation upon an anion increases with increasing charge on the cation, and decreasing cation radius. The ratio (charge)/(radius) is often used as a rough measure of polarising power. However, the electron affinity of the cation is also important. For example, Mg^{2+} and Zn^{2+} are comparable in size, but the latter appears to be much more polarising. This may be attributed to the higher ionisation potentials of zinc, giving Zn^{2+} a higher electron affinity (see also Sections 5.3 and 5.5).

Polarisation in 'ionic' crystals accounts qualitatively for many of the discrepancies between calculated and experimental lattice energies, and is tantamount to an admission that the ionic model is in many cases inappropriate. Significant polarisation means a breakdown of the 'hard sphere' model of an ionic crystal, and implies some degree of overlap between the electron clouds of cation and anion, i.e. covalent bonding. Even in a crystal of MgF_2 – where the Kapustinskii equation reproduces the cycle lattice energy most impressively – there may be appreciable covalency. It is possible that the 'real' ionic contribution to the bonding is less than that obtained by a calculation which assumes 100% ionic character, but that the sum of the ionic and covalent contributions happens fortuitously to coincide with the ionic result. We have already noted (Chapter 1) that the ionic model 'works' almost quantitatively in calculations of the bond energies in some molecular species such as BF_3 and SiF_4 (see also Section 3.3).

Section 5.2 is devoted to an attempt to rationalise the stabilities of solids which may be viewed – albeit with some circumspection – as ionic. Of particular interest is the stabilisation of oxidation states of elements which (at least formally) are represented as cations in the substances under scrutiny. We will have to present a number of thermochemical calculations in order to demonstrate the stabilities or otherwise of some representative compounds; these will necessarily involve the estimation of lattice energies. The Kapustinskii equation is not sufficiently reliable for general use. It does, however, offer a useful guide and may be used to interpret or predict a number of trends. For example, in the case of halides MX_n, the lattice energy is easily shown to be proportional to $n(n + 1)$. Thus for a given M and X, the lattice energies of MX, MX_2, MX_3 and MX_4 should follow the ratios $1:3:6:10$. This takes no account of the decreasing ionic radius of M^{n+} with increasing n, whose effect upon the denominator in the principal term of the Kapustinskii equation will be to increase the ratios. Table 5.3 cites some examples of fluorides (which are most likely to be ionic) in justification of these ratios. Allowing for the contraction in ionic radius with increasing charge, which will be most important where the cation charge differs by two units in the two fluorides considered, the agreement between experiment and theory is quite gratifying. Given an element which forms MF_3, but no other fluoride, we should be able to estimate the lattice energies of MF, MF_2 or MF_4 by reference to the experimental value for MF_3.

Table 5.3 *Experimental lattice energies (in kJ mol^{-1}) of selected fluorides MF_n where M exhibits different oxidation states*

Fluoride	Lattice energy	Ratio
AgF	-972	
AgF$_2$	-2951	3.04
TlF	-848	
TlF$_3$	-5447	6.42
FeF$_2$	-2951	
FeF$_3$	-5932	2.01
EuF$_2$	-2491	
EuF$_3$	-5032	2.02
PbF$_2$	-2527	
PbF$_4$	-9473	3.75
CeF$_3$	-4915	
CeF$_4$	-8391	1.71

5.2   Thermochemical treatments of some ionic solids

We first look at the fluorides of barium. Only BaF$_2$ is known, a typically ionic solid having the fluorite (8:4) structure. From Table 5.2, we see that the calculated lattice energy is very close to the experimental value; in other words, we can calculate the enthalpy of formation of BaF$_2$(s) almost within the limits of experimental uncertainty. Why have BaF$_3$ and BaF not been prepared? Presumably they are thermodynamically unstable with respect to other species. In order to verify this supposition, let us estimate the enthalpies of formation ΔH_f^0 of BaF(s) and BaF$_3$(s), assuming these to be ionic.

The formation of BaF$_3$(s) from the elemental substances can be broken down as follows:

		ΔH^0 (kJ mol^{-1})
Ba(s)	$\rightarrow$ Ba(g)	180
Ba(g)	$\rightarrow$ Ba$^+$(g) + e	503
Ba$^+$(g)	$\rightarrow$ Ba^{2+}(g) + e	965
Ba^{2+}(g)	$\rightarrow$ Ba^{3+}(g) + e	3454
$\frac{3}{2}$F$_2$(g)	$\rightarrow$ 3F(g)	237
3F(g) + 3e	$\rightarrow$ 3F$^-$(g)	-984
Ba^{3+}(g) + 3F$^-$(g)	$\rightarrow$ BaF$_3$(s)	-4800 ± 100
Ba(s) + $\frac{3}{2}$F$_2$(g)	$\rightarrow$ BaF$_3$(s)	-445 ± 100

All the required data can be obtained from standard tabulations except for the lattice energy of $BaF_3(s)$, which has to be calculated or estimated.

The Kapustinskii equation is unreliable for trifluorides (Table 5.2) but the lattice energy of BaF_3 can be estimated to be a little more than twice that of BaF_2, for which an experimental value is given in Table 5.2. Our estimate of $-4800 \pm 100\,\text{kJ}\,\text{mol}^{-1}$ is in line with the experimental value for LaF_3 (Table 5.2).

The formation of $BaF_3(s)$ is predicted to be exothermic to an extent which should override any entropy considerations. This result may surprise some readers, who might think the noble gas core of Ba^{2+} to be inviolable. Why, then, is BaF_3 nonexistent? Although it is evidently stable with respect to the elemental substances, BaF_3 is unstable to decomposition, to BaF_2 and fluorine gas. This may be demonstrated by the following thermochemical breakdown (rounding enthalpy changes to the nearest $10\,\text{kJ}\,\text{mol}^{-1}$):

		$\Delta H^0\,(\text{kJ}\,\text{mol}^{-1})$
$BaF_3(s)$	$\rightarrow Ba^{3+}(g) + 3F^-(g)$	$+4800 \pm 100$
$Ba^{3+}(g) + e$	$\rightarrow Ba^{2+}(g)$	-3450
$F^-(g)$	$\rightarrow F(g) + e$	$+330$
$F(g)$	$\rightarrow \frac{1}{2}F_2(g)$	-80
$Ba^{2+}(g) + 2F^-(g)$	$\rightarrow BaF_2(s)$	-2350
$BaF_3(s)$	$\rightarrow BaF_2(s) + \frac{1}{2}F_2(g)$	-750 ± 100

Thus the decomposition of BaF_3 is certainly favourable thermodynamically, and the proximity of the ions in a three-dimensional crystal should facilitate an easy electron-transfer mechanism. The instability of BaF_3 clearly arises from the fact that the gain in lattice energy in going from BaF_2 to BaF_3 is more than outweighed by the large third ionisation energy of Ba.

We now look at the unknown $BaF(s)$. An obvious route to this substance would be to treat barium metal with a deficiency of elemental fluorine or some other fluorinating agent:

$$2Ba(s) + F_2(g) \rightarrow 2BaF(s)$$

But in practice we find:

$$2Ba(s) + F_2(g) \rightarrow BaF_2(s) + Ba(s)$$

This suggests that BaF is unstable with respect to disproportionation, i.e. the reaction:

$$2BaF(s) \rightarrow Ba(s) + BaF_2(s)$$

is thermodynamically favourable. The disproportionation breaks down as follows:

$$\Delta H^0$$

2BaF(s)	$\rightarrow 2\text{Ba}^+(\text{g}) + 2\text{F}^-(\text{g})$	$-2U_L(\text{BaF,s})$
$\text{Ba}^+(\text{g})$	$\rightarrow \text{Ba}^{2+}(\text{g}) + e$	I_2
$\text{Ba}^+(\text{g}) + e$	$\rightarrow \text{Ba(g)}$	$-I_1$
Ba(g)	$\rightarrow \text{Ba(s)}$	$-\Delta H^0_{\text{atom}}$
$\text{Ba}^{2+}(\text{g}) + 2\text{F}^-(\text{g}) \rightarrow \text{BaF}_2(\text{s})$		$U_L(\text{BaF}_2,\text{s})$

2BaF(s)	$\rightarrow \text{Ba(s)} + \text{BaF}_2(\text{s})$

The enthalpy change for the reaction will therefore be given by:

$$\Delta H^0 = -2U_L(\text{BaF,s}) + I_2 - I_1 - \Delta H^0_{\text{atom}} + U_L(\text{BaF}_2,\text{s})$$

We need an estimate for the lattice energy of BaF(s). From the Kapustinskii equation, this should be about one-third of the lattice energy for $\text{BaF}_2(\text{s})$. Alternatively, we might argue that it should be close to the lattice energy of CsF, Cs^+ and Ba^+ being presumably of similar size. From the experimental lattice energies of CsF ($-750\,\text{kJ mol}^{-1}$) and BaF_2 ($-2352\,\text{kJ mol}^{-1}$) we estimate the lattice energy of BaF to be about $-765 \pm 20\,\text{kJ mol}^{-1}$. We can then calculate ΔH^0 for the disproportionation as written above to be about $-540\,\text{kJ mol}^{-1}$. The enthalpy of formation of BaF(s) at 298 K is estimated to be about $-340\,\text{kJ mol}^{-1}$. Thus although the formation of BaF from the elemental substances is exothermic, it is unstable with respect to disproportionation. Analysis of the underlying factors shows that the high lattice energy of BaF_2 compared with BaF, together with the relatively small difference $(I_2 - I_1)$ which is the major endothermic term, can be held responsible for the instability of BaF.

Could any barium(I) halide be stable as an ionic solid? With increasing anion size, the difference in lattice energy between BaX and BaX_2 becomes less advantageous to the latter. Even BaI, however, can be shown to be unstable, albeit less so than BaF:

$$2\text{BaI(s)} \rightarrow \text{Ba(s)} + \text{BaI}_2(\text{s}) \quad \Delta H^0 = -360\,\text{kJ mol}^{-1}\ (\text{approx.})$$

Now look at the fluorides of lanthanum, next-door to Ba in the Periodic Table. Only LaF_3 is known. Given that its enthalpy of formation is $-1732\,\text{kJ mol}^{-1}$, that the atomisation enthalpy of lanthanum metal is $423\,\text{kJ mol}^{-1}$ (Table 7.1), and that the first four ionisation energies of lanthanum are 538, 1067, 1850 and $4820\,\text{kJ mol}^{-1}$, the reader should have little difficulty in estimating the enthalpies of formation of LaF, LaF_2 and LaF_4. The lower fluorides can be shown to be quite unstable with respect to disproportionation to LaF_3 and lanthanum metal, while LaF_4 is unstable to decomposition to LaF_3 and fluorine gas, although its formation from the elemental substances is exothermic. Similarly, given the enthalpy of formation of $\text{LaI}_3(\text{s})$ ($-700\,\text{kJ mol}^{-1}$), the reader can estimate the enthalpies of formation of LaI, LaI$_2$ and LaI$_4$. The results are conveniently summarised in Fig. 5.1, on which we plot the enthalpies of formation (experimental values for BaX$_2$ and LaX$_3$, otherwise estimated values) against the oxidation number of the more electropositive atom. Diagrams of this type can be interpreted in much the same way as the free energy diagrams for ions in solution which the reader may have encountered (see Section 5.4). The slope of a line joining two points represents the degree of difficulty in going from the lower to the higher oxidation state. Thus we see that Ba(s) is more readily oxidised to BaX$_2$ than to BaX, while La(s) is more easily oxidised to LaX$_3$ than to LaX or LaX$_2$. The oxidation of BaX$_2$ to BaX$_3$, or LaX$_3$ to LaX$_4$, is clearly 'uphill'

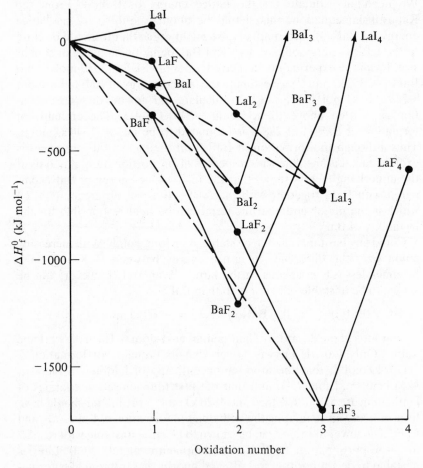

Fig. 5.1. Enthalpies of formation ΔH_f^0 for barium and lanthanum fluorides and iodides; experimental values for BaX$_2$ and LaX$_3$, otherwise calculated values as explained in text.

and thermodynamically unfavourable. BaX, LaX and LaX_2 appear as convex – as opposed to concave – points on the diagram, which shows at a glance their instability to disproportionation. For example, in order that LaF_2 be stable to the disproportionation:

$$3LaF_2(s) \rightarrow 2LaF_3(s) + La(s)$$

its enthalpy of formation must be at least two-thirds of that of LaF_3, in which case it would lie below the line joining La(s) – the zero oxidation state – and LaF_3.

It will be apparent that although no BaX_3 or LaX_4 can be stable, the fluorides are less unstable than the iodides, as a consequence of the much greater gain in lattice energy for the former which accompanies an increase in oxidation state. Likewise, the lower oxidation states Ba(I), La(I) and La(II), though also unstable, are less so for the iodides than for the fluorides. It will also be apparent to the reader who has performed the calculations necessary to draw up Fig. 5.1 that the high atomisation enthalpy of La discourages the lower oxidation states and assists their disproportionation.

Evidently the most – practically the only – stable oxidation state of La in ionic compounds is III. Does this hold for the later members of the lanthanide series? Fig. 5.1 suggests that the I oxidation state has little prospect of stability, given the high atomisation enthalpies and the relatively low second and third ionisation energies. The II oxidation state has better prospects, however. Consider the disproportionation:

$$3LnX_2(s) \rightarrow 2LnX_3(s) + Ln(s)$$

where Ln is one of the 4f elements, and X is a halogen. The thermochemical breakdown of this reaction can be performed as follows:

		ΔH^0
$3LnX_2(s)$	$\rightarrow 3Ln^{2+}(g) + 6X^-(g)$	$-3U_L(LnX_2,s)$
$2Ln^{2+}(g)$	$\rightarrow 2Ln^{3+}(g) + 2e$	$2I_3$
$Ln^{2+}(g) + 2e$	$\rightarrow Ln(g)$	$-(I_1 + I_2)$
$Ln(g)$	$\rightarrow Ln(s)$	$-\Delta H^0_{atom}$
$2Ln^{3+}(g) + 6X^-(g)$	$\rightarrow 2LnX_3(s)$	$2U_L(LnX_3,s)$

Thus the standard enthalpy change for the disproportionation can be written as:

$$\Delta H^0 = 2U_L(LnX_3,s) - 3U_L(LnX_2,s)$$
$$+2I_3 - I_1 - I_2$$
$$-\Delta H^0_{atom}$$

As we traverse the series, the lattice energies of LnX_2 and LnX_3 increase steadily in magnitude, a consequence of the lanthanide contraction; since U_L is always a little more than twice as great for LnX_3 as for LnX_2 (as indicated by the Kapustinskii equation), this trend should favour the III oxidation state. But this is more than offset by the steady increase in the third ionisation energy, subject to discontinuities at Gd and Lu; in the case of Gd, the third electron removed is from a 5d orbital instead of a 4f orbital for Ce–Eu and Tb–Yb, while in Lu the 5d electron is removed first and the third ionisation energy corresponds to the removal of a 6s electron. The thermochemical terms relevant to the disproportionation of LnF_2 (Ln = La, Eu) are compared below. (All energies in kJ mol^{-1}.)

	La	Eu
$-3U_L(LnF_2,s)$	7200 (est.)	7472
$2U_L(LnF_3,s)$	-9714	-10064
$2I_3 - I_1 - I_2$	2095	3176
$-\Delta H^0_{atom}$	-423	-175
	-842	$+409$

Thus the higher I_3 of Eu is mainly responsible for the limited stability of its II oxidation state, although the lower atomisation enthalpy helps as well. As shown on Fig. 5.2, the third ionisation energy follows the sequence:

$$La < Ce < Pr < Nd < Pm < Sm < Eu > Gd < Tb$$
$$< Dy \sim Ho \sim Er < Tm < Yb > Lu$$

This correlates very well with the stability of the II oxidation state in ionic crystals. Thus only Sm, Eu and Yb form difluorides. The dichlorides should be more stable to disproportionation since the lattice energies of $LnCl_2$ and $LnCl_3$ must be considerably lower than those of the corresponding fluorides. $NdCl_2$, $SmCl_2$, $EuCl_2$, $DyCl_2$, $TmCl_2$ and $YbCl_2$ have all been prepared. $PmCl_2$ is probably stable but the chemistry of promethium (which has no long-lived isotopes) has been inadequately explored. The nonexistence of $HoCl_2$ and $ErCl_2$ may be attributed to their anomalous third ionisation energies – their atomisation enthalpies, too, are a little higher than might be expected (Table 7.1).

We now turn to the 3d series elements. The dihalides and trihalides can be treated as ionic solids, although the chlorides, bromides and iodides adopt layer structures which might be better viewed as polymeric covalent crystals. In Fig. 5.2 the third ionisation energies of the 3d atoms are plotted alongside those of the lanthanides. These all involve the removal of an electron from a 3d orbital; from Fe onwards, the orbital concerned is doubly occupied so that spin-pairing energy assists the ionisation. This accounts for the 'break' between Mn and Fe, as previously discussed (Section 4.3). The increase from Sc to Mn, and from Fe to Zn, is much sharper than the corresponding increases in the lanthanide series. However, the 'break' at the half-filled shell is less abrupt for the 3d series. This explains why the II oxidation state – which is

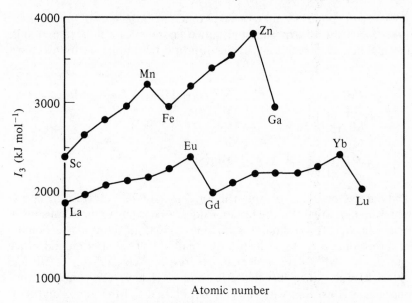

Fig. 5.2. Third ionisation energies I_3 for 3d and 4f atomic substances.

quite unstable for Sc – appears earlier in the 3d series than in the 4f series, and does not vanish at the half-filled shell. For the lanthanides, the II oxidation state has limited stability towards the end of each half of the series; III is the most important state for *all* the lanthanides (EuI_3 is the only unstable trihalide, and EuO is the only stable monoxide). However, among simple binary compounds and salts, the III oxidation state becomes strongly oxidising towards the right-hand side of the 3d series. Thus all the 3d elements (save scandium) form stable monoxides, and triiodides are unstable from Mn onwards.

The ionic model is of limited applicability for the heavier transition series (4d and 5d). Halides and oxides in the lower oxidation states tend to disproportionate, chiefly because of the very high atomisation enthalpies of the elemental substances. Many of the lower halides turn out to be cluster compounds, containing 'metal–metal' bonds (see Section 8.5). However, the ionic model does help to rationalise the tendency for high oxidation states to dominate in the 4d and 5d series. As an example, we look at the fluorides MF_3 and MF_4 of the triad Ti, Zr and Hf. As might be expected, the reaction between fluorine gas and the elemental substances leads to the formation of the tetrafluorides MF_4. We now investigate the stabilities of the trifluorides MF_3 with respect to the disproportionation:

$$4MF_3(s) \rightarrow 3MF_4(s) + M(s)$$

As a starting point, we take the experimental enthalpies of formation of MF_4. From these, experimental lattice energies can be obtained. The thermochemical breakdown of the disproportionation is as follows:

$$\Delta H^0$$

$4MF_3(s)$	$\rightarrow 4M^{3+}(g) + 12F^-(g)$	$-4U_L(MF_3,s)$
$3M^{3+}(g)$	$\rightarrow 3M^{4+}(g) + 3e$	$3I_4$
$M^{3+}(g) + 3e$	$\rightarrow M(g)$	$-(I_1 + I_2 + I_3)$
$M(g)$	$\rightarrow M(s)$	$-\Delta H^0_{atom}$
$3M^{4+}(g) + 12F^-(g)$	$\rightarrow 3MF_4(s)$	$3U_L(MF_4,s)$

The lattice energies of the trifluorides may be inferred from the experimental values for the tetrafluorides. According to the Kapustinskii equation, $U_L(MF_3,s)$ should be 0.6 times $U_L(MF_4,s)$ – this may err on the generous side, since M^{3+} will be larger than M^{4+}. The experimental lattice energies of TiF_4, ZrF_4 and HfF_4 are respectively -9910, -8970 and $-9110\,kJ\,mol^{-1}$. The fact that ZrF_4 and HfF_4 have very similar lattice energies is consistent with the comparable sizes of the atoms and their cations, as a consequence of the lanthanide contraction. The relevant thermochemical data can be collected thus (all energies in $kJ\,mol^{-1}$):

	Ti	Zr	Hf
$-4U_L(MF_3,s)$	23784	21528	21864
$3U_L(MF_4,s)$	-29730	-26910	-27330
$3I_4$	12525	9939	9645
$-(I_1 + I_2 + I_3)$	-4621	-4145	-4370
$-\Delta H^0_{atom}$	-470	-609	-619
	$+1488$	-197	-810

Thus we may predict that TiF_3 should be stable, HfF_3 unstable and ZrF_3 rather marginal, but probably unstable given that our estimated lattice energy for ZrF_3 is generous. This is in agreement with the experimental facts. TiF_3 is known; its experimental lattice energy is $-5690\,kJ\,mol^{-1}$, compared with the estimated value of $-5946\,kJ\,mol^{-1}$ in the above calculation. This, of course, reflects the smaller size of Ti^{4+} compared with Ti^{3+}, which was not taken into account in estimating the ratio $U_L(TiF_3,s)/U_L(TiF_4,s)$ from the Kapustinskii equation. The 'true' enthalpies of disproportionation for ZrF_3 and HfF_3 are likely to be several hundred $kJ\,mol^{-1}$ more negative than the calculated values, reinforcing our insistence that ZrF_3 and HfF_3 are unstable. (Reference can be found in the literature to ZrF_3 as a known substance, but this is probably $ZrOF_2$; crystallographers have difficulty in distinguishing between F and O atoms, especially in a lattice where O^{2-} and F^- ions are randomly distributed.) But let us not concern ourselves unduly about the absolute validity of the calculated enthalpies of disproportionation. Let us concentrate on the underlying factors which determine the trend of decreasing stability down the triad. The difference in lattice energy between $4MF_3$ and $3MF_4$ is favourable to disproportionation. Going from Ti to Zr, the increase in atomic size makes this term less important. But the lower ionisation energies of Zr compared with Ti counteract the lattice energy term; and the atomisation enthalpy of Zr is greater. The net result is to the disadvantage of the trifluoride. Going now from Zr to Hf, there is little change in atomic/ionic radii and hence lattice energies. But although the first three ionisation energies of Hf are greater than those of Zr, its fourth ionisation energy is lower. As noted in Section 4.3, successive ionisations from 5d orbitals require less energy than from 4d or 3d orbitals, as a consequence of the greater penetration of nd orbitals with increasing n and their lower sensitivity to the increasing charge on the atom. The result is that the ionisation energy term ($3I_4 - I_1 - I_2 - I_3$) is some $500\,kJ\,mol^{-1}$ smaller for Hf compared with Zr.

Having examined some ionic compounds of s, d and f block elements, we turn now to the p block. The formation of cations as chemical entities by atoms of the p block elements is somewhat restricted. As examples, we compare the halides of Al and Tl. Aluminium forms only one fluoride, AlF_3; this may be fairly described as an ionic solid. Thallium forms TlF and TlF_3. The other halides of Al are best regarded as covalent solids; $AlBr_3$ and AlI_3 are molecular solids, containing Al_2X_6 molecules, while $AlCl_3$ has a layer structure. All the thallium monohalides are known and have at least some ionic characteristics. $TlCl_3$ and $TlBr_3$ have layer structures and are only marginally stable with respect to decomposition to TlX and X_2. Thallium(III) iodide is unknown. The decreasing stability of the higher oxidation state down the Group is in marked contrast to the d block, as discussed above. We look at the factors which determine the stability or otherwise of MF (M = Al, Tl) with respect to the disproportionation:

$$3MF(s) \rightarrow MF_3(s) + 2M(s)$$

The usual thermochemical analysis gives the standard enthalpy change for this reaction as:

$$\Delta H^0 = -3U_L(MF,s) + U_L(MF_3,s) + I_2 + I_3 - 2I_1 - 2\Delta H^0_{atom}$$

Experimental data are available for all of these quantities except for the lattice energy of AlF. This we estimate as $-950\,kJ\,mol^{-1}$, which is consistent with the experimental lattice energies of NaF and MgF_2 (-923 and $-2957\,kJ\,mol^{-1}$ respectively). It also leads to about equal ratios

$U_L(MF_3,s)/U_L(MF,s)$ for the aluminium and thallium fluorides. The relevant energies (in kJ mol^{-1}) are given below:

	Al	Tl
$-3U_L(MF,s)$	2850	2535
$U_L(MF_3,s)$	-6215	-5427
$I_2 + I_3$	4562	4849
$-2I_1$	-1156	-1178
$-2\Delta H^0_{atom}$	-652	-364
	-611	$+415$

All three factors – lattice energies, ionisation energies and atomisation enthalpies – work in the same direction, to stabilise Tl(I) relative to Al(I). For the smaller Al, the gain in lattice energy in going from 3AlF to AlF$_3$ is greater than the corresponding term for thallium. But this is not compensated by lower ionisation energies for thallium (cf. Ti and Hf). The first ionisation energies of Al and Tl are very comparable, but the second and third ionisation energies of Tl are significantly greater, notwithstanding the larger size of the Tl atom compared with Al. The strongly-penetrating 6s orbitals are poorly shielded from the nucleus by the filled 4f and 5d orbitals; this is the 'inert pair' effect, mentioned in Section 4.3. The lower atomisation enthalpy of thallium metal also contributes to the stability of its I oxidation state.

It is left to the reader as an exercise to estimate the enthalpies of formation of AlF$_2$ and TlF$_2$ and to show that these are unstable to disproportionation. It need hardly be added that AlF$_4$ and TlF$_4$ are out of the question. The fourth ionisation energies of Al and Tl are respectively 11 578 and c. 5000 kJ mol^{-1}, compared with the estimated gains in lattice energy in going from MF$_3$ to MF$_4$ of 4400 and 3800 kJ mol^{-1} respectively.

It is instructive to compare the p block elements of Group 13 with the 3d transition elements with respect to the stabilisation of oxidation states in ionic fluorides. Why are the difluorides of Al, Ga, In and Tl all unstable, while V, Cr, Mn, Fe and Co all form both difluorides and trifluorides? The higher atomisation enthalpies of the 3d elemental substances should be to the disadvantage of the lower oxidation state by favouring disproportionation. The 3d series M^{3+} ions are not very different in size from Ga^{3+}, and the lattice energies of their trifluorides lie between those of GaF$_3$ and InF$_3$. The decisive difference rests in the ionisation energies. The disproportionation of MF$_2$ to MF$_3$ and the elemental substance involves the algebraic sum $(2I_3 - I_1 - I_2)$. A large value for this quantity will favour the stability of MF$_2$ against such disproportionation. A comparison is shown in Table 5.4. Apart from Sc and Ti (ScF$_2$ is unknown and TiF$_2$ is doubtful), the ionisation energy term is much less favourable to the II oxidation state for the Group 13 elements than for the 3d elements. The jump in ionisation energy between I_2 and I_3 is always greater in the 3d series; the third ionisation always involves the removal of a 3d electron, while the second usually (though not always) involves a 4s electron. In Group 13, the biggest jump occurs between the first and second ionisations, which respectively involve the removal of an np and an ns electron. This has the effect of stabilising the II state for the 3d atoms and the I state for the Group 13 atoms.

The data in Table 5.4 suggest that gallium, of all the Group 13 elements, is most likely to exhibit the II oxidation state. A thermochemical calculation suggests that $GaCl_2$ might be on the brink of stability with respect to disproportionation to $GaCl_3$ and the elemental substance, and that $GaBr_2$ and GaI_2 should certainly be stable. A substance having the stoichiometry $GaCl_2$ is known, but (as noted in Section 3.2) its structure shows that it is best formulated as $Ga^+[GaCl_4]^-$. In other words, Ga(II) may be stable in some circumstances with respect to disproportionation to Ga(III) and Ga(0), but it is unstable with respect to Ga(I) + Ga(III) (note, however, the existence of the $[Cl_3Ga—GaCl_3]^{2-}$ ion; Ga(II) can be stabilised by M–M bond formation).

We can now summarise the factors which determine the stability or otherwise of M^{n+} in crystals. We are talking about stability with respect to oxidation and/or reduction; a stable cation is one which is neither easily reduced nor easily oxidised by other ions in the lattice. M^{n+} is likely to be

Table 5.4 *Values of $(2I_3–I_1–I_2)$ for selected atomic substances (in $kJ\ mol^{-1}$)*

Element	$(2I_3–I_1–I_2)$
Sc	2912
Ti	3338
V	3592
Cr	3729
Mn	4272
Fe	3596
Co	4060
Ni	4298
Cu	4404
Zn	5026
Al	3095
Ga	3368
In	3031
Tl	3196

stable against oxidation if the ionisation energy I_{n+1} is large, and if the atom M and its cations are large, so that the difference in lattice energy between, say, MX_3 and MX_4 does not compensate for the investment in ionisation energy. Small anions (especially F^- or O^{2-}) lead to the greatest lattice energies, and hence tend to stabilise high oxidation states in ionic crystals. Low oxidation states tend to be unstable with respect to disproportionation. Such instability is likely if, for an M^{n+} ion, the ionisation energy I_{n+1} is not much greater than I_n (or I_{n-1} etc., where appropriate). It is also assisted if the elemental substance M(s) has a high atomisation enthalpy. In other words, cations which are easily oxidised (low I_{n+1}) and easily reduced (high I_n, high $\Delta H^\circ_{\text{atom}}$) may be subject to disproportionation, especially if the associated anion is relatively small (e.g. F^-, O^{2-}), in which case the difference in lattice energy between the lower and the higher oxidation state will favour disproportionation. A cation M^{n+} is likely to be stable, other things being equal, if there is a large jump in ionisation energy between I_n and I_{n+1}.

These points are well illustrated by comparing Cu, Ag and Au with respect to the relative stabilities of their oxidation states. Although few compounds formed by these elements can properly be described as ionic, the model can quite successfully rationalise the basic facts. The copper Group 11d is perhaps the untidiest in the Periodic Table. For Cu, II is the most common oxidation state; Cu(I) compounds are quite numerous but have some tendency towards oxidation or disproportionation, and Cu(III) compounds are rare, being easily reduced. With silver, I is the dominant oxidation state; the II oxidation state tends to disproportionate to I and III. For gold, III is the dominant state; I tends to disproportionate and II is very rare. No clear trend can be discerned. The relevant quantities are the ionization energies I_1, I_2 and I_3; the atomisation enthalpies of the metallic substances; and the relative sizes of the atoms and their cations. These are collected below: I_n and the atomisation enthalpies ΔH^0_{atom} are in kJ mol^{-1} and r, the metallic radii, are in pm.

	I_1	I_2	I_3	r	ΔH^0_{atom}
Cu	746	1958	3554	128	339
Ag	731	2074	3361	144	284
Au	890	1980	2900	144	366

Comparing the data for Cu and Ag, it is not difficult to see why Ag(I) is more stable than Cu(I) with respect to disproportionation. As might be expected, I_1 is slightly lower for Ag than for Cu; but I_2 is unexpectedly higher for Ag. The larger difference $(I_2 - I_1)$ for silver is the dominant factor in stabilising Ag(I); the larger size of the Ag atom, and the smaller atomisation enthalpy of the elemental substance, also discourage disproportionation. Why is I_2 for Ag higher than for Cu? The second ionisation involves the removal of an electron from the filled nd subshell. The process is assisted by interelectron repulsion, but to a lesser extent for 4d electrons compared with 3d; the greater radial extension of 4d orbitals allows the electrons to keep their distance from each other. However, I_3 for Ag is smaller than for Cu; as noted in discussing the data in Table 4.3, higher ionisation energies involving the removal of 4d electrons tend to be lower than those which involve the removal of 3d electrons. Given relatively low values of I_1 and I_3, and an unexpectedly high I_2, we can see why Ag(II) compounds tend to disproportionate. For example, AgO is not silver(II) oxide; it is best formulated as Ag(I)Ag(III)O$_2$. Turning now to gold, we note the higher I_1, the lower I_3 and the higher atomisation enthalpy compared with silver. These clearly render the I state less stable to disproportionation. The relatively small jump between I_2 and I_3 is to the disadvantage of the II state. Thus III is the dominant state for gold. The high I_1 – which, of course, involves the removal of a 6s electron – arises from the stability of the 6s orbital following the filling of the 4f and 5d subshells. The low I_3 is in keeping with the observation that successive ionisations from d orbitals increase less sharply with increasing n in the sequence 3d > 4d > 5d.

This section has been largely devoted to ionic halides. The same considerations are valid for oxides, *mutatis mutandis*. As noted in Section 4.3, the electron attachment energy of O$^-$ is highly endothermic and the reader may think O^{2-} to be rather improbable as a viable chemical species. But the lattice energies of oxides are large enough to overcome this. From the Kapustinskii equation, and ignoring the small difference in radius between F$^-$ and O^{2-}, it can be predicted that the lattice energy of MO will be 1.33 times that of MF$_2$, and the lattice energy of MO$_2$ will be 1.2 times that of MF$_4$. This ensures that the formation of many ionic oxides is exothermic. The Kapustinskii equation predicts also that the lattice energies of M$_2$O, MO, M$_2$O$_3$ and MO$_2$ should follow the ratios 1:1.33:5:4. These are seen from Table 5.5 to be approximately valid.

Ionic solids containing polyatomic anions can also be subjected to thermochemical analysis. Their lattice energies can be estimated from the Kapustinskii equation by use of thermochemical radii for the anions; tabulations of such radii will be found in the monographs cited at the end of this chapter. It should be noted that these bear no relation to the actual sizes of the anions, and are purely empirical quantities. For example, the thermochemical radius of IO$_3^-$ is less than that of ClO$_3^-$, despite the larger size of the I atom compared with Cl. All this means is that iodates tend to have larger lattice energies than the corresponding chlorates; the electrostatic attraction depends on the fractional charge on the O atoms in the anions, and will increase with decreasing electronegativity of the central atom.

Table 5.5 *Experimental lattice energies for some oxides, comparing calculated (Kapustinskii) and experimental ratios for different oxidation states of elements (all lattice energies in kJ mol⁻¹)*

| | | Ratio | |
Oxide	Lattice energy	expt.	calc.
Cu_2O	$-3\,273$	1.24	1.33
CuO	$-4\,050$		
Tl_2O	$-2\,659$	5.53	5.00
Tl_2O_3	$-14\,702$		
FeO	$-3\,865$	3.82	3.75
Fe_2O_3	$-14\,774$		
Ce_2O_3	$-12\,574$	1.20	1.25
CeO_2	$-10\,518$		
PbO	$-3\,520$	3.19	3.00
PbO_2	$-11\,217$		

5.3   Ionic or covalent bonding?

Here we consider the factors which determine whether a given compound prefers an ionic structure or a covalent one. We may imagine that for any binary compound – e.g. a halide or an oxide – either an ionic or a covalent structure can be envisaged, and these alternatives are in thermochemical competition. Bear in mind that there may be appreciable covalency in 'ionic' substances, and that there may be some ionic contribution to the bonding in 'covalent' substances. Since there is no simple means – short of a rigorous MO treatment – of calculating covalent bond energies, and since quantitative calculations based upon the ionic model are subject to some uncertainties, the question of whether an ionic or a covalent structure is the more favourable thermodynamically cannot be answered in absolute terms. We can, however, rationalise the situation to some extent.

As an example, let us pose the question: why does BF_3 adopt a molecular structure, while AlF_3 is apparently ionic? As shown in Table 5.2, the ionic model (using the Kapustinskii equation) gives a fair approximation to the thermochemistry of formation of AlF_3. Let us estimate the enthalpy of formation of a hypothetical ionic substance $BF_3(s)$, having a structure similar to that of AlF_3. The lattice energy can be estimated by means of the Kapustinskii equation. We require the radius of six-coordinate B^{3+}; in the absence of any empirical data, this is estimated by Pauling's method to be 27 pm, about half the radius of Al^{3+}. Does the more negative lattice energy of $BF_3(s)$ compared with that of $AlF_3(s)$ compensate for the higher ionisation energies of $B(g)$? The analysis is as follows:

	$\Delta H^0 (\text{kJ mol}^{-1})$
$B(s) \rightarrow B(g)$	573
$B(g) \rightarrow B^{3+}(g) + 3e$	6888
$\frac{3}{2}F_2(g) \rightarrow 3F(g)$	233
$3F(g) + 3e \rightarrow 3F^-(g)$	-984
$B^{3+}(g) + 3F^-(g) \rightarrow BF_3(s)$	-7142
$B(s) + \frac{3}{2}(g) \rightarrow BF_3(s)$	-432

The formation of $BF_3(s)$ from the elemental substances is apparently exothermic; but the experimental enthalpy of formation of $BF_3(g)$ is $-1137\,\text{kJ mol}^{-1}$.

Bearing in mind that the Kapustinskii equation tends to overestimate the lattice energies of trifluorides (Table 5.2), we can be confident that ionic BF_3 is not thermodynamically competitive with molecular BF_3. We may even have been generous in our assessment of $BF_3(s)$. The Kapustinskii lattice energy is based upon an ionic radius sum of 160 pm $(133 + 27)$. This is unrealistically small; the Kapustinskii equation demands ionic radii appropriate to octahedral six-coordination (never observed for boron) and an ionic radius sum of 160 pm for $BF_3(s)$ implies a distance of 226 pm between nearest F^- ions in the lattice. This seems unreasonably short, in view of the ionic radius of 133 pm for F^- (but remember that absolute ionic radii are subject to considerable doubt). Taking this value for the radius, the distance between two F^- ions in a crystal should be not less than 266 pm, so that the ionic radius sum where the cation is octahedrally coordinated should be not less than 188 pm. In crystalline AlF_3, the shortest Al–F distances are actually 179 pm, with shortest F–F distances of 253 pm. No shorter distances than these are to be found in any other ionic fluoride, and AlF_3 may well represent the most compact array of F^- ions in a crystal lattice, in which case its lattice energy $(-6215\,\text{kJ mol}^{-1})$ represents the limit attainable in an ionic trifluoride. Thus although there are difficulties in estimating the enthalpy of formation of ionic $BF_3(s)$, the reader should be in no doubt that it is unstable with respect to molecular $BF_3(g)$. Compared with AlF_3, the higher ionisation energies required to attain B^{3+} are not compensated by any proportionate advantage in lattice energy. For chlorides, bromides and especially iodides, these restrictions will become increasingly severe as the anion size increases.

Between the extremes of ionic and molecular substances, we have polymeric crystals (e.g. CdI_2, $AlCl_3$, SiO_2) which could be described as assemblies of cations and anions but which may be better viewed in terms of covalent bonding. In such crystals there are doubtless contributions from both ionic and covalent bonding. Where the latter assumes some importance, the structure adopted may reflect the need to maximise it, even at the expense of the ionic component in the bonding. For example, nearly all crystalline difluorides have the rutile ($6:3$) or fluorite ($8:4$) structures. In most cases where MF_2 has the rutile structure, the other dihalides adopt layer structures of the $CdCl_2$ or CdI_2 type. In the rutile structure, each F^- ion is surrounded in trigonal planar fashion by three cations. A covalent description of the bonding, in terms of VB theory, would require sp^2 hybridisation at the F atom with the lone pair in a pure 2p orbital. Such a situation has never been observed in any covalently-bonded molecule or discrete polyatomic ion containing fluorine. In the $CdCl_2$ and CdI_2 structures, the halogen atom has pyramidal three-coordination, as in discrete species such as ClO_3^- and IO_3^- whose bonding can be described in much the same way (see Chapter 6).

In more refined treatments, the lattice energy is proportional to the Madelung constant (Table 5.1). From the viewpoint of maximising the ionic bonding, the rutile structure would be preferable to the $CdCl_2$ or CdI_2 structures for an AB_2 solid. A factor which favours the two-dimensional layer structure ($CdCl_2/CdI_2$) over the three-dimensional rutile structure for AB_2 solids is the stronger London attraction between B atoms in adjacent layers. This will be particularly advantageous where B is large and polarisable. This factor – aside from any arguments about the extent of ionic character in the bonding – may well determine which type is adopted by a given dihalide.

Simple MO arguments (Chapter 7) are helpful in assessing qualitatively the extent of the covalent contribution. The dominant interaction, leading to the formation of strongly-bonding MOs, will involve the highest occupied AOs on the respective atoms. For example, in $ZnCl_2(s)$, covalent bonding will mainly involve the combination of zinc 4s and chlorine 3p orbitals. Such combination is most effective if the AOs concerned are comparable in energy. The energies of the principal valence orbitals of atoms are closely paralleled by their electronegativities, so that the electronegativity difference between M and X in MX_n may determine (roughly) the relative magnitudes of ionic and covalent contributions to the bonding. Where this difference is relatively small (e.g. between zinc and the heavier halogens), the covalent contribution is large enough to favour a structure which optimises this component. Where the difference is large (e.g. between Ca and F), the structure adopted is that which maximises the ionic bonding. The variation of electronegativity with oxidation number should not be forgotten.

5.4 Energetics of simple ions in solution

The remainder of this chapter is concerned with the stabilities of ions (mainly cations) in aqueous solution, with respect to oxidation, reduction and disproportionation. Ions in solution are surrounded by solvent molecules, oriented so as to maximise ion–dipole attraction (although there may be appreciable covalency as well). The hydration number of an ion in aqueous solution is not always easy to determine experimentally; it is known to be six for most cations, but may be as low as four for small cations of low charge (e.g. Li^+) or as high as eight or nine for larger cations (e.g. La^{3+}).

An important thermochemical quantity associated with the formation and stability of an ion in aqueous solution is its hydration enthalpy ΔH^0_{hyd}, the enthalpy change under standard conditions for the process:

$$X^{n\pm}(g) \rightarrow X^{n\pm}(aq)$$

This, of course, is always negative, and plays the same role in aqueous thermochemistry as the lattice energy does in the energetics of ionic solids. The hydration enthalpy cannot be measured directly, and many thermodynamicists frown upon this or any other single-ion quantity. For example, the enthalpy of solution of sodium chloride can be measured and subjected to the following analysis:

$$(1) \quad NaCl(s) \rightarrow Na^+(g) + Cl^-(g)$$
$$(2) \quad Na^+(g) \rightarrow Na^+(aq)$$
$$(3) \quad Cl^-(g) \rightarrow Cl^-(aq)$$

$$\overline{NaCl(s) \rightarrow Na^+(aq) + Cl^-(aq)}$$

The enthalpy change for the first step is simply the negative of the lattice energy, which can be obtained experimentally (though indirectly). From this and the enthalpy of solution of NaCl, we can obtain the sum of the hydration enthalpies of Na^+ and Cl^-, but not the individual values. The problem is analogous to the determination of ionic radii from internuclear distances. From solubility and other thermochemical data, it is possible to draw up a table of hydration enthalpies relative to the proton. There is now general agreement that the hydration enthalpy ΔH^0_{hyd} for H^+ can be assigned the absolute value of $-1091 \pm 10 \, kJ \, mol^{-1}$, so that absolute hydration enthalpies for other ions can be obtained. Some representative values are given in Table 5.6. For cations having noble gas configurations, the hydration enthalpies are in excellent agreement with the empirical formula:

$$\Delta H^0_{hyd}(M^{n+}) = -\frac{70\,000n^2}{r+80} \quad (kJ\,mol^{-1})$$

Table 5.6 *Ion hydration enthalpies* (*kJ* mol^{-1})

H^+	-1091	Zn^{2+}	-2046	Pm^{3+}	-3427
		Cd^{2+}	-1807	Sm^{3+}	-3449
Li^+	-519	Hg^{2+}	-1824	Eu^{3+}	-3501
Na^+	-409			Gd^{3+}	-3517
K^+	-322	Sn^{2+}	-1556	Tb^{3+}	-3559
Rb^+	-293	Pb^{2+}	-1481	Dy^{3+}	-3567
Cs^+	-264			Ho^{3+}	-3623
		Sm^{2+}	-1444	Er^{3+}	-3637
Cu^+	-593	Eu^{2+}	-1458	Tm^{3+}	-3664
Ag^+	-473	Yb^{2+}	-1594	Yb^{3+}	-3706
				Lu^{3+}	-3722
In^+	-344	Al^{3+}	-4665		
Tl^+	-326	Ga^{3+}	-4700	Zr^{4+}	-6593
		In^{3+}	-4112	Hf^{4+}	-7120
Be^{2+}	-2494	Tl^{3+}	-4105	Ce^{4+}	-6309
Mg^{2+}	-1921			Th^{4+}	-6136
Ca^{2+}	-1577	Sc^{3+}	-3897	U^{4+}	-6470
Sr^{2+}	-1443	Ti^{3+}	-4154		
Ba^{2+}	-1305	V^{3+}	-4375	F^-	-515
		Cr^{3+}	-4560	Cl^-	-381
Ti^{2+}	-1862	Mn^{3+}	-4544	Br^-	-347
V^{2+}	-1918	Fe^{3+}	-4430	I^-	-305
Cr^{2+}	-1904	Co^{3+}	-4651	OH^-	-460
Mn^{2+}	-1841			NO_3^-	-314
Fe^{2+}	-1946	Y^{3+}	-3583		
Co^{2+}	-1996	La^{3+}	-3278	S^{2-}	-1495
Ni^{2+}	-2105	Ce^{3+}	-3326	SO_4^{2-}	-1059
Cu^{2+}	-2100	Pr^{3+}	-3373	CO_3^{2-}	-1314
Ag^{2+}	-1931	Nd^{3+}	-3403		

where r is the Pauling crystal radius of the ion in pm. The equation is less satisfactory if Shannon–Prewitt radii are used. An expression of this general form can be derived theoretically from purely electrostatic considerations of ion–dipole attraction. For cations having the d^{10} configuration, the experimental values of $-\Delta H^0_{hyd}$ are always well in excess of those calculated. This may be taken to indicate a substantial covalent contribution to the bonding between these strongly-polarising cations and water molecules. Ions of the transition elements having partly-filled d subshells fall between these extremes, and crystal field stabilisation energy (CFSE) has to be considered in order to explain the trends observed across the series.

For simple anions such as halide and S^{2-}, the experimental hydration enthalpies are in good agreement with the empirical formula:

$$\Delta H^0(X^{n-}) = -\frac{70\,000 n^2}{r} \quad (\text{kJ mol}^{-1})$$

where r is the Pauling radius in pm.

Notice that the denominator does not include the addition of 80 pm, as is required for cations. This may be taken to reflect the different orientations of water molecules in the two cases:

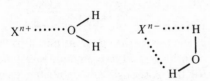

The distance between the fractionally-positive H atom and the anion will be rather smaller than that between the O atom and a cation, other things being equal. A similar expression works reasonably well for polyatomic anions, with the denominator set equal to $(r + 30)$ pm, r in this case being the thermochemical radius of the anion.

We now look at some examples of redox reactions involving simple cations in aqueous solution. Electrochemical terminology will often be encountered, since e.m.f. measurements on electrochemical cells are important sources of thermodynamic data in this area. For example, the reduction potential E^0 for the half-reaction:

$$Zn^{2+}(aq) + 2e \rightarrow Zn(s)$$

is quoted as -0.77 V, measured against the standard hydrogen electrode. What this means is that ΔG^0 for the cell reaction:

$$Zn^{2+}(aq) + H_2(g) \rightarrow Zn(s) + 2H^+(aq)$$

is given by the equation:

$$\Delta G^0 = -nFE^0$$

where n is the number of electrons involved (2 in this case) and F is the Faraday constant (9.65×10^4 C mol^{-1}). Thus ΔG^0 for this reaction is 149 kJ mol^{-1}. For the reverse reaction, ΔG^0 is, of course, equal to -149 kJ mol^{-1}, consistent with the fact that zinc metal dissolves in dilute acids with evolution of hydrogen gas; i.e. zinc is oxidised by proton under standard conditions to give $Zn^{2+}(aq)$. Let us now explore the factors which determine the magnitude of ΔH^0 for this reaction. The thermochemical analysis is as follows:

$$
\begin{array}{lr}
 & \Delta H^0 \,(\text{kJ mol}^{-1}) \\
\text{Zn(s)} \rightarrow \text{Zn(g)} & 129 \\
\text{Zn(g)} \rightarrow \text{Zn}^{2+}(\text{g}) + 2e & 2640 \\
\text{Zn}^{2+}(\text{g}) \rightarrow \text{Zn}^{2+}(\text{aq}) & -2046 \\
2\text{H}^+(\text{aq}) \rightarrow 2\text{H}^+(\text{g}) & 2182 \\
2\text{H}^+(\text{g}) + 2e \rightarrow 2\text{H(g)} & -2624 \\
2\text{H(g)} \rightarrow \text{H}_2(\text{g}) & -436 \\
\hline
\text{Zn(s)} + 2\text{H}^+(\text{aq}) \rightarrow \text{Zn}^{2+}(\text{aq}) + \text{H}_2(\text{g}) & -155
\end{array}
$$

Since ΔH^0 is not very different from ΔG^0, the entropy term is evidently quite small, and we are justified in concentrating on the enthalpy terms in our analysis. The atomisation enthalpy of the elemental substance, the relevant ionisation energies of the gaseous atomic substance and the hydration enthalpy of the cation are obviously the quantities to be compared when looking at different species. The last three steps in the analysis above amount to $-439n$ kJ mol^{-1}.

Let us then compare zinc with other metallic elemental substances with respect to the formation of $M^{2+}(\text{aq})$ in acid solution at a pH of zero. The relevant data are summarised in Table 5.7. Note that the E^0 values given refer (in accordance with the European Convention) to reduction potentials for the half-reactions:

$$M^{2+}(\text{aq}) + 2e \rightarrow M(\text{s})$$

The excellent correlation between the total ΔH^0 values for the oxidation of $M(\text{s})$ by protons to $M^{2+}(\text{aq})$ and the E^0 values shows that the entropy terms cancel almost to zero and we can interpret variations in E^0 from

Table 5.7 *Thermodynamic data relevant to the formation of* M^{2+} *(aq) for selected elements (all energies in kJ mol^{-1})*

M	Ca	Fe	Cu	Zn	Sn
ΔH^0_{atom}	177	418	339	129	302
$I_1 + I_2$	1736	2320	2704	2640	2120
$\Delta H^0_{\text{hyd}}(M^{2+})$	-1577	-1946	-2100	-2046	-1556
Sum[a]	-542	-86	65	-155	-22
$E^0(\text{V})$	-2.87	-0.44	+0.34	-0.77	-0.14

[a] Including -878 kJ mol^{-1} in respect of the process:
$$2\text{H}^+(\text{aq}) + 2e \rightarrow \text{H}_2(\text{g})$$

enthalpy considerations. Comparing copper with zinc, we see that the slightly higher ionisation energy sum for copper is very largely balanced by the more negative hydration enthalpy of Cu^{2+}. The decisive difference which makes elemental zinc the more electropositive metal is its lower atomisation enthalpy. The highly electropositive character of calcium metal can be attributed to the low ionisation energy sum for the gaseous atomic substance. Comparing calcium with tin, the hydration enthalpies of M^{2+} are very similar, but the higher ionisation energies of tin lead to a much less negative E^0 value (with some contribution from the higher atomisation enthalpy).

To summarise: E^0 values for redox couples of the type $M(s)/M^{n+}(aq)$ can largely be rationalised in terms of the atomisation enthalpies, ionisation energies and hydration enthalpies. The entropy terms can be neglected in most cases.

We now turn to the relative stabilities of $M^{2+}(aq)$ and $M^{3+}(aq)$, as measured by E^0 for the half-reaction:

$$M^{3+}(aq) + e \rightarrow M^{2+}(aq)$$

These can be rationalised by analysis of the enthalpy changes for the cell reaction:

$$
\begin{array}{ll}
 & \Delta H^0 \\
M^{3+}(aq) \rightarrow M^{3+}(g) & -\Delta H^0_{hyd}(M^{3+}) \\
M^{3+}(g) + e \rightarrow M^{2+}(g) & -I_3 \\
M^{2+}(g) \rightarrow M^{2+}(aq) & \Delta H^0_{hyd}(M^{2+}) \\
\tfrac{1}{2}H_2(g) \rightarrow H^+(aq) & +439\,kJ\,mol^{-1} \\
\hline
\end{array}
$$

$$M^{3+}(aq) + \tfrac{1}{2}H_2(g) \rightarrow M^{2+}(aq) + H^+(aq)$$

Thus for the overall reaction:

$$\Delta H^0 = -[\Delta H^0_{hyd}(M^{3+}) - \Delta H^0_{hyd}(M^{2+})] - I_3 + 439\,kJ\,mol^{-1}$$

and

$$\Delta G^0 = -FE^0$$

where E^0 is the $M^{3+}(aq)/M^{2+}(aq)$ reduction potential in volts. The reduction of $M^{3+}(aq)$ to $M^{2+}(aq)$ will be favoured by a large third ionisation energy I_3, and by a small difference between the hydration enthalpies of M^{2+} and M^{3+}. Let us now look at the relevant data for the 3d and 4f series. Fig. 5.3 shows the variation across each series of E^0 and ΔH^0 for the reaction under scrutiny. The E^0 values are taken from standard tabulations of experimental data; the ΔH^0 values are obtained from the ionisation energies and the hydration enthalpies in Table 5.6. The excellent correlation between these reflects the fact that the entropy term is small and approximately constant within each series. Thus the variation

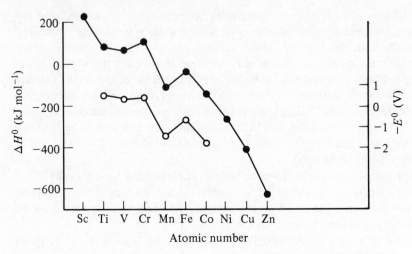

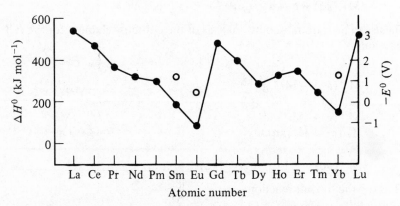

Fig. 5.3. Values of ΔH^0 (kJ mol^{-1}) for the reduction of M^{3+}(aq) to M^{2+}(aq) by H_2(g), plotted against atomic number for 3d and 4f elements. Open circles represent experimental values of $-E^0$ (V), where available.

of E^0 can be interpreted in terms of the enthalpy changes. For unknown ions, the hydration enthalpies have to be estimated, by interpolation and extrapolation. Thus the hydration enthalpy of Sc^{2+} is estimated to be the mean of the values for Ca^{2+} and Ti^{2+}, and the near-linear dependence on atomic number for the lanthanide M^{2+} ions (including Ba^{2+}) permits easy interpolation for unknown Ln^{2+} ions. The hydration enthalpies of Ni^{3+}, Cu^{3+} and Zn^{3+} were estimated – by extrapolation from Fe^{3+} and Co^{3+} taking into account the value for Ga^{3+} – to be all equal to -4800 kJ mol^{-1}.

The ΔH^0 and E^0 values correlate fairly well with the variation of I_3, the third ionisation energy (see Fig. 5.2). Along the 3d series, the most conspicuous discontinuity in E^0 – between Mn and Fe – matches the fall in I_3 between the atomic d^5 and d^6 configurations, as discussed in Section 4.3. There is a small 'break' in going from V to Cr. This may be attributed to crystal field effects. The $M^{n+}(aq)$ ions are actually $[M(H_2O)_6]^{n+}$ octahedral ions, and CFSE reaches a maximum at the d^3 configuration represented by V^{2+} and Cr^{3+}. This makes V^{2+} rather more difficult and Cr^{2+} rather easier to oxidise than would otherwise have been expected.

One of the factors which governs the viability of a cation in aqueous solution is the susceptibility of water to oxidation:

$$\tfrac{1}{2}O_2(g) + 2H^+(aq) + 2e \rightarrow H_2O(l) \quad E^0 = 1.23\,V$$

This means that at a pH of 0, the oxidising component of any couple whose reduction potential is greater than 1.23 V should oxidise water, with liberation of oxygen, e.g.:

$$2M^{3+}(aq) + H_2O(l) \rightarrow 2M^{2+}(aq) + 2H^+(aq) + \tfrac{1}{2}O_2(g)$$

Conversely, in the presence of oxygen M^{2+} may be susceptible to oxidation. In practice, species such as $Mn^{3+}(aq)$ – which ought to oxidise water – can survive in low concentrations in acid solution. But $Ni^{3+}(aq)$, $Cu^{3+}(aq)$ and $Zn^{3+}(aq)$ cannot.

If the reduction potential of the $M^{3+}(aq)/M^{2+}(aq)$ couple is negative, $M^{2+}(aq)$ is thermodynamically unstable with respect to oxidation by $H^+(aq)$ at a pH of 0. But at a lower concentration than one mole per litre, and at higher pH values, species such as $V^{2+}(aq)$ are moderately stable, provided that oxygen is rigorously excluded. The estimated reduction potential for $Sc^{3+}(aq)/Sc^{2+}(aq)$ precludes the stabilisation of $Sc^{2+}(aq)$.

The variation of the $M^{3+}(aq)/M^{2+}(aq)$ reduction potential across the lanthanide series can be easily understood, without the need to consider crystal field effects (which are very small for 4f orbitals). The slow increase in I_3 across the series means that the stability of $M^{2+}(aq)$ relative to $M^{3+}(aq)$ is much slower to emerge than across the 3d series. The abrupt drop in I_3 at Gd causes the disappearance of the II oxidation state (as for crystalline halides and oxides, discussed in Section 5.2), which then has to resume its uphill struggle and finally achieves a limited viability at ytterbium, only to lose it when we move to lutetium, where the drop in I_3 favours the III oxidation state. Only $Eu^{2+}(aq)$ among all $Ln^{2+}(aq)$ can be obtained and studied at high concentrations; it is often kinetically inert towards oxidising agents (e.g. ClO_4^-) which, thermodynamically, ought to oxidise it. Otherwise, apart from $Ce^{4+}(aq)$ the aqueous chemistry of the lanthanides is restricted to $Ln^{3+}(aq)$ and their hydrolysis products.

As with all cations of low charge, the $M^{2+}(aq)$ ions considered above may be vulnerable to disproportionation as well as to oxidation. The reaction:

$$3M^{2+}(aq) \rightarrow 2M^{3+}(aq) + M(s)$$

can be analysed in the usual way to yield:

$$\Delta H^0 = 2[\Delta H^0_{hyd}(M^{3+})] - 3[\Delta H^0_{hyd}(M^{2+})] +$$
$$2I_3 - I_1 - I_2 - \Delta H^0_{atom}$$

The reaction is therefore favoured if:

(a)   The difference in hydration enthalpy between M^{3+} and M^{2+} is large.

(b)   The third ionisation energy I_3 is not very much larger than I_1 and I_2.

(c)   The atomisation enthalpy of M(s) is large.

The relevant quantities are collected below for Sc, Fe, La and Eu. (Estimated hydration enthalpies for Sc^{2+} and La^{2+} are given in brackets; all energies in $kJ\,mol^{-1}$.)

	Sc	Fe	La	Eu
$2\Delta H^0_{hyd}(M^{3+})$	−7794	−8860	−6556	−7002
$-3\Delta H^0_{hyd}(M^{2+})$	(5100)	5838	(3980)	4374
$2I_3$	4778	5914	3700	4800
$-(I_1 + I_2)$	−1866	−2320	−1605	−1632
$-\Delta H^0_{atom}$	−378	−418	−423	−175
Sum	−160	154	−904	365

Thus $Sc^{2+}(aq)$ and $La^{2+}(aq)$ are unstable to disproportionation. It is of interest to note that CFSE plays a part in stabilising $Fe^{2+}(aq)$ with respect to disproportionation. The CFSE is estimated (from spectroscopic data) to be about $50\,kJ\,mol^{-1}$. If this could be 'switched off', the enthalpy change for the disproportionation of $Fe^{2+}(aq)$ would be close to zero. The entropy change, however, is always negative for such a reaction; the entropy of a metallic solid is low, and the entropies of aqueous ions decrease with increasing charge – presumably on account of the greater ordering of the solvent produced by highly-charged cations (see Section 3.3). Thus ΔG^0 for the disproportionation:

$$3Fe^{2+}(aq) \rightarrow 2Fe^{3+}(aq) + Fe(s)$$

is $233\,kJ\,mol^{-1}$, compared with $154\,kJ\,mol^{-1}$ for ΔH^0. Entropy tends to discourage disproportionation.

The dramatic improvement in the fortunes of the lanthanide $M^{2+}(aq)$ ions from La to Eu arises mainly from the relatively static ionisation energy sum $(I_1 + I_2)$ alongside the steadily increasing I_3; the low atomisation enthalpy of Eu also plays a part. Only Sm^{2+}(aq), Eu^{2+}(aq) and Yb^{2+}(aq) among the lanthanide M^{2+}(aq) ions are stable to disproportionation.

The scarcity of M^+(aq) ions (apart from those formed by the Group 1 elements) can be attributed to their susceptibility to disproportionation, giving M^{n+}(aq) ($n = 2$ or 3 usually) and the elemental substance $M(s)$. Consider the relative stabilities of Ca^+(aq), Cu^+(aq) and Ag^+(aq) with respect to:

$$2M^+(aq) \rightarrow M^{2+}(aq) + M(s)$$

From the usual thermochemical breakdown, we obtain for this reaction:

$$\Delta H^0 = \Delta H^0_{hyd}(M^{2+}) - 2\Delta H^0_{hyd}(M^+) + I_2 - I_1 - \Delta H^0_{atom}$$

These are tabulated below (all energies in $kJ\,mol^{-1}$; an estimated value for Ca^+ is in brackets):

	Ca	Cu	Ag
$\Delta H^0_{hyd}(M^{2+})$	-1577	-2100	-1931
$-2\Delta H^0_{hyd}(M^+)$	(700)	1186	946
$I_2 - I_1$	555	1212	1343
$-\Delta H^0_{atom}$	-177	-339	-284
Sum	-499	-41	74

Remembering that entropy will tend to oppose disproportionation we can predict that Ca^+(aq) is hopelessly unstable, Ag^+(aq) is quite stable and Cu^+(aq) marginal. This is in accordance with observation. Ca^+(aq) is unknown, while Ag^+(aq) is familiar and stable. Cu^+(aq) undergoes disproportionation, but the equilibrium constant:

$$K = [Cu^{2+}(aq)]/[Cu^+(aq)]^2 = 1 \times 10^6$$

shows that Cu^+(aq) can exist in millimolar concentrations. Even higher concentrations can be obtained at low temperatures in certain media where the disproportionation reaction is slow. Thus experimental electrochemical data are available for Cu^+(aq), but not for Ca^+(aq), whose hydration enthalpy is estimated to be $c. -350\,kJ\,mol^{-1}$ (cf. $-322\,kJ\,mol^{-1}$ for K^+).

The instability of Cu^+(aq) reflects the preference of copper for the II oxidation state, while silver prefers the I state. This is perhaps a little surprising, since other elements of the 4d series tend to favour higher oxidation states compared with their congeners in the 3d series. The enthalpy data given above help to rationalise this apparent anomaly. It might have been expected that the hydration enthalpy terms would favour the higher oxidation state for the smaller atom; the difference in hydration enthalpy between M^+ and M^{2+} should be greater for copper than for silver. But this is not the case. The hydration enthalpy of Ag^{2+} seems rather higher (i.e. more negative) than might have been expected; it is $124\,kJ\,mol^{-1}$ more negative than for Cd^{2+}, which compares with a difference of only $54\,kJ\,mol^{-1}$ between Cu^{2+} and Zn^{2+}. This may be attributed to crystal field effects. The CFSE of $Cu^{2+}(aq)$ – which is enhanced by the Jahn–Teller effect – is estimated to be about $100\,kJ\,mol^{-1}$. (This figure is derived by fitting the hydration enthalpies of Ca^{2+}, Mn^{2+} and Zn^{2+} to a quadratic function of the atomic number, and interpolating the intermediate M^{2+} ions to obtain their hydration enthalpies without CFSE.) Since crystal field splittings of 4d orbitals are much greater than for 3d orbitals, the CFSE enjoyed by Ag^{2+} may well be $150–180\,kJ\,mol^{-1}$, which would account for the large hydration enthalpy. This should be to the advantage of the II oxidation state for silver. However, the ionisation energy terms also work in the opposite direction to that expected. I_1 for silver is slightly smaller than for copper, but I_2 is larger for silver (for reasons explained in Section 5.2). Thus the greater $(I_2 - I_1)$ for silver is mainly responsible for the stabilisation of $Ag^+(aq)$.

The relative stabilities of aqueous species can be usefully displayed on a free energy diagram, in which we plot the free energies of formation against the oxidation number. Sometimes 'volt equivalents' are plotted instead of explicit free energies; since $\Delta G^0 = -nFE^0$ for a cell reaction, the volt-equivalent nE^0 is proportional to ΔG^0. Fig. 5.4 shows a free energy diagram for manganese, at a pH of zero. Thus the free energy of formation of $Mn^{2+}(aq)$ is equal to ΔG^0 for the reaction:

$$Mn(s) + 2H^+(aq) \rightarrow Mn^{2+}(aq) + H_2(g)$$

while for $MnO_4^-(aq)$ the relevant quantity is ΔG^0 for:

$$Mn(s) + 4H_2O \rightarrow MnO_4^-(aq) + H^+(aq) + \tfrac{7}{2}H_2(g)$$

No experimental data are available for $Mn^+(aq)$; it may easily be shown that this is unstable to disproportionation, to $Mn(s)$ and $Mn^{2+}(aq)$. The negative slope of the line joining $Mn(s)$ and $Mn^{2+}(aq)$ indicates that the former should react spontaneously with protons to give the latter. The position of $Mn^{2+}(aq)$ as having the most negative free energy of formation shows II to be the most stable oxidation state for Mn in acid solution. The large slope of the line joining $Mn^{2+}(aq)$ and $Mn^{3+}(aq)$ indicates that the oxidation of the former to the latter is difficult. However, the oxidation of $Mn^{3+}(aq)$ to $MnO_2(s)$ is easier than the oxidation of $Mn^{2+}(aq)$ to $Mn^{3+}(aq)$. Thus any agent strong enough to take manganese from the II to the III oxidation state in acid solution will also take the III to the IV state. This is another way of saying that $Mn^{3+}(aq)$ is unstable to disproportionation; the disproportionation is rather slow, however. The V state, represented by MnO_4^{3-}, is on the borderline of stability with respect to MnO_2 and $MnO_4^-(aq)$, while $MnO_4^{2-}(aq)$, in acid solution, is

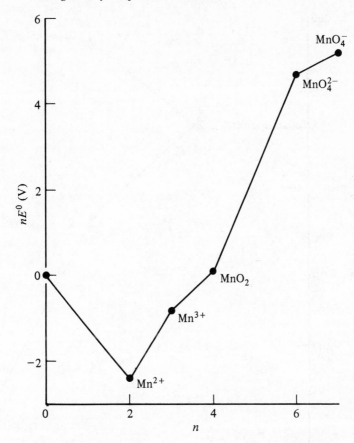

Fig. 5.4. Free energy diagram for manganese in aqueous solution, at pH = 0.

unstable to disproportionation. At acid pH, $MnO_4^-(aq)$ ought to oxidise water, with evolution of oxygen. This process is rather slow but readers who have done some volumetric analysis may recall that permanganate solutions should be standardised immediately before use.

Much the same information can be displayed on an enthalpy diagram, in which we plot the enthalpies of formation (relative to the proton) of aqueous species instead of free energies. This facilitates the inclusion on the diagram of unknown and presumably unstable species, making reasonable assumptions concerning their hydration enthalpies. As an example, Fig. 5.5 presents enthalpy diagrams for $M^{n+}(aq)$ (M = Al, Tl) at a pH of zero. The enthalpy changes are ΔH^0 values for the processes:

$$M(s) + nH^+(aq) \rightarrow M^{n+}(aq) + \tfrac{1}{2}nH_2(g)$$

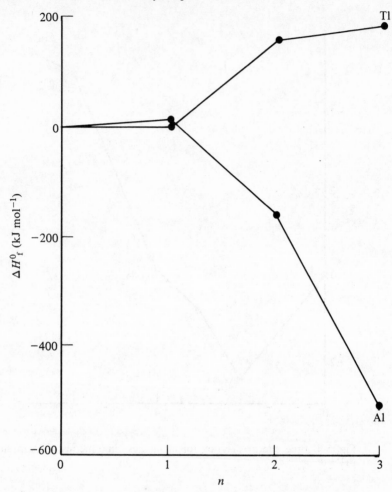

Fig. 5.5. Enthalpy diagram for M^{n+}(aq) (M = Al, Tl) at pH = 0.

and can readily be obtained from the relevant atomisation enthalpies, ionisation energies and hydration enthalpies. The hydration enthalpies of the unknown cations Al^{+}(aq), Al^{2+}(aq) and Tl^{2+}(aq) were estimated to be -450, -2000 and $-1700\,kJ\,mol^{-1}$ respectively, by reference to the experimental values for neighbouring ions, or ions of about the same size, having the same charge. Al^{+}(aq) and Al^{2+}(aq) are both unstable with respect to disproportionation to Al(s) and Al^{3+}(aq), while Tl^{2+}(aq) should disproportionate to Tl^{+}(aq) and Tl^{3+}(aq). The underlying reasons for this behaviour are closely analogous to those which govern the stabilities of Al and Tl fluorides, discussed in Section 5.2, with hydration enthalpies playing the same role in aqueous chemistry as lattice energies in the crystalline state. Tl^{+}(aq) is the most stable species for thallium, and is not oxidised to Tl^{3+}(aq) by protons. However, the oxidation can be accomplished by means of a stronger agent; the reduction potential of the $Tl^{3+}(aq)/Tl^{+}(aq)$ couple is $+1.25$ V.

This section has been devoted to simple aquo-cations in acid solution, in the absence of complexing agents. If such agents are present, or if a solvent other than water is used, the relative stabilities of oxidation states may change dramatically. Thus in acetonitrile, $Cu(CH_3CN)_n^{+}$ is stable to disproportionation and the II state is strongly oxidising. In the presence of ammonia, Co(III) is greatly stabilised relative to Co(II).

We have looked only at acid solutions since relatively few simple cations $M^{n+}(aq)$ can exist in appreciable concentrations at alkaline or even neutral pH values. Note that in Fig. 5.4, Mn(IV) is represented by $MnO_2(s)$ and not by $Mn^{4+}(aq)$, while the higher oxidation states are represented by anionic species. The relative stabilities of the oxidation states are strongly pH-dependent. Thus $Mn^{2+}(aq)$ is quite difficult to oxidise to $Mn^{3+}(aq)$ in acid solution. When a manganous salt is dissolved in water and the pH adjusted to about 10, $Mn(OH)_2$ is precipitated. If allowed to stand in contact with air, this undergoes oxidation to an Mn(III) hydrated oxide usually (but erroneously) formulated as $Mn(OH)_3$:

$$Mn(OH)_3(s) + e \rightarrow Mn(OH)_2(s) + OH^{-}(aq) \quad E^0 = 0.15 \text{ V}$$

It is now necessary to consider the hydrolysis of ions in aqueous solution.

5.5 The hydrolysis of cations in aqueous solution

The term 'hydrolysis' implies a reaction in which the O–H bonds of water molecules are ruptured. Consider the following equilibria and associated equilibrium constants K which are set up in aqueous solution:

$[Fe(H_2O)_6]^{3+} \rightleftharpoons [Fe(H_2O)_5OH]^{2+} + H^{+}(aq)$
$K = 8.9 \times 10^{-4}$

$[Fe(H_2O)_5OH]^{2+} \rightleftharpoons [Fe(H_2O)_4(OH)_2]^{+} + H^{+}(aq)$
$K = 5.5 \times 10^{-4}$

$2[Fe(H_2O)_5OH]^{2+} \rightleftharpoons [Fe_2(H_2O)_8(OH)_2]^{4+} + 2H_2O$
$K = 1.5 \times 10^{3}$

The above species are conveniently abbreviated as $Fe^{3+}(aq)$, $Fe(OH)^{2+}(aq)$, $Fe(OH)_2^{+}(aq)$ and $Fe_2(OH)_2^{4+}(aq)$. The last of these is a hydroxo-bridged dimer:

These equilibria are, of course, responsible for the fact that the dissolution of a ferric salt in water produces an acidic solution. If the total iron concentration in such a solution is one mole per litre, with the pH adjusted to zero, it is easy to show that 99.8% of the iron is present in the form of $Fe^{3+}(aq)$, with $Fe(OH)^{2+}(aq)$ and $Fe_2(OH)_2^{4+}(aq)$ each contributing roughly 0.1%. At a pH of 2, these proportions become respectively about 24%, 2% and 73%, with $Fe(OH)_2^+$ contributing only 0.1%. The dimer is now the dominant species; in more dilute solutions, it assumes less prominence. Thus at a total iron concentration of $0.01\,mol\,l^{-1}$ and a pH of 2, the proportions become about 84% $Fe^{3+}(aq)$, 7% $Fe(OH)^{2+}(aq)$ and 8% $Fe_2(OH)_2^{4+}(aq)$. If the pH is raised above about 2, further complex polynuclear hydroxo-bridged cations are formed. At a pH of around 3, hydrous ferric oxide $Fe_2O_3.nH_2O$ ('$Fe(OH)_3$') comes down as a gelatinous precipitate. Over the pH range approximately 3–14 this is virtually insoluble, but in strongly alkaline solutions it becomes slightly soluble with formation of anionic hydroxo-complexes, such as $Fe(OH)_4^-(aq)$.

Most $M^{3+}(aq)$ and a number of $M^{2+}(aq)$ cations behave in much the same way as $Fe^{3+}(aq)$. Evidently the viabilities of simple aquo-cations are restricted by the possibilities for hydrolysis, as well as the redox processes discussed in Section 5.4. The simplest explanation for the hydrolysis of $Fe^{3+}(aq)$ acknowledges the considerable polarising power of Fe^{3+}. When surrounded by six water molecules, the cation will tend to attract electron density from the O atoms. This makes the coordinated O atoms less attractive to protons than those in the bulk solvent, and encourages the transfer of protons from the coordination sphere. If the bonding in the complex $[Fe(H_2O)_6]^{3+}$ is regarded as covalent, with the formation of coordinate bonds, we can arrive at the same conclusion; the delocalisation of the positive charge over the complex ion $[Fe(H_2O)_5OH]^{2+}$ will tend to discourage recombination of the coordinated OH^- ion with a proton.

Where the polarising power of a cation is very great, no simple aquocation – or even no cationic species whatever – may be stable to hydrolysis, even at extremely acid pH. For example, let us contemplate the viability of $B^{3+}(aq)$. The hydration enthalpy of B^{3+} is estimated to be about $-6000\,kJ\,mol^{-1}$. From this and the other relevant data given in the treatment of $BF_3(s)$ in Section 5.3, we can estimate ΔH^0 for the reaction:

$$B(s) + 3H^+(aq) \rightarrow B^{3+}(aq) + \tfrac{3}{2}H_2(g)$$

to be about $140\,kJ\,mol^{-1}$ (though with considerable uncertainty, given the likely error in ΔH^0_{hyd}). This is comparable in magnitude with the values for some well-known $M^{3+}(aq)$ ions (e.g. Tl^{3+}) and implies that, while $H^+(aq)$ cannot oxidise $B(s)$ to $B^{3+}(aq)$, stronger oxidising agents should. But $B^{3+}(aq)$ does not exist at any pH, nor does any cationic B(III) species in aqueous solution. B(III) is represented at neutral and acid pH by $B(OH)_3(aq)$, i.e. $[B(OH)_3H_2O]$, with tetrahedral coordination about the B atom. This is a weak acid:

$$[B(OH)_3H_2O] \rightleftharpoons [B(OH)_4]^- + H^+(aq) \quad K = 6.3 \times 10^{-10}$$

so that the anionic species does not predominate until the pH is raised above about 9. For the reaction:

$$B(s) + 3H_2O(l) \rightarrow B(OH)_3(aq) + \tfrac{3}{2}H_2(g)$$

ΔH^0 is $-228\,kJ\,mol^{-1}$ (although elemental boron is kinetically inert to water under any conditions). Thus for the reaction:

$$[B(H_2O)_4]^{3+} \rightarrow [B(H_2O)(OH)_3] + 3H^+(aq)$$

ΔH^0 is estimated to be about $-370\,kJ\,mol^{-1}$, i.e. $B^{3+}(aq)$ – written above as $[B(H_2O)_4]^{3+}$ – is a very strong acid, too strong to exist in aqueous solution. $[B(OH)_3(H_2O)]$ is a very weak base; the OH oxygens have no measurable tendency to accept protons, presumably because of the withdrawal of electron density towards the boron atom.

The stabilities of cations $M^{n+}(aq)$ towards hydrolysis can be rationalised in terms of the polarising power of M^{n+}. This is often measured by the ratio (charge/radius) but this takes no account of the electron affinity of the cation. Thus, as noted in Section 5.1, Mg^{2+} and Zn^{2+} are comparable in size but the latter is more polarising, as might be expected from the higher ionisation energies of Zn. A better measure is the ratio I_n/r where I_n, as usual, is the nth ionisation energy in $kJ\,mol^{-1}$ and r is the radius of M^{n+} in pm (for six-coordination, taken from Table 4.2). The aqueous chemistry of the elements in various oxidation states can then be summarised as follows:

$I_n/r = 2$–5 Examples: Na(I), K(I), Rb(I), Cs(I)

Simple cations $M^{n+}(aq)$ formed over entire accessible pH range. Hydroxides very soluble.

$I_n/r = 5$–20 Examples: Ca(II), La(III), Li(I)

Simple cations $M^{n+}(aq)$ at acid pH. For $n > 1$, some hydrolysis around neutral pH. Sparingly-soluble hydroxide precipitated in strongly alkaline solution.

$I_n/r = 20$–55 Examples: Be(II), Zn(II), Al(III), Ce(IV)

Simple cations $M^{n+}(aq)$ stable only at strongly acid pH. Polynuclear hydroxo-cations usually present at weakly acid pH.

Insoluble hydroxide or oxide precipitates around neutral pH. Hydroxo-anions such as $[Zn(OH)_4]^{2-}$ stable in strongly alkaline solutions.

$I_n/r = 55–110$ Examples: Sn(IV), V(IV)

No simple cation $M^{n+}(aq)$ exists. Hydroxo- or oxocations present at acid pH; anionic species stable in alkaline solutions. Otherwise insoluble oxide precipitates.

$I_n/r > 110$ Examples: B(III), Si(IV), P(V), Cr(VI)

No cationic species whatsoever exist. Neutral hydroxide (often insoluble) or oxohydroxide obtained in acid solution. Hydroxo-, oxohydroxo- or oxoanions present in alkaline solution (e.g. $B(OH)_4^-$, $SiO_2(OH)_2^{2-}$, PO_4^{3-}). In extreme cases (e.g. Cl(VII)), oxoanion dominant over entire pH range.

The summaries given above are to be taken as broad generalisations. There are numerous exceptions. For example, V^{5+} has $I_n/r = 117$, but $VO_2^+(aq)$ exists in strongly acid solutions. Cations having an 'inert pair' electron configuration (e.g. Tl^+, Pb^{2+}, As^{3+}) are rather more polarising than their I_n/r values would suggest.

It should be apparent that the aqueous species formed by an element in a given oxidation state and the oxide (or hydroxide) are intimately connected. The formation of aqueous species – ions or molecules – depends on the solubility of the oxide (or hydroxide) in water at a given pH, and this is governed in part by the stabilities of the aqueous species with respect to hydrolysis.

Binary oxides are often classified as follows:

(i)   Acidic oxides dissolve in water to give acidic solutions, or an acidic oxide may be insoluble in water but dissolves in excess alkali.

(ii)   Basic oxides dissolve in water to give alkaline solutions, or if insoluble in water will dissolve in excess acid.

(iii)   Amphoteric oxides are usually insoluble in water but dissolve in both acids and alkalies.

(iv)   Neutral oxides dissolve in water to give neutral solutions, or are insoluble at any pH.

An oxide $EO_{n/2}$ (n even) or E_2O_n (n odd) will be acidic if E when carrying the oxidation number n forms anionic species in aqueous solution at alkaline pH, and no cationic species even at acid pH. The oxide will be basic if E forms only cationic species in aqueous solution, or amphoteric if E forms cationic species at acid pH and anionic species at alkaline pH. These are best illustrated by means of examples.

Consider first the reaction:

$$Na_2O(s) \xrightarrow{\text{H}_2\text{O}} 2Na^+(aq) + 2OH^-(aq)$$

That this reaction proceeds is related to the fact that $Na^+(aq)$ can exist at strongly alkaline pH; the high solubility in water of NaOH is likewise related to the stability of $Na^+(aq)$. Sodium(I) forms no anionic species, even in strongly alkaline solution.

Magnesium(II) oxide reacts with the stoichiometric amount of water to form $Mg(OH)_2$; this is insoluble at neutral or alkaline pH. It does, however, dissolve in excess acid:

$$Mg(OH)_2(s) + 2H^+(aq) \rightarrow Mg^{2+}(aq) + 2H_2O$$

The appearance of proton on the left-hand side of the equation requires a cationic species on the right-hand side, and the solubility of $Mg(OH)_2$ in acids depends upon the stability of $Mg^{2+}(aq)$ in acid solution. If the pH is raised to about 7, $Mg(OH)_2$ is precipitated and this does not redissolve at alkaline pH; Mg(II) forms no anionic species such as $Mg(OH)_4^{2-}(aq)$. Sodium oxide and magnesium oxide are classified as basic oxides.

Zinc(II) oxide is insoluble in neutral water, but soluble in dilute acids; $Zn^{2+}(aq)$ is stable at low pH values. If an acid solution containing Zn^{2+} is treated with an alkali, $Zn(OH)_2$ precipitates around neutral pH but redissolves at a pH of about 14, with formation of $Zn(OH)_3^-(aq)$ and $Zn(OH)_4^{2-}(aq)$. The fact that Zn(II) forms cationic aqueous species in acid solution and anionic species at alkaline pH determines the amphoteric character of ZnO.

Silicon dioxide is insoluble except at strongly alkaline pH, at which $SiO(OH)_3^-$ and $SiO_2(OH)_2^{2-}$ are stable as aqueous species. Silicon forms no cationic species in aqueous solution, even at very low pH values; SiO_2 is classed as a (relatively weakly) acidic oxide.

Nitrogen(V) oxide N_2O_5 dissolves rather violently in water to give a solution of nitric acid. The nitrate ion – unlike $SiO(OH)_3^-(aq)$ or $SiO_2(OH)_2^{2-}(aq)$ – is stable over the whole accessible pH range in aqueous solution, and N_2O_5 is classed as a strongly-acidic oxide.

This discussion should not be seen as 'explaining' the acid/base character of oxides, i.e. their solubilities in water at various pH values. We are emphasising the close relationship between the acid/base behaviour of oxides and the nature of aqueous species. The dissolution of an oxide (other than a neutral oxide) in water, or in acids/alkalies, is an acid–base process, a chemical reaction rather than a mere separation of ions. The relative acid/base strengths of oxides are further discussed in Section 9.2.

5.6   Further reading

The books listed in Section A.8 in the Appendix should be consulted for more detail on lattice energies etc. These are also valuable for ions in solution; see also Burgess, J. (1978). *Metal Ions in Solution.* Chichester: Ellis Horwood. Among the standard inorganic texts (see Section A.3), Phillips and Williams (1965) and Sharpe (1981) are most useful for the material discussed in this chapter.

6

Covalent bonding in Main Group chemistry: a VB approach

6.1 Valence bond theory and the 'octet rule': obedience and violation

This chapter seeks to rationalise the occurrence of molecules and polyatomic ions (finite and infinite) formed by atoms of the Main Group elements, and the structures adopted. VB theory is used because it provides a fairly simple classification of the types of bonding exhibited by Main Group atoms and helps to rationalise (if not explain) the shapes of many molecules and ions. MO theory is covered in Chapter 7. For the reasons set out in Section 1.4, VB theory is unsatisfactory for open-shell systems, and is therefore of limited value in d block chemistry. However, it can be applied in compounds where the atom has the d^0 or d^{10} configuration. Thus, Zn, Cd and Hg can be considered alongside Be and Mg since their nd subshells are invariably filled, and compounds of Ti(IV), V(V) and Cr(VI) often have close analogies with their Main Group analogues, Si, P and S.

The strategy for devising a VB description of a molecule or polyatomic ion has been outlined in Section 1.4. Such a description can be regarded as satisfactory – giving the species a 'right to exist' – if:

(*a*) A hybridisation scheme consistent with the observed (or predicted) geometry can be devised.

(*b*) The promotion energy required to attain the valence state is not excessive, and is recompensed by bond formation.

(*c*) There are enough valence electrons to form the required electron-pair bonds.

(*d*) All valence electrons are accounted for, either in bonding pairs or in stable lone-pair orbitals.

'Stable' and 'unstable' species are further discussed in Section 6.4. The important types of hybridisation and the associated directional properties are summarised in Table 6.1. Given also is the 'intrinsic strength' of a

Table 6.1 *Principal types of hybridisation*

Hybridisation	Directional properties	Relative strength[a]
sp	linear	3.73
sp^2	trigonal planar	3.96
sp^3	tetrahedral	4.00
sp^2d	square planar	7.25
sp^3d	trigonal bipyramidal	8.63 (axial)
		5.06 (equatorial)
sp^3d^2	octahedral	8.54
sp^3d^3	pentagonal bipyramidal	8.86 (equatorial)
		8.52 (axial)

[a] A crude measure of the intrinsic strength of a bond formed using a hybrid orbital, relative to that obtainable using a pure s orbital: for a pure p orbital, the figure is 3.00

bond, relative to that obtained using a pure s orbital. The intrinsic strength of a bond is the stabilisation that occurs when the bond is formed with the atoms in their respective valence states. This is not the same as the thermochemical bond energy (Section 6.2), which measures the strength of a bond in terms of the energy of the molecule relative to the separated atoms in their ground states. If promotion energy has to be invested to attain the valence state, the intrinsic bond strength will be greater than the thermochemical bond energy.

For σ bonding, sp^n hybrids are superior to pure s or p orbitals; they are strongly directional and give better overlap. For example, the intrinsic strength of a bond formed by use of an sp^3 hybrid is 1.33 times greater than that obtained by use of a pure np orbital. However, sp^n hybrids are inferior to pure p orbitals for π bonding; since an s orbital cannot overlap with anything in π fashion, the s component of a hybrid is 'wasted'. To minimise the promotion energy, lone pairs are best placed in hybrid orbitals having at least some s character. For example, consider the NH_3 molecule. If this were planar, the valence state of the N atom would be $(sp^2)^1(sp^2)^1(sp^2)^1p^2$, i.e. the three bonds are formed by overlap of singly-occupied sp^2 hybrids on the N atom with singly-occupied hydrogen 1s orbitals, with the lone pair in the remaining pure 2p orbital (remember that an sp^2 hybrid is $\frac{1}{3}$ s and $\frac{2}{3}$ p in character; three sp^2 hybrids use up one s orbital and two p orbitals, leaving one 2p orbital). This valence state corresponds to the configuration s^1p^4 (three electrons are $\frac{1}{3}$ s and $\frac{2}{3}$ p, and two are pure 2p), compared with the ground state s^2p^3. Thus one electron has to be promoted from 2s to 2p. If we now distort the molecule to a pyramidal configuration, with interbond angles of 109.5° as in a tetrahedron, we require sp^3 hybrids. The valence state is now $(sp^3)^1(sp^3)^1(sp^3)^1(sp^3)^2$, which corresponds to the configuration $s^{1.25}p^{3.75}$ and thus requires the promotion of 0.75 of an electron from 2s to 2p. Thus in performing this distortion, we have saved some promotion energy, and sp^3 hybrids are slightly better than sp^2 for bonding. If the interbond angle is further reduced to 90°, we can use pure 2p orbitals for bonding, with the lone pair in the 2s orbital. No promotion energy is required at all, but the intrinsic bond strengths are much lower than those obtained by use of hybrid orbitals. Experimentally, we find an interbond angle of 107° for NH_3. With other pyramidal AB_3 molecules, the angle is nearly always between 90° and 109.5°, and is partly determined by the compromise struck between the need to minimise the promotion energy while maximising the overlap and hence the intrinsic bond strength. Thus the shape of the molecule can be understood qualitatively in terms of VB theory, although the VSEPR approach leads to the same result rather more simply. The important point here is that the promotion energy decreases as the s component in the lone pair orbital increases.

The electron-pair bond theory of Lewis – forerunner of VB theory – emphasised the 'octet' or 'noble gas' rule as the primary criterion for stability; this was thought to be related to the chemical inertness of the noble gases. The VB theory places this rule on a firmer foundation. A Main Group atom having just four valence orbitals – one ns and three np – will necessarily have an octet configuration if all of these orbitals are fully engaged, either in bond formation or in accommodating nonbonding electrons. Thus a carbon atom which forms four covalent bonds makes full use of its valence orbitals and electrons. A nitrogen atom, with five valence electrons and four valence orbitals, has exhausted its bonding capacity if it forms three 'ordinary' covalent bonds (contributing one electron to each) plus one dative or coordinate bond; it is often content to form only three bonds, leaving the two remaining valence electrons as a lone pair. In either case, it can be portrayed in terms of a Lewis structure in which it has an octet configuration. A boron atom, with three valence electrons, can form three bonds, with one valence orbital left empty and unemployed. A molecule BX_3 is not necessarily unstable; but the B atom can better itself by making use of its unused orbital in the formation of a coordinate bond, e.g. $Cl_3B \leftarrow NR_3$. Thus the octet rule (or duet rule in the case of hydrogen, with only one valence orbital) simply states that atoms tend to adopt a policy of full employment for their valence orbitals.

Molecules and polyatomic ions in which the octet is exceeded – sometimes called 'hypervalent' species – present problems. Many well-known molecules such as PF_5, SF_6 and ClF_3 cannot be represented by Lewis structures which obey the octet rule. VB descriptions of hypervalent species can be devised by postulating hybridisation schemes which make use of nd orbitals as well as ns and np. Thus the hybridisation of ns, $n\mathrm{p}_x$, $n\mathrm{p}_y$, $n\mathrm{p}_z$ and $n\mathrm{d}_{z^2}$ leads to five (non-equivalent) orbitals which point to the corners of a trigonal bipyramid. These sp^{3}d hybrids can be invoked for PF_5, SF_4 and ClF_3, with lone pairs in the latter two occupying equatorial positions in the trigonal bipyramid. For SF_6 (and the isoelectronic species SiF_6^{2-}, PF_6^- and ClF_6^+), sp^{3}d^2 hybrids can be constructed, pointing to the corners of an octahedron. The same scheme can be used for BrF_5 and ICl_4^-, with respectively one and two lone pairs. The postulate of d orbital participation rationalised the formation of hypervalent species by heavier atoms such as P and S; analogous species are not formed by their lighter congeners N and O, since there are no 2d orbitals to supplement their 2s and 2p valence orbitals.

If the heavier p block atoms can use nd orbitals in hybridisation schemes for σ bonding, it seems reasonable to postulate their availability for π bonding as well. Lewis could describe the bonding in molecules such as $POCl_3$, SO_2F_2 and $HClO_4$ in terms of coordinate bonds to O atoms, thus:

or:

These descriptions preserve octet configurations about the central atoms. However, alternative descriptions are possible if we promote electrons to nd orbitals, and use these to form d_π–p_π bonds:

Experimental measurements of bond lengths support the formulation E=O in oxohalides EO_mX_n. In oxofluorides such as POF_3, SO_2F_2 and

IO_2F_3, the E–O distances are always shorter than E–F. Since F has a smaller covalent radius than O, we would expect an E–O single bond to be longer than E–F. Thermochemical measurements of E–O bond strengths in these molecules are consistent with doubly-bonded oxygen atoms, as are the high E–O stretching frequencies and force constants shown in their infra-red spectra. Moreover, the formulation of E–O bonds as 'semi-polar', with donation from E to O, can lead to rather unrealistic positive charges on E unless there is some degree of 'back-bonding', via π overlap, from O to E which results effectively in a double bond. For example, the formulation of XeO_4 as $Xe^{4+}(O^-)_4$ seems absurd.

But serious objections have been raised to the involvement of nd orbitals, which are rather diffuse (at least in the free, gaseous atoms) and have poor overlap properties. Spectroscopic data for the free atoms suggest that promotion energies to nd orbitals (typically about $1000\,kJ\,mol^{-1}$ from np, and $1500\,kJ\,mol^{-1}$ from ns) are prohibitive. On the other hand, the nd orbitals can be greatly contracted, and thus made more suitable for bonding, if the central Main Group atom is bonded to atoms or groups of high electronegativity (e.g. O, F, CF_3, OCF_3); a positive charge on the central atom diminishes the otherwise effective shielding of the nd orbitals from the nucleus. This is consistent with the fact that fluorides, oxides, oxoanions and oxofluorides are particularly well represented among hypervalent species, and dominate the known chemistry of the noble gases. It may also be argued that spectroscopic data for free atoms are not necessarily valid in estimating promotion energies for atoms in molecules.

These arguments have been debated for many years. Chemists who reject the use of nd orbitals in hybridisation schemes prefer three-centre bonds to describe hypervalent species. These are best portrayed in MO language (see Section 7.4); translated into VB terminology, they correspond to polar (or ionic) structures, e.g.:

Similarly, the bonding in POX_3, SO_2X_2, XeO_4 etc. is portrayed in terms of semi-polar (or semi-ionic) bonds between E and O, as in Lewis's original structures. The fact that P, S, Cl, Xe etc. form such bonds is attributed to their lower ionisation potentials compared with their lighter congeners, while the high electronegativities of F, O etc. favour structures in which they appear bearing negative charges.

The results of MO calculations (insofar as they can be transalated into

VB language) should help us to decide between these conflicting viewpoints; but the results are inconclusive, and often provide succour for both factions. A good example is PH_3F_2. This gaseous molecular substance (isostructural with PCl_3F_2 – see Section 2.3) is unstable with respect to disproportionation:

$$2PH_3F_2 \rightarrow PH_2F_3 + PH_3 + HF$$

However, the above reaction (which presumably requires a bimolecular mechanism) is rather slow under low pressure and the substance can be isolated. A fairly rigorous MO calculation on the PH_3F_2 molecule suggests that the degree of phosphorus 3d orbital involvement is much less than would be expected if the sp^3d structure were dominant; translating into VB language, the sp^3d structure contributes some 20–25% of the total, the polar structures $H_3FP^+F^-$ being the major contributors. But if phosphorus 3d orbitals are completely neglected in the calculation, the total binding energy is reduced by about $170\,kJ\,mol^{-1}$, or about 10% of total. This is very significant, since the enthalpy change for the decomposition:

$$PH_3F_2(g) \rightarrow PH_2F(g) + HF(g)$$

is about $170\,kJ\,mol^{-1}$. Such a reaction would proceed rapidly, even at low pressures (via a unimolecular mechanism since pentacoordinate phosphorus compounds are known to be very flexible) if it were thermodynamically favourable – entropy will certainly favour such a decomposition. It is thus fair to say that PH_3F_2 owes its existence, both as a molecule and as a substance, to the availability of phosphorus 3d orbitals for bonding. In the series of molecules PH_nF_{5-n}, the extent of phosphorus 3d orbital participation increases with decreasing n – the more F atoms, the higher the fractional positive charge on the P atom, resulting in contraction of the 3d orbitals and improving their bonding capability. Thus the charge on the P atom in PH_3F_2 is calculated to be +1.5, but for the PF_5 this increases to +2.1. The nonexistence of PH_5 and PH_4F may be attributed to the lack of phosphorus 3d orbital involvement.

Consider now the E–O bonds in species such as X_3PO and X_2SO_2. The bonding in X_3PO can be described in terms of resonance:

$$X_3P^+\!\!-\!O^- \leftrightarrow X_3P^-\!\!=\!\!O^+ \quad (or\ X_3P\!\!\equiv\!\!O)$$

The structure $X_3P{=}O$ should, strictly speaking, be rejected; there are two phosphorus 3d orbitals, at right angles to one another, capable of forming d_π–p_π bonds with oxygen 2p orbitals, and symmetry demands that both carry equal weight. In MO terminology, a twofold-degenerate pair of π bonding MOs is formed and a bond order of 2 – a double bond – would imply two unpaired electrons. A rigorous MO calculation on H_3PO (which is unknown) suggests that the triple-bonded structure

$H_3P^-\!\!=\!\!O^+$ contributes only about 15% of the total, in VB language. However, even this modest degree of d_π–p_π bonding leads to a shortening of the P–O bond by 13 pm and contributes largely to its 'double-bond' characteristics. Considerably more d_π–p_π bonding may be expected for F_3PO and Cl_3PO: well-known molecules for which MO calculations comparable to that described for H_3PO have yet to be performed.

To summarise, it seems that VB structures which depend upon the involvement of Main Group atom nd orbitals are not wholly satisfactory, and structures based on polar or semi-polar bonds are often dominant. But nd orbital participation can make a significant *and often decisive* contribution to the bonding, and it seems reasonable to give VB structures involving central atom nd orbitals some prominence, especially where we are primarily concerned with the right of a molecule or substance to exist. We shall portray the E–O bonds in X_3PO etc. as double bonds, notwithstanding the valid objections raised above to such a formulation; their properties indicate a bond order close to two, after all. And we should bear in mind the philosophical approach, that a theory which serves us well need not be summarily discarded on the grounds that some of its fundamental assumptions are called into question.

Finally, it may be noted that VSEPR theory – which is firmly based upon two-centre electron-pair bonds and may be viewed as a half-sibling to VB theory – works remarkably well for hypervalent molecules and polyatomic ions. In VSEPR treatments, E–O bonds in species such as SO_2F_2 and XeO_3F_2 are regarded as double bonds.

6.2 Covalent bond energies

A tabulation of bond energies is intended to convey numerically the strengths of covalent bonds. The total binding energy of a molecule is its energy relative to the separated atoms, i.e. the atomisation enthalpy of the gaseous substance. The idea is to partition this among the bonds in the molecule, so that the atomisation enthalpy is given by the sum of the bond energies. Many authors express bond energies as ΔU quantities measured at absolute zero temperature; here, all values are obtained from enthalpy data at 25 °C for conformity with the rest of the text, but we still use the more familiar term 'bond energy' rather than 'bond enthalpy'.

As well as enabling us to compare the strengths of different bonds, bond energies can be used to estimate the atomisation enthalpies of molecular substances in the absence of thermodynamic data, and hence enthalpy changes for chemical reactions involving gaseous species (ΔH is given by the appropriate algebraic sum of the atomisation enthalpies). For example, one might envisage the possibility of preparing $H_3Si—SiH_2OH$. In planning the synthesis, one would have to take note of the fact that this is certainly unstable with respect to its isomer H_3Si—O—SiH_3; this can be predicted with some confidence from a table of bond energies. Even where complete thermodynamic data are available for all relevant species, it is often worthwhile to make use of bond energies to rationalise the sign and magnitude of ΔH. For example, you may be asked to consider the possibility of preparing CF_4 by the reaction:

$$CH_4(g) + 4HF(g) \rightarrow CF_4(g) + 4H_2(g)$$

From the enthalpies of formation listed in standard tabulations, you will quickly find that ΔH^0 at 25 °C for the reaction is $+234\,kJ\,mol^{-1}$, so that it is out of the question as a realistic preparative route. From the table of bond energies, you will be able to offer the rationalisation that although the C–F bond is stronger than C–H, this does not compensate for the much weaker H–H compared with H–F bond. A more detailed explanation will require analysis of the factors which determine the relative strengths of E–H and E–F bonds.

The strength of a bond, thermochemically speaking, must be measured by the energy required to break it. If we are dealing with a diatomic molecule, this should be given directly by the dissociation enthalpy ΔH^0_{diss} for the process:

$$XY(g) \rightarrow X(g) + Y(g)$$

Apart from those molecules which are present in stable, familiar molecular substances such as H_2, HCl, CO etc., a great number of diatomic molecules have been observed by various means, and their dissociation energies determined to varying degrees of accuracy. Most of these are of little chemical interest. The bond energies thus obtained for, e.g., H—F, Cl—Cl, N≡N etc., are of great importance. But the dissociation energies of diatomic molecules such as CH, NBr or BO are of marginal relevance to stable molecules containing these kinds of bonds. The dissociation enthalpy of CH is $337\,kJ\,mol^{-1}$, but the C–H bond energy for practical purposes in stable molecular substances is much higher.

The C–H bond energy applicable to 'real' organic molecules is most simply obtained by taking the atomisation enthalpy of $CH_4(g)$, and dividing it by four to give the mean C–H bond energy. This gives $416\,kJ\,mol^{-1}$. It is not the same as the bond dissociation energy of CH_4:

$$CH_4(g) \rightarrow CH_3(g) + H(g) \quad \Delta H^0 = 439\,kJ\,mol^{-1}$$

Remembering that we want bond energies which, when added together, reproduce the experimental atomisation enthalpy of the molecular substance, 416 rather than $439\,kJ\,mol^{-1}$ is appropriate for the C–H bonds in methane. However, some caution must be exercised in transferring this value to other situations. Consider, for example, CH_3Cl. From the atomisation enthalpy of $CCl_4(g)$, the mean C–Cl bond energy is found to be $326\,kJ\,mol^{-1}$. Thus the atomisation enthalpy of $CH_3Cl(g)$ is predicted to be $(3 \times 416) + (1 \times 326)$, i.e. $1574\,kJ\,mol^{-1}$. The experimental value is $1573\,kJ\,mol^{-1}$, and the agreement is as admirable as one could hope for. But in many cases, serious discrepancies occur. For example, the Si–H and Si–F bond energies are found to be respectively 324 and $597\,kJ\,mol^{-1}$ from the experimental atomisation enthalpies of $SiH_4(g)$ and $SiF_4(g)$. Hence we might estimate the atomisation enthalpy of SiH_3F to be $1569\,kJ\,mol^{-1}$; but the experimental value is $1628\,kJ\,mol^{-1}$. An error of about $60\,kJ\,mol^{-1}$ is far from trivial in thermodynamic arguments. Evidently – and there is evidence from independent sources – the Si–H bonds are appreciably stronger when one of them is replaced by F. Variability of bond energy from one molecule to another is to be expected where one of the atoms involved in the bond is in a different state of hybridisation; for example, the $C = O$ bond in CO_2 (sp carbon) appears to be rather stronger than $C = O$ where the C atom is sp^2. But even within a range of molecules where the hybridisation remains the same, there is evidence for considerable variability, and the X–Y bond energy obtained directly from data for XY_n may not be universally applicable.

Many bonds occur only in conjunction with other bonds, and indirect means have to be employed to obtain bond energies in these cases. A good example is the important C–C bond. We might try to determine the C–C bond energy from the atomisation enthalpy of diamond, whose structure reveals sp^3 carbon atoms with two C–C bonds per atom (i.e. each C atom forms four tetrahedral bonds, but we have to divide by two because it takes two atoms to form each bond). This gives a C–C bond energy of $357\,kJ\,mol^{-1}$. It might be argued, however, that bond energies should be based on data for gaseous molecular substances; in the diamond lattice, there is evidence for appreciable interaction between next-nearest neighbours (see also Section 7.5). The C–C bond energy might thus be determined from data such as these:

$$H_3C\!-\!CH_3(g) \rightarrow 2CH_3(g) \qquad \Delta H^0 = 376\,kJ\,mol^{-1}$$
$$F_3C\!-\!CF_3(g) \rightarrow 2CF_3(g) \qquad \Delta H^0 = 343\,kJ\,mol^{-1}$$
$$Cl_3C\!-\!CCl_3(g) \rightarrow 2CCl_3(g) \qquad \Delta H^0 = 260\,kJ\,mol^{-1}$$

The large variation looks discouraging; but these are bond dissociation energies, not mean bond energies; if we could be sure that the C–X bond strengths remained the same in CX_3 as in C_2X_6, these values would indeed represent the C–C bond strengths. But, as in the case of the bond dissociation energies of CH_4, or of CH, thermochemical data for reactions involving radicals are inappropriate for determining bond energies if we wish ultimately to estimate the atomisation enthalpies of stable molecular substances. However, you will find in the literature values for, e.g., C–C and Si–Si bond energies which have been taken from bond dissociation data.

A better C–C bond energy is obtained from the atomisation enthalpies

Table 6.2 *Mean bond energies from atomisation enthalpies at 25 °C of gaseous molecular substances, in kJ mol⁻¹. Ranges of values are not given in cases where the relevant data are available only for one or two molecules*

H—H	436	$C(sp^3)$—$C(sp^3)$	325–360
H—B	365–385	$C(sp)$=$C(sp)$	567
H—C(sp)	400–430	$C(sp)$=$C(sp^2)$	585–630
H—$C(sp^2)$	380–410	$C(sp^2)$—$C(sp^2)$	630–675
H—$C(sp^3)$	410–420	$C(sp^2)$....$C(sp^2)$	520–525ᵃ
H—N	385–405	C≡C	795–835
H—O	450–485	$C(sp^2)$—N	275–360
H—F	568	$C(sp^3)$—N	260–300
H—Si	320–340	C=N	555–595
H—P	320	C≡N	835–865
H—S	360–380	$C(sp^2)$—O	375–430
H—Cl	432	$C(sp^3)$—O	335–380
H—Ge	290	$C(sp)$=O	790–815
H—As	297	$C(sp^2)$=O	720–780
H—Se	317	C(sp)—F	485–500
H—Br	366	$C(sp^2)$—F	430–455
H—Sn	250–260	$C(sp^3)$—F	465–490
H—Sb	257	C—Si	290–305
H—Te	267	C=Si	460
H—I	298	C—P(PR₃)	230–265
H—Pb	204	C—P(O=PR₃)	275–350
H—Bi	194	C—S	270–290
Be(sp)—F	638	C=S	565–600
Be(sp)—Cl	463	C—Cl	315–340
Be(sp)—Br	394	C—Ge	230–245
Be(sp)—I	311	C—As	190–225
B—B	305–325	C—Br	260–285
B—C	330–325	C—Sn	190–210
B(sp²)—N	440–455	C—Sb	175–205
B—O	515–530	C—I	210–220
B=O	800–840	C—Pb	125–140
B—F	635–655	C—Bi	105–135
B—S	360–375	N—N	120–210
B=S	522	N=N	450–485
B—Cl	425–460	N≡N	946
B—Br	360–370	N—O	205–245
B—I	270–280	N=O	560–620
C(sp)—C(sp)	370–385	N...O(RNO₂)	420–450
C(sp)—$C(sp^2)$	335–375	N—F	260–280
C(sp)—$C(sp^3)$	350–395	N—Si	290–310
$C(sp^2)$—$C(sp^2)$	270–400	N—Cl	185–200
$C(sp^2)$—$C(sp^3)$	325–360	N—Br	163

N—I	140	Si—Br	330
O—O	120–175	Si—I	271
O=O	496	P—P	200
O—F	190	P—S	258
O—Si	420–435	P=S	355–370
O=Si	640	P—Cl	310–330
O—P(PX$_3$)	350–370	P—Br	262
O—P(O=PX$_3$)	370–400	S—S	250–280
O=P	480–550	S=S	425
O—S	275–350	S—Cl	255–285
O=S(SOR$_2$)	360–380	S—Br	230–250
O=S(SO(OR)$_2$)	390–445	Cl—Cl	243
O=S(SO$_2$R$_2$)	420–435	Cl—Ge	340
O=S(SO$_2$(OR)$_2$)	435–475	Cl—As	310
O—Cl	195–215	Cl—Se	180–240
O=Cl	215–225	Cl—Br(BrCl)	219
O—Ge	360	Cl—Sn	316
O=Ge	500	Cl—Sb	314
O—As	305–325	Cl—I(ICl)	211
O=Se	330–450	Ge—Ge	135–195
O—Br	190	Ge—Br	281
O=Te	330	Ge—I	215
O—I	195	As—As	142
F—F	158	As—Br	256
F—Si	565–625	As—I	208
F—P	485–500	Se—Se	195–250
F—S(VI)	315–325	Se—Br	205–230
F—Cl(ClF)	255	Br—Br	194
F—As	442	Br—Sn	235–300
F—Se	303	Br—Sb	264
F—Br(BrF)	285	Br—I	179
F—Te	332	Sn—Sn	140–185
F—I(IF)	282	I—I	153
Si—Si	180–220		
Si—S	314		
Si—Cl	390–410		

^{a} C–C bonds in aromatic rings

of C$_2$X$_6$ if we assume that the C–X bond energies in C$_2$X$_6$ are the same as in CX$_4$. The bond energies presented in Table 6.2 (apart from those for bonds like H–F, Br–Br etc. which occur in only one diatomic molecule) were obtained by means of a least-squares fitting procedure to reproduce as closely as possible the atomisation enthalpies of about 500 gaseous molecular substances. Rather than quote a single value, the variability of bond energy is emphasised by giving a range of values. Statistically, a given bond energy can be placed within the range quoted in any molecule with a confidence level of about 70%. In estimating the atomisation enthalpy of any molecular substance, the average of the extreme values should be used. Among the most variable bond energies are $P=O$ and $S=O$. These tend to increase with increasing electronegativities of the other atoms bonded to P or S, consistent with the view that $d_\pi-p_\pi$ bonding is important in molecules like POX_3 and SO_2X_2 and the 3d orbitals of P and S come most strongly into play when the atom bears a large fractional charge.

As far as the data permit, the dependence of bond energies on the hybridisations of the atoms concerned is allowed for. In the case of F—P, Cl—P and Br—P, the bond energies are based upon data for molecules in which the P atom has (approximately) sp^3 hybridisation, e.g. PX_3, PX_2Y, POX_3. These values are inappropriate for PX_5 species, or for PX_3Y_2 etc. The mean bond energies found for, e.g., PF_5 are considerably lower than for trivalent phosphorus species.

The classification of bonds as single, double or triple is not always clear-cut. For example, the Si–F bond may have some double-bond character, depending on the extent of overlap between empty silicon 3d orbitals and filled F 2p orbitals, and this will vary from one situation to another. The large range quoted for the $C(sp^2)-C(sp^2)$ bond energy reflects the variable amount of $p_\pi-p_\pi$ overlap which may be present.

Homopolar single bond energies E–E might be expected to show some correlation with the E–E bond length, i.e. with the covalent radii of the atoms, since short bonds are likely to afford better overlap. Thus it is not surprising to find that H–H is the strongest homopolar single bond. But along the Period Li $\rightarrow$ F, we find the strongest E–E bond to be C–C, with N–N, O–O and F–F about equally weak. Several explanations have been put forward for this apparent anomaly and doubtless several factors are at work. The most favoured interpretation involves consideration of repulsions between lone pairs on the E atoms in the E–E bond; molecules X_2N—NX_2, XO—OX and F—F have respectively one, two and three lone pairs on each atom forming the E–E bond, and their mutual repulsion largely offsets the expected increase in bond strength across the Period. It is significant that P–P, S–S and Cl–Cl bonds are stronger than N–N, O–O and F–F respectively; repulsion between lone pairs is less important at the greater internuclear distances found for single bonds between the larger atoms. The bond energies for N—N, $N=N$ and $N\equiv N$ increase dramatically as the hybridisation of the N atom goes from sp^3, to sp^2 and finally to sp. In each case, there is one lone pair on each N atom. If the direction along which the nitrogen lone pair points is represented by a vector, the angle between this vector and the N–N axis increases from about 109° in X_2N—NX_2 to 120° in $XN=NX$ and 180° in $N\equiv N$. This enables the lone pairs to keep further away from each other and reduces their mutual repulsion. In the case of carbon, where lone pair repulsions play no part, the $C=C$ bond is about 1.9 times stronger than C—C while $C≡C$ is only 2.4 times stronger than C—C. Evidently π bonds between C atoms are weaker than σ bonds, and compounds containing such bonds will be thermodynamically unstable with respect to addition reactions. However, multiple bonds between carbon atoms tend to be kinetically rather stable, especially if delocalised via resonance.

For the heavier atoms Si, P and S, the π component in double and triple bonds is much weaker than for the lighter atoms C, N and O. This arises from the greater internuclear distances which prevent effective $p_\pi-p_\pi$ overlap. Experimental data are sparse, but the $P=P$ bond in the P_2 molecule (observable in the gas phase at high temperatures) is about 2.4 times as strong as P—P ($490\,kJ\,mol^{-1}$ versus *c*. $200\,kJ\,mol^{-1}$). This ratio is actually the same as for $C≡C$ versus C—C, but the fact that a P atom would rather form three P–P single bonds than one triple bond means that an assembly of P atoms will form tetrahedral P_4 molecules (each containing six P–P bonds) in preference to P_2 – except when the temperature is high enough for entropy to favour $2P_2(g)$ over $P_4(g)$. On the other hand, a collection of N atoms will form N_2 molecules.

Heteropolar bond energies are conveniently rationalised in terms of electronegativities, according to Pauling's interpretation (Section 4.4). The fact that Pauling's original scale was based upon bond energy data might suggest a circular argument if we use electronegativities to rationalise bond energies; however, independent methods (as discussed in Section 4.4) lead to electronegativity values very similar to those of Pauling. As an example of the use of electronegativity arguments, it may be noted that the Si–H bond is weaker than C–H, while Si–F is stronger than C–F. For C–H and Si–H, the polar contributions are both very small and the fact that Si–Si is weaker than C–C is the dominant factor. But for Si–F, there is a much larger electronegativity difference (squared) compared with C–F and this is now dominant. In general, Si forms stronger single bonds than carbon with atoms of high electronegativity (F, Cl, Br, O, N). But carbon invariably forms stronger multiple bonds than silicon. The $C=O$ bond is more than twice as strong as C—O, there being a significant polar contribution. Thus carbon would rather form one $C=O$ bond than two C—O bonds and molecules of the type $R_2C(OH)_2$ are quite unstable with respect to $R_2C=O + H_2O$. However, $R_2C(OR')_2$ is kinetically stabilised since the elimination of R'OR' may be mechanistically difficult. The $Si=O$ bond energy (obtained from data for molecular SiO_2 in the gas phase) is only 1.5 times the Si–O single bond energy, so that Si would rather form two Si—O bonds than one $Si=O$. In general, we can say that a silicon atom is wasting one of its valence orbitals by devoting it to formation of a $p_\pi-p_\pi$ bond. But, as we shall see in Section 10.6, such bonds can be kinetically stabilised.

The comparison between C and Si in the previous paragraph also holds good for N and P, and for O and S. However, a greater tendency to form multiple bonds can be discerned as we move towards the right-hand side of the Period Na $\rightarrow$ Cl. As the atoms contract in size across the Period, p_π–p_π overlap is improved. C=S bonds are far more common than C=Si or C=P. Table 6.2 shows that C=S is about twice as strong as C—S.

6.3   Covalent bonding schemes for Main Group atoms

This section attempts a systematic classification of the covalent bonding schemes, in terms of VB theory, exhibited by atoms of the Main Group elements. The aim is to enable the reader to 'fit together' the structure of a molecule or polyatomic ion by identifying the appropriate bonding scheme for each atom. You should try not to memorise by rote the tables in this section; the material should be studied in conjunction with a text which covers the descriptive chemistry in some detail. The lists of bonding schemes are not exhaustive, but they contain the most important ones.

It must be remembered that many molecules/polyatomic ions cannot be adequately represented by a single VB structure, and resonance is often called for. Many of the structures and associated types of hybridisation have been idealised for simplicity. Deviations from regular geometries (which can be so neatly explained in terms of VSEPR theory) need not concern us since they are of little importance in understanding the broad principles of molecular architecture. Thus the bonding in molecules such as H_2O, NH_3 and PF_3 is described in terms of sp^3 hybridisation. The fact that the interbond angles are invariably less than the tetrahedral angle of 109.5° need not worry the reader. Again, for the sake of simplicity, we consider only the types of hybridisation listed in Table 6.1, although intermediate hybridisation may be more appropriate in many cases.

In the following tables, the valence state of the central atom is described in terms of orbital occupancy. Thus, for example, $h^1h^1h^2p^1$ denotes a valence state in which two hybrid orbitals (of specified type) are singly occupied (and are used in bond formation) while a third accommodates a lone pair. The singly-occupied pure np orbital forms a p_π–p_π bond. Empty hybrid orbitals (denoted h^0) always function as acceptor orbitals in the formation of coordinate (or dative) bonds. Doubly-occupied hybrid orbitals (h^2) may be lone pairs, or may function as donor orbitals in coordinate bonds, in which case the h is underlined. A doubly-occupied np orbital always forms a dative π bond, while a singly-occupied np orbital always forms an 'ordinary' π bond. Singly-occupied nd orbitals form d_π–p_π bonds. For the reasons discussed in Section 6.1, you will find few stable species where np orbitals are left empty.

Group 1 (including hydrogen)

Hydrogen, having one valence orbital and one electron, may be expected to form just one σ bond. In fact, the covalent bonding in many hydrides is much more complicated than this, with H atoms often being bonded to two other atoms, forming a three-centre two-electron bond. Such 'electron-deficient' species are discussed more fully in Section 7.4.

The other Group 1 atoms have little tendency to form covalent bonds, and their chemistry is dominated by compounds which are best thought of as containing M^+ cations. Apart from the gaseous M_2 molecules (obtainable by vaporisation of the elemental substances), covalent species are largely restricted to the organometallic chemistry of lithium (see Section 7.4). The low dissociation energies of $M_2(g)$, which range from $105 \, kJ \, mol^{-1}$ for Li_2 to only $42 \, kJ \, mol^{-1}$ for Cs_2, suggest that the valence orbitals are too diffuse to allow effective overlap. The dissociation energies of the monohalides in the gas phase are much greater (in the range $300–600 \, kJ \, mol^{-1}$) but it is generally agreed that these are to be regarded as ion pairs, rather than covalent molecules.

Group 2 (also zinc, cadmium and mercury, Group 12d)

The heavier alkaline earths – from Ca downwards in Group 2 – have little covalent chemistry and are almost invariably found (except in high-temperature gaseous species) as M^{2+} ions. However, the chemistry of Be is almost entirely covalent with Mg being intermediate. Table 6.3 lists the principal types of covalent bonding situations for Be and Mg.

Table 6.3 *Covalent bonding of Be, Mg*

	Hybridisation	Valence state	Examples
(1) —M—	sp	$h^1h^1p^0p^0$	$BeCl_2(g)$
(2) $\overset{\backslash}{\underset{/}{M}}\leftarrow$	sp^2	$h^1h^1h^0p^0$	$Be_2Et_4(N_2Me_4)_2$
(3) $\overset{\downarrow}{\underset{/}{M}}\searrow$	sp^3	$h^1h^1h^0h^0$	$BeCl_2(s)$ $BeO(s)$ $MgCl(CH_3)(OEt_2)_2$
(4) $\overset{\downarrow}{\underset{\uparrow}{M}}$	sp^3d^2	$h^1h^1h^0h^0h^0h^0$	$[Mg(NH_3)_6]^{2+}$ $MgI_2(s)$

Restricted to the use of only four valence orbitals, Be very rarely attains coordination numbers greater than four; higher coordination numbers are more common for Mg, whose 3d orbitals may be invoked.

Having only two ns electrons in their valence shell, Be and Mg can form only two covalent bonds unless donor groups can furnish the electron pairs necessary for the formation of coordinate bonds. Scheme (1) in Table 6.3 is rare since two 2p orbitals are unemployed and the atom has not fulfilled its bonding potential. Thus molecular $BeCl_2$ is (at room temperature) unstable with respect to polymerisation to $BeCl_2(s)$:

$$BeCl_2(g) \rightarrow BeCl_2(s) \quad \Delta H^0 = -137 \, kJ \, mol^{-1}$$

The structure of beryllium(II) chloride in its various phases has been described, together with the relevant thermochemistry, in Section 3.3. The dimer Be_2Cl_4 is an example of scheme (2) bonding, while the chain structure of $BeCl_2(s)$ represents scheme (3). Scheme (2) is rather unusual, and in compounds stable at room temperature it is mainly restricted to some organometallic compounds of Be. The example given in Table 6.3 has the structure:

$$
\begin{array}{ccc}
& CH_3 \quad CH_3 & \\
CH_3CH_2 & | \qquad | & CH_2CH_3 \\
\diagdown & | \qquad | & \diagup \\
Be \leftarrow N \text{---} N \rightarrow Be \\
\diagup & | \qquad | & \diagdown \\
CH_3CH_2 & | \qquad | & CH_2CH_3 \\
& CH_3 \quad CH_3 &
\end{array}
$$

Scheme (3) is by far the most important for Be, and dominates the extensive organometallic chemistry of Mg. The example given of $MgCl(CH_3)(OEt_2)$ is one of the well-known Grignard reagents, much loved by organic chemists. Crystalline BeO has the wurtzite structure with tetrahedral 4:4 coordination, and can be depicted as a polymeric covalent structure in which both Be and O atoms form two 'ordinary' and two coordinate bonds:

$$
\begin{array}{ccc}
O & & Be \\
\downarrow & & \uparrow \\
Be & & O \\
\diagup \diagdown & & \diagup \diagdown \\
O \quad O & & Be \quad Be \\
O & & Be
\end{array}
$$

Scheme (4) is restricted to Mg, since nd orbitals are called for. MgI_2 – cited as an example in Table 6.3 – has the two-dimensional CdI_2 layer structure described in Section 3.3.

Can Be and Mg form p_π–p_π bonds? In $BeCl_2(g)$, we can envisage π overlap between filled Cl 3p orbitals and empty Be 2p orbitals, but the latter are likely to be too diffuse for effective overlap. The gas-phase molecules BeO and MgO can be written as $E=O$ but their properties (bond lengths, dissociation energies, force constants etc.) suggest that the π bond is rather weak. The high boiling point of BeO(s) (3850 °C) clearly indicates that four Be—O bonds are very much better than one $Be=O$ bond.

Zn, Cd and Hg may be admitted as honorary members of Group 2 and as Main Group elements inasmuch as their filled nd subshells remain intact in all their compounds and apparently play little or no part in bonding. They exhibit all the bonding schemes in Table 6.3, plus five-coordination which is rare in Group 2. Scheme (1) is quite common for Hg; HgO(s) has a polymeric chain structure, described in Section 3.3. Mercury(II) chloride has a structure in which linear $HgCl_2$ molecules can be identified. From spectroscopic data for the free atoms, the $ns \rightarrow np$ promotion energies for Zn, Cd and Hg are found to be respectively about 450, 400 and 520 kJ mol^{-1} (for Be and Mg, the corresponding figures are 320 and 280 kJ mol^{-1}). The high figure for mercury presumably reflects the poor shielding of the 6s electrons from the nucleus. Since Hg–X bond energies are expected to be intrinsically rather weak (given the large size of the Hg atom) and since the $6s \rightarrow 6p$ promotion energy is rather large, it is not surprising that mercury should often be content with linear (sp) structures which demand less promotion energy than sp^2 or sp^3 schemes.

The chemistry of zinc has much in common with that of Be. The atoms are of comparable electronegativity, and although Zn is somewhat larger than Be, analogous compounds of Zn and Be are often isostructural.

Group 13

Covalent bonding schemes for the Group 13 atoms are summarised in Table 6.4. The formation of M^{3+} and M^+ ions has been discussed in Chapter 5.

The formation of only three covalent bonds, using the three valence electrons (ns^2np^1), is rare. Scheme (1) leaves one np orbital unemployed, but it is difficult to see how a trialkyl boron compound BR_3 can do any better. BH_3 is unstable with respect to dimerisation, as discussed in Section 7.4. Scheme (2) is perhaps the most common for boron, and is found for all the other Group 13 atoms; full use is made of all the ns and np valence orbitals. The bonding in BH_4^-, cited in Table 6.4, can be described in terms of resonance:

Table 6.4 *Covalent bonding of B, Al, Ga, In and Tl*

		Hybridisation	Valence state	Examples
(1)	$\diagdown$ M $-$ $\diagup$	sp^2	$h^1h^1h^1p^0$	$B(CH_3)_3$ BO_3^{3-}
(2)	M	sp^3	$h^1h^1h^1h^0$	BH_4^- Al_2I_6 $GaCl_4^-$
(3)	$\diagdown$ M $=\!\!=$ $\diagup$	sp^2	$h^1h^1h^1p^0$	BF_3 Me_2BNR_2 $B_3N_3H_6$
(4)	$-$ M $\diagup$ $\diagdown$	sp^3d	$h^1h^1h^1h^0h^0$	$AlCl_3(NR_3)_2$
(5)	M	sp^3d^2	$h^1h^1h^1h^1h^0h^0h^0$	$Al(acac)_3^a$ $TlCl_6^{3-}$

a The ligand acac⁻ is depicted in Fig. 8.5

The anion is a regular tetrahedron (isoelectronic with CH_4) and might alternatively (and perhaps preferably?) be described as $B^-(H)_4$; since B and H are of about the same electronegativity, the negative charge should be about equally shared over all five atoms. Many molecules and ions containing tetrahedrally-coordinated boron can be regarded as Lewis acid–Lewis base adducts $X_3B \leftarrow L$.

Scheme (3) affords an alternative (though less satisfactory) means whereby a B atom can make full use of its valence orbitals (and attain the octet configuration). The 2p orbitals of boron are (like those of Be) rather too diffuse for effective p_π–p_π overlap. Molecules $X\!-\!B\!=\!Y$ (X = H, F, Cl etc.; Y = O, S) are known, but they are unstable with respect to polymerisation to give complex O- or S-bridged species. However, the situation depicted in scheme (3) may make a significant contribution to the bonding in molecules such as BF_3, thus:

The short B–F distance in BF_3 (130 pm compared with 143 pm in BF_4^-) and the large B–F bond energy are suggestive of partial double bond character. BF_3 is a weaker Lewis acid than the other boron trihalides, which may indicate that the four boron valence orbitals in BF_3 are more fully engaged than in BCl_3 etc. Molecules of the type R_2BNR_2 are believed to have substantial p_π–p_π bonding:

$$R_2B\!\!-\!\!NR_2 \leftrightarrow R_2\overset{-}{B}\!\!=\!\!\overset{+}{N}R_2 \quad \text{(i.e. } R_2B \leftarrow NR_2)$$

These molecules (isoelectronic with ethylene derivatives) are planar and have short B–N bonds. Borazine $B_3N_3H_6$ is isoelectronic with benzene and has a similar planar, hexagonal structure. It is probably best formulated as:

The heavier atoms of Group 13 appear not to form p_π–p_π bonds. Five- and six-coordinate structures (schemes (4) and (5)) are almost entirely restricted to the heavier atoms, although at least one complex containing five-coordinate boron is known. According to VSEPR theory, scheme (4) structures should be trigonal bipyramidal. However, $InCl_5^{2-}$ and $TlCl_5^{2-}$ are found to be square pyramidal in crystalline solids. As noted in Section 8.2, this shape (also adopted by $MnCl_5^{2-}$) is possibly favoured by crystal packing requirements.

Group 14

Table 6.5 summarises the principal covalent bonding schemes exhibited by atoms of the Group 14 elements. Carbon is restricted to schemes (1)–(9), having just four valence orbitals and four valence electrons; its chemistry can largely be rationalised by assigning it a valency of four. Molecules/polyatomic ions having lone pairs on the Group 14 atoms (schemes (7), (9), (10)) become increasingly common as we go down the Group. This 'inert pair effect' arises from the fact that diminishing bond energies down the Group are not recompensed by diminishing $ns \rightarrow np$ promotion energies, which are needed in order to

Table 6.5 *Covalent bonding of C, Si, Ge, Sn, Pb*

		Hybridisation	Valence state	Examples
(1)	$\overset{\displaystyle\mid}{\underset{\diagup\,\diagdown}{E}}$	sp^3	$h^1h^1h^1h^1$	CH_4
(2)	$\rangle E =\!=$	sp^2	$h^1h^1h^1p^1$	C_2H_4 $COCl_2$
(3)	$=\!= E =\!=$	sp	$h^1h^1p^1p^1$	CO_2
(4)	$- E \equiv$	sp	$h^1h^1p^1p^1$	HCN
(5)	$E \equiv\!\!\equiv$	sp	$h^1h^2p^1p^0$	CO
(6)	$:\!\overline{E} \equiv$	sp	$h^1h^2p^1p^1$	CN^-
(7)	$\overset{\displaystyle\cdot\cdot^-}{\underset{\diagup\,\diagdown}{E}}$	sp^3	$h^1h^1h^1h^2$	$GeCl_3^-$
(8)	$\overset{+}{\underset{\displaystyle\mid}{E}}$	sp^2	$h^1h^1h^1h^0$	$C(C_6H_5)_3^+$
(9)	$\overset{\displaystyle\cdot\cdot}{\underset{\diagup\,\diagdown}{E}}$	sp^2	$h^1h^1h^2p^0$	$SnCl_2(g)$
(10)	$\overset{\displaystyle\cdot\cdot}{\underset{\diagup\,\diagdown}{E}}$	sp^3	$h^1h^1h^2h^0$	$SnCl_2(s)$
(11)	$-\underset{\displaystyle\mid}{E}\diagdown^{\diagup}$	sp^3d	$h^1h^1h^1h^1h^0$	SiF_5^-
(12)	$\rangle\underset{\displaystyle}{E}\langle$	sp^3d^2	$h^1h^1h^1h^1h^0h^0$	$SnCl_6^{2-}$

attain 'tetravalency'. The promotion energies for Ge and Pb are likely to be particularly high, since their ns orbitals are very stable. Thus for the heavier atoms, structures which preserve a lone pair, having substantial s character, are favoured.

As noted in the previous section, the heavier members of the Group have much less tendency than carbon to form $p_\pi–p_\pi$ bonds. However, a number of disilenes $R_2Si=SiR_2$ and distannenes $R_2Sn=SnR_2$ have been prepared, where R is a very bulky organic group such as 2,4,6-trimethyl-phenyl. Attempts to prepare $H_2Si=SiH_2$ by obvious methods produce a polymer $(SiH_2)_n$. But with bulky substituents, such polymerisation is sterically inhibited. Thus the otherwise kinetically vulnerable $Si=Si$ bond can be stabilised, as discussed further in Section 10.6. We have already noted in Section 1.5 that even ethylene $H_2C=CH_2$ is thermodynamically unstable.

Only the heavier members of Group 14 exhibit the schemes (11) and (12), which require nd orbitals. A number of complex coordination geometries for Sn and Pb have been omitted from Table 6.5.

Group 15

Scheme (1) in Table 6.6 is the most important. Schemes (2), (3) and (5)–(8) involve $p_\pi–p_\pi$ bonding and are commonly encountered in nitrogen chemistry. The heavier members of Group 15 rarely form such bonds; however, a fair number of compounds containing $P=P$, $P=N$, $As=As$, $As=N$ etc. bonds have been characterised. These are stabilised (as for $Si=Si$ bonds) by bulky groups in molecules such as $RP=PR$, where R is, e.g., $CH(SiMe_3)_2$ (see also Section 10.6).

Some of the schemes in Table 6.6 may be difficult to distinguish. Thus NH_4^+ could be described either by (4) or (10), and molecules such as X_3PO are best described (as discussed in Section 6.1) in terms of resonance between structures (4) and a triple-bonded structure $X_3P^-\!\equiv\!\overset{+}{O}$; however, the properties of the molecule are consistent with a P–O bond order of about two, consistent with scheme (9). Schemes (9) and (11)–(13) require nd orbitals on the Group 15 atom and are found only for the heavier atoms, from P downwards. Arsenic and bismuth show some reluctance to adopt (9), (11) and (12), probably because their ns electrons are quite strongly bound as 'inert pairs'. Thus $AsCl_5$, which defied preparative attempts for many decades, is stable only at low temperatures, and H_3AsO_3 has a different structure from that of H_3PO_3:

Table 6.6 *Covalent bonding of N, P, As, Sb, Bi*

		Hybridisation	Valence state	Examples
(1)	$\overset{\cdot\cdot}{E}$ (two single bonds)	sp^3	$h^1h^1h^1h^2$	NH_3, $AsCl_3$
(2)	E (double + single)	sp^2	$h^1h^1h^2p^1$	$NOCl$
(3)	$:E\equiv$	sp	$h^1h^2p^1p^1$	N_2, CN^-
(4)	$\overset{\uparrow}{E}$	sp^3	$h^1h^1h^1\underline{h}^2$	R_3NO
(5)	$\overset{\uparrow}{E}$	sp^2	$h^1h^1\underline{h}^2p^1$	HNO_3
(6)	$\leftarrow E\equiv$	sp	$h^1\underline{h}^2p^1p^1$	N_2O
(7)	$:\overset{\cdot\cdot}{E}=$	sp^2	$h^1h^2h^2p^1$	NCO^-, CN_2^{2-}
(8)	$=\overset{+}{E}=$	sp	$h^1h^1p^1p^1$	NO_2^+
(9)	$\overset{\|}{E}$	sp^3	$h^1h^1h^1h^1d^1$	POF_3
(10)	$\overset{\|+}{E}$	sp^3	$h^1h^1h^1h^1$	NH_4^+, PCl_4^+
(11)	$-E$	sp^3d	$h^1h^1h^1h^1h^1$	PF_5
(12)	$\overset{\downarrow}{E}$	sp^3d^2	$h^1h^1h^1h^1h^1h^0$	$SbCl_6^-$
(13)	$\overset{\cdot\cdot}{E}$	sp^3d^2	$h^1h^1h^1h^0h^0h^2$	$BiCl_5^{2-}$

Scheme (13), exemplified by $BiCl_5^{2-}$, can also be described in terms of p^3d^2 hybrids, with the lone pair in the 'inert' 6s orbital.

Group 16

The formation of two covalent bonds – scheme (1) in Table 6.7 – is the most obvious option for Group 16 atoms, having two unpaired electrons in their ground states. The lone pairs can form coordinate bonds, as in schemes (4)–(7). Sulphur is less willing than oxygen to commit one of its precious p orbitals to formation of a p_π–p_π bond, but it does so more readily than Si or P. Thus CS_2 and $R_2C=S$ unequivocally contain such bonds and have been known since the dawn of chemistry. The main difference between oxygen's covalent chemistry and that of the other members of the Group arises from the availability of schemes (8)–(15) for the heavier atoms. Like its neighbour As, Se exhibits some 'anomalies' compared with S and Te. Thus SeO_4^{2-} is a stronger oxidising agent than SO_4^{2-} (just as AsO_4^{3-} is a stronger oxidising agent than PO_4^{3-}) and the formation of SeF_6 is less exothermic than for SF_6 or TeF_6. This 'anomaly' stems from the stability of the filled 4s orbital in Se. As befits its greater size, Te tends to prefer higher coordination numbers than the lighter members of the Group. Telluric acid is H_6TeO_6 – i.e. $Te(OH)_6$ – rather than H_2TeO_4.

The discussion of nd orbital participation in Section 6.1 is relevant to the bonding schemes (8)–(15). The examples given for (9)–(13) can all be described in terms of semi-polar bonds, i.e. donation from S to O, and SO_2 might be placed alongside ozone O_3 as an example of scheme (6). A VB description of the SO_2 molecule would require resonance among the structures:

whose relative contributions are uncertain. It may be noted, however, that the S–O bond length in SO_2 (143 pm) is shorter than that in the unstable SO (148 pm), whereas the O–O bond length in ozone (128 pm) is longer than in O_2 (121 pm). The mean bond energy in O_3 is $300\,kJ\,mol^{-1}$, compared with $494\,kJ\,mol^{-1}$ for O_2: but the mean bond energy in SO_2 is $548\,kJ\,mol^{-1}$, compared with $520\,kJ\,mol^{-1}$ for SO. These observations suggest that a bond order of about two is appropriate for SO_2; there may even be some triple bond character, via resonance structures such as O—S≡O. If SO_2 is to be described in terms of a single structure, for simplicity, scheme (12) seems most appropriate.

Table 6.7 *Covalent bonding of O, S, Se, Te, Po*

		Hybridisation	Valence state	Examples
(1)	$\diagdown\!\overset{\cdots}{\underset{}{E}}\!\diagup$	sp^3	$h^1h^1h^2h^2$	H_2O, S_8
(2)	$:\!E\!=$	sp^2	$h^1h^2h^2p^1$	$COCl_2$, CS_2
(3)	$:\!\overset{\cdot\cdot}{\underset{\cdot\cdot}{E}}\!\leftarrow$	sp^3	$h^2h^2h^2h^0$	$(CH_3)_3NO$
(4)	$:E\equiv$	sp	$h^1h^2p^1\underline{p}^2$	CO
(5)	$\diagup\!\overset{\cdot\cdot}{\underset{}{E}}\!\searrow$	sp^3	$h^1h^1h^2\underline{h}^2$	H_3O^+
(6)	$\swarrow\!\overset{\cdot\cdot}{\underset{}{E}}\!=$	sp^2	$h^1h^2\underline{h}^2p^1$	O_3
(7)	$\diagup\!\overset{\uparrow}{\underset{}{E}}\!\searrow$	sp^3	$h^1h^1\underline{h}^2\underline{h}^2$	BeO, ZnS
(8)	$\diagdown\!\overset{\mid}{\underset{\mid}{E}}\!:$	sp^3d	$h^1h^1h^1h^1h^2$	SF_4, $TeCl_4$
(9)	$\diagdown\!\overset{\mid}{\underset{\diagup}{E}}\!=$	sp^3d	$h^1h^1h^1h^1h^1d^1$	$F_4S{=}CH_2$, SF_4O
(10)	$\overset{\parallel}{\underset{\diagup\diagdown}{E}}$	sp^3	$h^1h^1h^1h^1d^1d^1$	SO_2Cl_2, SeO_4^{2-}
(11)	$\overset{\cdot\cdot}{\underset{\diagup\diagdown}{E}}$	sp^3	$h^1h^1h^1h^2d^1$	SO_3^{2-}, $SeO_2(s)$
(12)	$=\!\overset{\cdot\cdot}{\underset{}{E}}\!\diagdown$	sp^2	$h^1h^1h^2d^1d^1$	SO_2
(13)	$\overset{\parallel}{\underset{\diagup}{E}}\!\diagdown$	sp^2	$h^1h^1h^1d^1d^1d^1$	$SO_3(g)$
(14)	$\diagdown\!\overset{\downarrow}{\underset{\cdot\cdot}{E}}\!\diagup$	sp^3d^2	$h^0h^1h^1h^1h^1h^2$	TeF_5^-
(15)	$\diagdown\!\overset{\mid}{\underset{\mid}{E}}\!\diagup$	sp^3d^2	$h^1h^1h^1h^1h^1h^1$	SF_6, TeO_6^{6-}

Group 17

A halogen atom can obviously form one covalent bond by use of its singly-occupied np orbital, without the need to invoke any hybridisation. In scheme (1) of Table 6.8, sp^3 hybridisation has been used as a formality, in order to illustrate its relationship to schemes (3)–(5) where lone pairs are used to form coordinate bonds. Schemes (1) and (2) may be difficult to distinguish. In the example given for scheme (2), either may be used. BCl_4^- may be described for book-keeping purposes either as B^-Cl_4 or as $BCl_3(Cl^-)$; in the latter case, the negatively-charged Cl is bonding in accordance with scheme (2). It seems more sensible to place the negative charge on the more electronegative (Cl) atom. Since BCl_4^- is a regular tetrahedron (isoelectronic with CCl_4), resonance is called for among scheme (2) structures with the coordinate bond alternating among the four B–Cl bonds.

Scheme (3) implies a bent halogen bridge between two other atoms, as in $BeCl_2$(s). Fluorine often forms linear bridges, in which case sp hybridisation is required and two pure 2p orbitals are available for π bonding. Fluorine is the most likely halogen to employ its np orbitals in this way. Many compounds of the d block elements exhibit linear M–F–M bridges (see Section 9.2); an example from Main Group chemistry is $(Et_3Al{-}F{-}AlEt_3)^-$, where we might postulate some π overlap between filled F 2p orbitals and empty Al 3d orbitals. Scheme (9) appears to be important as a contributing structure for fluorides, of which BF_3 is given as an example. The extent of p_π–p_π overlap in the boron trihalides is probably in the order $BF_3 > BCl_3 > BBr_3 > BI_3$, although this has aroused some controversy.

Group 18

Apart from XeF^+ (isoelectronic with IF and identifiable in a number of crystalline solids), the bonding in all noble gas compounds requires nd orbitals, subject to the reservations expressed in Section 6.1. The known chemistry of the noble gases is almost entirely that of xenon. KrF_2 is the only well-characterised compound formed by krypton, and radon chemistry has been little explored as yet. We have seen in Section 6.1 that the use of nd orbitals in bonding schemes for Main Group atoms is justified only if the atom in question is bonded to groups of high electronegativity; a substantial partial positive charge on the atom is needed if its nd orbitals are to be sufficiently contracted to be of any use for bonding. The electronegativities of the noble gas atoms are difficult to determine but they may be expected to decrease, roughly in parallel with those of the halogens, down the Group. Thus the most likely noble gas compounds should be the fluorides, oxides and oxofluorides of

Table 6.8 *Covalent bonding of F, Cl, Br, I*

		Hybridisation	Valence state	Examples
(1)	$:\!\ddot{E}\!-$	sp^3	$h^1h^2h^2h^2$	HF, Cl_2
(2)	$:\!\ddot{E}\!\rightarrow$	sp^3	$h^2h^2h^2\underline{h}^2$	BCl_4^-
(3)	$\overset{\cdot\dot{E}\cdot}{\diagdown}$	sp^3	$h^1h^2h^2\underline{h}^2$	$BeCl_2(s)$
(4)	$\ddot{E}$	sp^3	$h^1h^2\underline{h}^2\underline{h}^2$	$CdI_2(s)$
(5)	E	sp^3	$h^1\underline{h}^2\underline{h}^2\underline{h}^2$	$CuI(s)$
(6)	$\overset{\cdot\ddot{E}\cdot}{\diagdown\!\!\!\equiv}$	sp^3	$h^1h^1h^2h^2d^1$	ClO_2^-
(7)	$\ddot{E}$	sp^3	$h^1h^1h^1h^2d^1d^1$	BrO_3^-
(8)	E	sp^3	$h^1h^1h^1h^1d^1d^1d^1$	IO_4^-
(9)	$:\!E\!\Rrightarrow$	sp^2	$h^1h^2h^2\underline{p}^2$	BF_3
(10)	$-\!\ddot{E}\!-$	sp^3d	$h^1h^1h^1h^2h^2$	ClF_3
(11)	$\overset{\diagup}{\underset{\diagup}{\ddot{E}}}$	sp^3d^2	$h^1h^1h^1h^1h^1h^2$	BrF_5
(12)	$\overset{\diagup}{\underset{\diagup}{E}}$	sp^3d^3	$h^1h^1h^1h^1h^1h^1h^1$	IF_7
(13)	$\overset{\cdot\cdot+\cdot\cdot}{\ddot{E}}$	sp^3	$h^1h^1h^2h^2$	ClF_2^+

	Hybridisation	Valence state	Examples
(14)	sp^3d	$h^1h^1h^1h^1h^2$	BrF_4^+
(15)	sp^3d^2	$h^1h^1h^1h^1h^2h^2$	BrF_4^-
(16)	sp^3d	$h^1h^1h^2h^2h^2$	ICl_2^-
(17)	sp^3d^2	$h^1h^1h^1h^1h^1h^1$	ClF_6^+

the heaviest noble gases. (The same conclusion can be reached if we describe hypervalent species in terms of polar bonds, e.g. $F—Xe^+F^- \leftrightarrow F^-Xe^+—F$.) Practically all known compounds of xenon have isoelectronic iodine analogues, most of which were known (or at least suspected) before the genesis of xenon chemistry in the early 1960s.

Some of the bonding schemes set out in Table 6.9 should not be taken too literally. In scheme (11) for XeO_4, the promotion of four electrons to 5d orbitals must require a colossal amount of energy. The alternative semi-polar structure $Xe^{4+}(O^-)_4$, with coordinate/dative bonds between Xe and O, is equally far-fetched. A complete VB description of the bonding in XeO_4 would require resonance among a large number of structures such as:

XeO_4 as a molecular substance is thermodynamically unstable, and highly explosive (as is XeO_3). The bonds are thermochemically weak – the mean bond energy in XeO_4 is only $88\,kJ\,mol^{-1}$ – but the stretching force constant, obtainable from infra-red spectra and a measure of the intrinsic bond strength, is consistent with at least some double-bond character.

Table 6.9 *Covalent bonding of the heavier noble gases*

		Hybridisation	Valence state	Examples
(1)	$:\overset{..}{\underset{..}{E}}{}^{+}\!\!-$	sp^3	$h^1h^2h^2h^2$	XeF^+
(2)	$-\overset{..}{\underset{..}{E}}-$	sp^3d	$h^1h^1h^2h^2h^2$	XeF_2
(3)	$-\overset{\mid+}{\underset{\mid}{E}}\!\!\cdot\cdot$	sp^3d	$h^1h^1h^1h^2h^2$	$XeF_3{}^+$
(4)	$:E\!\!\underset{\mid}{\overset{/\!\!/}{\diagdown}}$	sp^3d	$h^1h^1h^1h^1h^2d^1d^1$	XeO_2F_2
(5)	$=\!\!E\!\!\underset{\mid}{\overset{/\!\!/}{\diagdown}}$	sp^3d	$h^1h^1h^1h^1h^1d^1d^1d^1$	XeO_3F_2
(6)	$\diagdown\overset{..}{\underset{..}{E}}\diagup$	sp^3d^2	$h^1h^1h^1h^1h^2h^2$	XeF_4
(7)	$\diagdown\overset{\mid+}{\underset{..}{E}}\diagup$	sp^3d^2	$h^1h^1h^1h^1h^1h^2$	$XeF_5{}^+$
(8)	$\diagdown\overset{\|\|}{\underset{..}{E}}\diagup$	sp^3d^2	$h^1h^1h^1h^1h^1h^2d^2$	$XeOF_4$
(9)	$\diagup\overset{\mid}{\underset{\mid}{E}}\!\!\diagup\!\!\!/$	sp^3d^2	$h^1h^1h^1h^1h^1h^1d^1d^1$	$XeO_6{}^{4-}$
(10)	$/\!\!/\!\!\overset{..}{E}\!\!\diagdown\!\!\diagdown$	sp^3	$h^1h^1h^1h^2d^1d^1d^1$	XeO_3
(11)	$/\!\!/\!\!\overset{\|\|}{E}\!\!\diagdown\!\!\diagdown$	sp^3	$h^1h^1h^1h^1d^1d^1d^1d^1$	XeO_4
(12)	$-\overset{..}{\underset{\diagup\ \diagdown}{E}}-$	sp^3d^3	$h^1h^1h^1h^1h^1h^1h^2$	XeF_6

6.4   Unstable covalent species

In Section 1.5, we emphasised the pertinence of the question: 'stable or unstable; with respect to what?'. So far in this chapter, we have sought to rationalise the 'right to exist' of known, stable covalent substances by devising plausible descriptions of their bonding, accounting properly for the available valence electrons and orbitals of the constituent atoms. We now turn to unstable species, with a view to understanding the factors which deny them a right to exist.

We noted in Section 1.5 that a *molecule* may 'exist' in the sense that it can be observed and its properties studied (e.g. spectroscopically) while a collection of such molecules does not constitute a stable, isolable substance. Thus we have to distinguish between unstable molecules and unstable molecular substances.

Many molecules are unstable with respect to decomposition or rearrangement via a plausible unimolecular mechanism on account of inherent weakness in their bonding. Examples of the former type – molecules liable to 'fly apart' – are:

$$NCl_4 \rightarrow NCl_3 + Cl$$
$$OF_4 \rightarrow OF_2 + F_2$$
$$PH_5 \rightarrow PH_3 + H_2$$
$$FO_3(OH) \rightarrow 2O_2 + HF$$

In the case of NCl_4, it is impossible to devise a satisfactory VB bonding scheme. We can certainly form four N–Cl bonds, using the four valence orbitals available to nitrogen. But we have one N valence electron left over. Where are we to put it? Presumably in the 3s orbital – the lowest empty valence orbital of N – at a cost in promotion energy of more than $1000 \, kJ \, mol^{-1}$. The relatively weak N–Cl bonds afford poor compensation for such a large investment. NCl_4 would surely decompose to NCl_3 and Cl, with the Cl atoms recombining to form Cl_2. Indeed, NCl_4 may well be unstable with respect to N + 4Cl, not to mention the elemental substances N_2 and Cl_2; even the well-known NCl_3 is unstable with respect to these.

The bonding in OF_4 would require two three-centre O–F bonds in MO language; VB theory would employ polar structures $F_3O^+F^-$. In either description, the two additional O–F bonds must be weaker than those in OF_2, which are themselves only marginally stronger than F–F. For any molecular decomposition, the entropy change will be positive so that the enthalpy change must be appreciably positive if OF_4 is to be thermodynamically stable; this seems improbable.

The bonding in PH_5 can be described in terms of sp^3d hybridisation or alternatively in terms of polar structures $H_4P^+H^-$, i.e. three-centre bonds in MO language. PH_5 has been the subject of much theoretical analysis but has not been found experimentally. A factor which is bound to discourage the formation of PH_5 – starting presumably from PH_3 – is the high dissociation energy of the H_2 molecule, which appears on the right-hand side of the equation given above for the decomposition of PH_5. Given that entropy will favour decomposition, the total bond energy in PH_5 would have to exceed that of $PH_3 + H_2$ by a fair margin if the molecule is to survive; thus its total thermochemical bond energy would have to be at least $1400 \, kJ \, mol^{-1}$.

The total thermochemical bond energy of PF_5 is about 1.5 times that of PF_3, but the corresponding ratio for the chlorides and bromides is about 1.3. This can be understood in terms of decreasing 3d orbital involvement in PX_5 molecules as the electronegativity of X decreases. Since H is less electronegative than any of the halogens, it seems unlikely that the total thermochemical bond energy in PH_5 will be enough to rescue the molecule from instability. (An alternative explanation, based upon polar structures $H_4P^+H^-$, can be put forward; this is suggested as an exercise for the reader.)

$FO_3(OH)$ is the analogue of $ClO_3(OH)$, well known as perchloric acid. The F–O bonds must be single, since fluorine has no d orbitals available for bonding. From the bond energies in Table 6.2, it can be predicted that ΔH^0 for the reaction:

$$FO_3(OH) \rightarrow 2O_2 + HF$$

should be about $-300 \, kJ \, mol^{-1}$. Entropy considerations will add some weight to the proposition that $FO_3(OH)$ has no right to exist.

Another category of unstable molecule is that which is unstable to rearrangement, to give a more stable isomer. An example is H_3NO, which is unstable with respect to H_2NOH (hydroxylamine). The exchange of an N–H bond for an O–H bond is worth about $70 \, kJ \, mol^{-1}$, and such a rearrangement should find an easy mechanistic pathway. In the case of F_3NO, the exchange of an N–F bond for an O–F bond leads to a loss of about $130 \, kJ \, mol^{-1}$ in binding energy. F_3NO is well known, both as a molecule and as a molecular, gaseous substance, but F_2NOF is not.

Molecules which are stable towards decomposition or rearrangement may not form stable substances if the molecules are collectively unstable with respect to dimerisation/oligomerisation/polymerisation, or to disproportionation. Such reactions must involve bimolecular mechanisms – requiring collisions between molecules – and these may be slow enough to permit the isolation of the substance. The case of PH_3F_2 has been discussed in Section 6.1. Even if the reaction is fast, the unstable molecule may survive for long enough to permit its spectroscopic detection, at a high temperature in the gas phase or trapped in an inert matrix at low temperatures, e.g. in solid argon.

Dimerisation or polymerisation necessarily involves the formation of additional bonds. In Section 3.3, we discussed the case of $BeCl_2(g)$, whose polymerisation is thermodynamically 'driven' by the prospect of forming more bonds. 'Free radicals' such as BX_2, CX_3 and NX_2 (where X is a 'univalent' atom or group) are obviously unstable to dimerisation because they have an unpaired electron in a valence orbital. Just as a collection of H atoms will readily pair up to form H_2 molecules, so a collection of CH_3 molecules might be expected to form C_2H_6 molecules. The methyl radical CH_3 can survive in the gas phase for long enough to measure its spectroscopic constants, but methyl is unstable as a substance. Some radicals can be stabilised if dimerisation can be sterically discouraged. For example, $C(C_6H_5)_3$ has been known as a solid molecular substance for several decades; the delocalisation, via resonance, of the odd electron over the phenyl groups is of little importance, because the rings are not all coplanar. The steric bulk of the phenyl groups is the most important factor. Nitric oxide NO is the best-known substance consisting of 'odd-electron' molecules. From Table 6.2, it is evident that N–O bond energies increase impressively with increasing bond order. The NO molecule has a bond order of 2.5 – it is best described in terms of MO theory – and its dimerisation to give $O=N—N=O$ is thermodynamically unfavourable because the weak N–N bond fails to compensate for the great loss in N–O bonding. The weakness of the N–N bond accounts for the 'existence' of the NO_2 molecule in the equilibrium:

$$N_2O_4 \rightleftharpoons 2NO_2$$

Many molecules which are individually stable with respect to decomposition or rearrangement are collectively unstable to disproportionation. In Section 6.1 we mentioned the nonexistent species H_3PO; this is probably unstable with respect to the disproportionation:

$$2H_3PO \rightarrow PH_3 + H_2PO(OH)$$

The exchange of a P–H bond for an O–H bond is certainly beneficial. Disproportionation is an important consideration in determining the stability of a 'subvalent' species, i.e. where the atom of primary interest has not attained its full bonding potential, and exhibits a valency number lower than that usually found. As an example, consider SiF_2, which has been extensively studied in the gas phase at high temperatures. Molecular SiF_2 is obviously unstable to dimerisation to give $F_2Si=SiF_2$; the $Si=Si$ bond is worth about $330\,kJ\,mol^{-1}$. It is even more unstable with respect to disproportionation:

$$2SiF_2(g) \rightarrow SiF_4(g) + Si(s) \quad \Delta H^0 = -377\,kJ\,mol^{-1}$$
$$\Delta S^0 = -204\,J\,K^{-1}\,mol^{-1}$$

But if the reader notes the unfavourable entropy term and considers a simple application of Le Chatelier's principle, it should be apparent that the reverse reaction might be thermodynamically favourable at high temperatures and under low pressure. Thus $SiF_2(g)$ can be prepared at about $1200\,^{\circ}C$ at pressures in the range 10^{-5}–10^{-6} atmosphere. On cooling, the vapour condenses to give a polymer $(SiF_2)_n$ which disproportionates on mild heating to give a mixture of fluorosilanes Si_nF_{2n+2} and elemental silicon. Note that C_2F_4 is thermodynamically unstable with respect to CF_4 and graphite; however, the $C{=}C$ bond is kinetically more robust than $Si{=}Si$, and the latter will always undergo rapid polymerisation unless protected by bulky groups as discussed in the previous section, and in Section 10.6.

The instability of most subvalent species towards disproportionation should be apparent without recourse to explicit thermodynamic data. Consider, for example, the disproportionation:

$$3PH_2(g) \rightarrow 2PH_3(g) + P(s)$$

The number of P–H bonds on the right-hand side of the equation is the same as on the left-hand side, and unless there is good reason to suppose that the bonds in PH_2 are significantly different from those in PH_3, we can safely assume that the P–H bonding is unaffected by the disproportionation. But there are bonds in elemental phosphorus, whose enthalpy of atomisation is $315\,kJ\,mol^{-1}$, and this will tend to drive the disproportionation of PH_2.

As a further example of instability with respect to disproportionation, consider the fluorides of bromine. BrF, BrF_3 and BrF_5 are stable molecular substances; BrF_2, BrF_4, BrF_6 and BrF_7 are not. The mean bond energies found in BrF, BrF_3 and BrF_5 are respectively 249, 201 and $187\,kJ\,mol^{-1}$. One 4d orbital is needed to describe the bonding in BrF_3 (scheme (10) in Table 6.8); two are required in BrF_5 (scheme 11)), and three would be needed for BrF_7 (scheme 12)). The steadily decreasing bond energies reflect the increasing promotion energies to the valence state of the Br atom. However, BrF_3 and BrF_5 are stable towards decomposition to a lower fluoride. By extrapolation, the mean bond energy in BrF_7 should be about $170\,kJ\,mol^{-1}$. Thus ΔH^0 for the decomposition:

$$BrF_7(g) \rightarrow BrF_5(g) + F_2(g)$$

should be about $+100\,kJ\,mol^{-1}$. Entropy will, of course, favour decomposition, but BrF_7 might be isolable at low temperatures. Why are BrF_2 and BrF_4 unstable? To form BrF_2, one bromine 4d orbital is required and it seems reasonable to assume that the mean bond energy for BrF_2 would be about the same as that for BrF_3. We can then predict that the disproportionation:

$$2BrF_2 \rightarrow BrF + BrF_3$$

will be exothermic to the extent of around $-50\,kJ\,mol^{-1}$. Similarly, BrF_4 should disproportionate to BrF_3 and BrF_5. In other words, if one or more nd orbitals are to be invoked, the cost in promotion energy makes it essential to make the best possible use of them.

Polyatomic ions (as opposed to neutral molecules) may also be unstable with respect to decomposition, polymerisation or disproportionation. However, ions cannot be scrutinised in isolation. In a crystalline solid, there are always counter-ions of opposite charge to be considered, and in solution an ion is surrounded by solvent molecules. The intimacy of the chemical environment of any ion must influence its viability. For example, redox reactions involving electron transfer between cation and anion, or between ion and solvent, may find easy kinetic pathways. We look here at some examples of unstable oxoanions.

The carbonate ion CO_3^{2-} is familiar, both in ionic solids and in solution. The orthocarbonate ion CO_4^{4-} is unknown, although organic derivatives $C(OR)_4$ can be prepared. From Table 6.2, it is apparent that one $C{=}O$ bond is better than two $C{-}O$ bonds, and carbonate is further stabilised by resonance:

But the analogous silicate ion SiO_3^{2-} is unknown. Since two $Si{-}O$ bonds are better than one $Si{=}O$, SiO_3^{2-} is unstable with respect to polymerisation. $MgSiO_3$ contains the polymeric anion $(SiO_3)_n^{2n-}$:

The PO_3^- ion – analogous to nitrate – is likewise unstable to polymerisation, although many salts having the stoichiometry MPO_3 contain the cyclic $P_3O_9^{3-}$ ion:

The P atoms are tetrahedrally bonded to four O atoms; the terminal P–O bonds have some 'double' character and it is easy to see that the P atoms are much better off than in PO_3^-. There may also be some d_π–p_π bonding in silicates.

Earlier in this section we looked at the instability of perfluoric acid $FO_3(OH)$. We now consider the nonexistence of the perfluorate ion FO_4^-. A solid such as $NaFO_4$ is likely to decompose according to the equation:

$$NaFO_4(s) \to NaF(s) + 2O_2(g)$$

The weakness of F–O bonds compared with the bond in the O_2 molecule is clearly important, as for $FO_3(OH)$. But an additional factor is likely to be the high lattice energy of NaF compared with $NaFO_4$; the thermochemical radius of FO_4^- is expected to be larger than the crystal radius of F^-. Note, however, that $NaClO_4$ – a well-known substance – is likewise thermodynamically unstable. For the decomposition:

$$NaClO_4(s) \to NaCl(s) + 2O_2(g)$$

ΔH^0 is $-28\,kJ\,mol^{-1}$ and ΔG^0 is $-130\,kJ\,mol^{-1}$. This should illustrate the dangers of purely thermodynamic arguments in assessing stability. (It also demonstrates the dangers which accompany the handling of perchlorates in the laboratory.)

This chapter has been devoted to bonding among atoms of the Main Group elements: we have used VB theory, which is of limited value in dealing with compounds of the transition elements. We conclude this chapter by looking at a case where VB theory does help to explain a fundamental difference between Main Group and transition element chemistry. We have seen how Main Group molecules containing unpaired electrons are mostly unstable towards dimerisation and/or disproportionation. Transition element chemistry is largely characterised by variability in oxidation number (or valency number, where appropriate) and molecules/ions having unpaired electrons are commonplace. As an example, consider VCl_4, well known as a liquid substance consisting of tetrahedral molecules. The bonding can be described in terms of sp^3 or sd^3 hybridisation; some mixture of the two must be postulated. In either case, the odd electron can be placed in a nonbonding 3d orbital where it can be kept out of mischief. The 3d orbitals of vanadium are low in energy (especially when V is in a high oxidation state, bonded to atoms of higher electronegativity). This 3d orbital has poor overlap properties, and VCl_4 shows no tendency to dimerise. Disproportionation to VCl_3 and VCl_5 is discouraged by the fact that VCl_5 demands more promotion energy than VCl_4. This situation is to be compared with that for the analogous Main Group molecules ECl_4, where E is N, P etc., E having five valence electrons like vanadium. We showed earlier in this section that NCl_4 is unstable because no convenient orbital can be found for the odd electron.

Atoms of the transition elements have, in the nd orbitals, a comfortable haven for unwanted electrons. In the case of PCl_4, we might envisage a square coplanar structure using sp^2d hybrids; the odd electron can then be placed in a nonbonding phosphorus 3p orbital. There would then be a tendency to dimerise to form the dimer Cl_4P—PCl_4, by overlap of the singly-occupied orbitals. More importantly (since this dimer is unknown) there would be disproportionation to give PCl_3 and PCl_5. The latter (as a gas-phase molecule; the solid consists of PCl_4^+ and PCl_6^- ions) requires one 3d orbital, but the former does not and there would be some saving in promotion energy.

6.5   Further reading

Good expositions of VB theory are given by Pauling (1960), Cartmell and Fowles (1977) (see Section A.7 of the Appendix) and Lagowski (1973) (see Section A.3). McWeeny's revisions of Coulson's classic book (1979 and 1982) (see Section A.7) emphasise the three-centre bond approach to hypervalent species. See also Dasent (1965) (Section A.8) for discussion of 'nonexistent' compounds.

7

Molecular orbital theory in inorganic chemistry

7.1 Introduction: orthogonality of wave functions

This chapter is not intended to provide a rigorous treatment of MO theory. Its purpose is to help the reader who has some elementary acquaintance with the subject to appreciate the MO arguments likely to be encountered in the study of descriptive inorganic chemistry, and to emphasise the points which the reader should look out for in more detailed expositions of MO theory.

In Section 1.4, we discussed the history and foundations of MO theory by comparison with VB theory. One of the important principles mentioned was the orthogonality of molecular wave functions. For a given system, we can write down the Hamiltonian H as the sum of several terms, one for each of the interactions which will determine the energy E of the system: the kinetic energies of the electrons, the electron–nucleus attraction, the electron–electron and nucleus–nucleus repulsion, plus sundry terms like spin–orbit coupling and, where appropriate, other perturbations such as an applied external magnetic or electric field. We now seek a set of wave functions $\Psi_1, \Psi_2, \ldots$ which satisfy the Schrödinger equation:

$$H\Psi_i = E_i\Psi_i$$

Analytic, exact solutions cannot be obtained except for the simplest systems, i.e. hydrogen-like atoms with just one electron and one nucleus. Good approximate solutions can be found by means of the self-consistent field (SCF) method, the details of which need not concern us. If all the electrons have been explicitly considered in the Hamiltonian, the wave functions Ψ_i will be many-electron functions; Ψ_i will contain the coordinates of all the electrons, and a complete electron density map can be obtained by plotting Ψ_i^2. The associated energies E_i are the energy *states* of the molecule (see Section 2.6); the lowest will be the ground state E_1 and the calculated energy differences $E_n - E_1$ should match the spectroscopic transitions in the electronic spectrum.

A many-electron function is usually expressed as a product (actually an antisymmetrised product, written in the form of a determinant) of one-electron functions, known as molecular orbitals; an orbital is, by definition, a function of the coordinates of one electron although two electrons, of opposite spin, can have the same wave function. Molecular orbitals must, *mutatis mutandis*, obey the same rules for good behaviour as atomic orbitals. To the practitioner of elaborate, *ab initio* SCF–MO calculations, the formation of the MOs *per se* is only a preliminary to the determination of the total molecular wave function for the ground state, and the total energy E_1 of the molecule in that state. The real physical significance of an individual MO and its associated energy is somewhat obscure, but – in contrast to the corresponding functions in VB theory – it is theoretically respectable from the viewpoint of quantum-mechanical manipulation. Inorganic chemists find much profit and pleasure in MO calculations at the one-electron approximation, determining the MOs and their energies (often in a purely qualitative manner) without explicit regard for inter-electron repulsion.

One of the constraints to be imposed (as for AOs) is that the set of MOs for a given system must be linearly independent, i.e. it should not be possible to express any member of the set as a linear combination of the others. The fact that MOs ψ_1, ψ_2, . . ., ψ_n satisfy the Schrödinger equation under the one-electron Hamiltonian leads to the conclusion that any linear combination ψ_a:

$$\psi_a = a_1\psi_1 + a_2\psi_2 + \ldots + a_n\psi_n$$

will also be a solution, subject to appropriate choice of the constants a_1, a_2, . . ., a_n. Thus ψ_a is not a new, genuine member of the set; it is a trivial solution. To ensure that all the members of the set are non-trivial solutions, we impose the constraint that they must all be mutually *orthogonal*, i.e.:

$$\int \psi_i\psi_j d\tau = 0$$

It is easy to show that if any member of the set is a linear combination of the others, it will not be orthogonal to any function which has a non-zero coefficient in the linear combination. If all the functions are mutually orthogonal, they must be linearly independent. Integrals of this type are often called *overlap integrals*, because they provide a numerical measure of the extent to which ψ_i and ψ_j overlap with each other in space. Two functions which overlap are non-orthogonal.

The construction of MOs is very much a matter of devising a set of convenient functions which are orthogonal to one another; all the stratagems and 'tricks' – from intuitive reasoning, through qualitative arguments based on group theory (Section 7.3) to quantitative calculations – can ultimately be seen as ploys to ensure orthogonality. There is an obvious appeal in basing MOs upon AOs; the 'rules of the game' are well established, and reliable (albeit approximate) wave functions for atomic orbitals are available for all atoms. An appropriate choice of AOs constitutes the *basis set* for an MO calculation. The larger the basis set chosen, the better will be the calculation, at the expense of more computation and less facility of interpretation. Atomic orbitals as such cannot qualify as MOs if there is any overlap between AOs on different atoms (AOs on the same atom are necessarily orthogonal). Thus wherever we find such overlap, measures have to be taken to remedy this affront to the requirement of orthogonality. A linear combination of several individually acceptable wave functions can be a solution to the Schrödinger equation, although it may be unacceptable as a member of the set on the grounds of non-orthogonality. It follows that if we have a set of functions which turn out not to be orthogonal, or indeed are in any way displeasing or inconvenient, we can remedy the situation by constructing linear combinations of the original functions provided that the new set of functions are all mutually orthogonal. Thus MOs are either pure AOs which do not overlap with any other AOs in the basis set, or they are orthogonal linear combinations of overlapping AOs. An MO treatment begins by choosing the basis set of AOs; we will see later how to decide which AOs to include and which to leave out. We then have to divide the AOs into sets such that all the AOs in a set are non-orthogonal to each other. We now construct MOs by finding a new set of linear combinations which are all mutually orthogonal. A 'Law of Conservation of Orbitals' applies; if you start with n AOs, you must finish up with n MOs. Writing an MO ψ_a as:

$$\psi_a = a_1\phi_1 + a_2\phi_2 + \ldots + a_n\phi_n$$

the coefficients $a_1, a_2, \ldots, a_n$ must be such that the function is *normalised*, i.e.:

$$\int \psi_a^2 d\tau = 1$$

This simply means that the total probability of finding an electron in the MO over all space is 1 (remember that an orbital is a one-electron wave function). Functions that are orthogonal to each other are not necessarily normalised and they may need to be multiplied by the appropriate factor to make them comply. The normalisation condition can be expanded as:

$$\sum_{i=1}^{n} a_i^2 + 2\sum_{i \neq j} a_i a_j S_{ij} = 1$$

where S_{ij} is the overlap integral between ϕ_i and ϕ_j. The fractional contribution c_i made by the AO ϕ_i to the MO ψ_a is given by:

$$c_i = a_i^2 + \sum_{i \neq j} a_i a_j S_{ij}$$

Often the second term is neglected on the grounds that the overlap integrals are small; in any case the squared coefficients give a measure of the relative contributions made by the AOs to an MO. The sum of all the c_i over all the MOs to which ϕ_1 contributes must be unity; in other words, the complete set of MOs must ultimately add up exactly to all the AOs from which they were constructed.

If two or more AOs are symmetrically equivalent (see also Section 7.3) they must make equal contributions – i.e. the coefficients must be equal in magnitude though not necessarily in sign – to all MOs in which they appear; there are some exceptions to this rule in the case of degenerate MOs, having exactly equal energies. The relative contributions made by non-equivalent AOs to any MO require numerical calculation, taking into account the magnitudes of the relevant overlap integrals, the relative energies of the AOs and the requirement of orthogonality. The calculation yields the energies of the MOs as well as the coefficients of the constituent AOs.

Having constructed the MOs and calculated their energies, the next step is to feed in the appropriate number of electrons, to give the configuration associated with the ground state (see also Section 2.6). The minimum electronic energy will be obtained by filling the MOs from the bottom on an energy scale (the *aufbau* principle). At this stage, many chemists are content to inspect the occupied MOs and thereby 'explain' the bonding in the molecule, account for its stability or otherwise and perhaps gain insights into its reactivity, physical properties etc. But the professional theoretician, performing a full SCF calculation, has more work to do. According to the *variation principle*, the calculated total energy associated with the ground state will always be greater than the experimental value, and the best approximation to the true molecular wave function is found by optimising all adjustable parameters to give the minimum total energy. The calculated energy of the ground state will involve explicit treatment of interelectron repulsion as well as one-electron orbital energy. The variational treatment will lead to some modification of the MOs, and the ground state may turn out to be associated with a different configuration from that implied by a one-electron approximation. But most MO descriptions, including many supposedly based on quantitative calculations, are at the one-electron level of approximation.

7.2 Diatomic molecules

The reader is likely to have made some acquaintance with elementary MO treatments of diatomic molecules; if not, good expositions are to be found in most contemporary texts on 'general' chemistry. In this section, we concentrate on some important points which are often glossed over.

The H_2 molecule

If we bring two hydrogen atoms into close proximity, their 1s orbitals begin to overlap quite strongly. A 'zeroth-order' approximation to the electronic structure of the H_2 molecule might be to identify the 1s AOs directly as MOs; these would each be singly occupied in the ground state and the electron density distribution would be simply that resulting from the superimposition of the two AOs. The fact that the molecule is stable relative to the separated atoms could be attributed to the attraction of the electron on one atom to the nucleus on the other. The fatal flaw in this description is the fact that the AOs overlap and are therefore not orthogonal; they are unacceptable as MOs. We must take orthogonal linear combinations (two, to satisfy the Law of Conservation of Orbitals) of the AOs. Remembering the symmetry requirement that equivalent AOs must contribute equally in all MOs, the MOs, with subscripts 1 and 2 to indicate the respective atoms, must be:

$$\psi_1 = N_1(1s_1 + 1s_2)$$
$$\psi_2 = N_2(1s_1 - 1s_2)$$

Readers who are unafraid of elementary algebra and calculus may care to prove that these MOs are orthogonal, and that the normalising constants N_1 and N_2 are given by:

$$N_1 = [2(1 + S)]^{-\frac{1}{2}}$$
$$N_2 = [2(1 - S)]^{-\frac{1}{2}}$$

where S is the overlap integral between the AOs $1s_1$ and $1s_2$.

The reader may wonder whether a third MO:

$$\psi_3 = N_3(1s_2 - 1s_1)(= -\psi_2)$$

should be considered. To have three MOs would, of course, violate the conservation of orbitals. In any case, ψ_3 is in fact identical to ψ_2 for all practical purposes. Because $(-1) \times (-1) = +1$, the electron density distribution arising from occupancy of ψ_3 is the same as for ψ_2. You may, of course, substitute ψ_3 for ψ_2 if you wish; this will make no difference to any calculations. Note also that ψ_2 and ψ_3 are not orthogonal.

It is easy to show (another little exercise for the reader) that the electron distribution resulting from occupancy of ψ_1 reveals a distinct build-up of electron density between the nuclei, more than would be expected from the 'zeroth-order' superimposition of the two atomic orbitals; this is at the expense of a diminution of electron density in regions of space where there is little orbital overlap. One might expect such an orbital to be very stable, since an electron in it spends much of its time under the strong attractive influence of both nuclei. (More detailed calculations show that the stability – low energy – of ψ_1 arises from a lowering in the kinetic energy of the electron (s) as well as from a lowering in potential energy.) Thus ψ_1 is a *bonding* MO whose occupancy will tend to stabilise the molecule.

On the other hand, the electron distribution arising from occupancy of ψ_2 leads to a transfer of electron density from the overlap (internuclear) region, compared with the 'zeroth-order' superimposition model, and a build-up elsewhere. The result is an orbital of high energy, partly because the nuclei are exposed to mutual repulsion as a result of the deshielding effect of removing electron density from the internuclear region. Hence we call ψ_2 an *antibonding* MO.

The above observations are, in part, obvious from simple pictorial representations of the MOs:

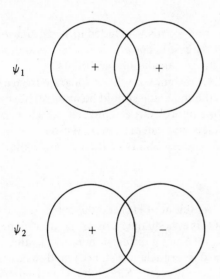

In the case of ψ_2, the presence of a nodal surface between the nuclei shows graphically the low electron density in this region.

In the case of ψ_1, we can say that there is positive (in-phase) overlap between the hydrogen 1s orbitals, because the product $\psi_1\psi_2$ is positive in the internuclear region (the amplitudes of both orbitals are positive); this leads to bonding between the two atoms. In the case of ψ_2, the amplitudes of the AOs in the overlap region are opposite in sign so that the product $\psi_1\psi_2$ is negative; negative overlap leads to an antibonding situation.

We can now draw up an MO energy diagram for H_2, as shown in Fig. 7.1. Clearly, in the ground state the two electrons will occupy ψ_1 with opposite spin in accordance with the Pauli principle. Note that in the energy diagram, ψ_2 is destabilised relative to the free hydrogen 1s orbital by a greater amount than ψ_1 is stabilised. This 'antibonding effect' only

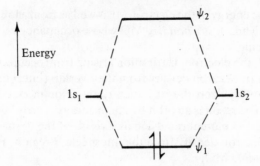

Fig. 7.1. MO energy level diagram for H_2.

appears if overlap integrals are rigorously included in all calculations and not ignored (as is frequently the case). The effect is moderated, however, if higher AOs are included in the calculation (see below).

The 'right to exist' of the H_2 molecule should be apparent from a glance at the MO energy diagram; the molecule should be more stable than the separated atoms (disregarding interelectron repulsion, which, of course, is not allowed for in 'one-electron' calculations). We now address the question of the nonexistence and instability of the He_2 molecule.

The He₂ molecule

The 'one-electron' treatment of He_2 and the determination of its MOs based on the He 1s AOs is exactly the same as the treatment of H_2, except that the overlap integral S has a different numerical value (which will depend upon the assumed internuclear distance), and the calculated energies of the MOs will be different. Since the He atom has two electrons, we have four electrons to accommodate in the two MOs ψ_1 and ψ_2. The He_2 molecule is evidently unstable, because the effect of the occupancy of ψ_2 will more than outweigh the bonding effect of filling ψ_1. In effect, the bond resulting from double occupancy of the bonding MO is cancelled out by double occupancy of its antibonding counterpart.

A caveat must be introduced at this point, which although not affecting our conclusions concerning the stability of He_2, does raise some points which we will return to later. What is the justification for considering only the 1s orbitals in the basis set? Should we not also include the 2s, 2p, 3s etc. orbitals in an MO calculation for H_2 or He_2? Let us investigate the effect upon the MOs of He_2 of including the 2s orbitals. The overlap between $2s_1$ and $2s_2$ leads to the construction of two new MOs, which we may label as ψ_3 (bonding) and ψ_4, taking the same forms as ψ_1 and ψ_2

respectively. A calculation which considers 1s–1s and 2s–2s interaction – but not, for the moment, 1s–2s interaction – leads to the same ψ_1 and ψ_2 as before. ψ_3 and ψ_4 lie much higher in energy; their separation is relatively small, because the 2s orbitals are very diffuse and high in energy (close to the zero energy at which an electron can escape). Such orbitals interact very weakly. The situation is illustrated in Fig. 7.2(a). Thus far, it would seem that the inclusion of 2s orbitals has made no difference at all to the calculation because neither ψ_3 nor ψ_4 will be occupied in the ground state.

Let us check, however, whether our four MOs are all orthogonal to each other. You should quickly be able to show that the integrals $\int\psi_1\psi_4 d\tau$ and $\int\psi_2\psi_3 d\tau$ are zero, bearing in mind that the AOs are normalised and that AOs based on the same atom are all orthogonal. But $\int\psi_1\psi_3 d\tau$ and $\int\psi_2\psi_4 d\tau$ are not zero because the overlap integrals between $1s_1$ and $2s_2$ and between $1s_2$ and $2s_1$ are non-zero. Thus we have to take linear combinations of ψ_1 and ψ_3, and of ψ_2 and ψ_4, which are properly orthogonal. When two AOs on different atoms overlap, they give rise to two MOs, one of which is of lower energy and one of higher energy than either of the

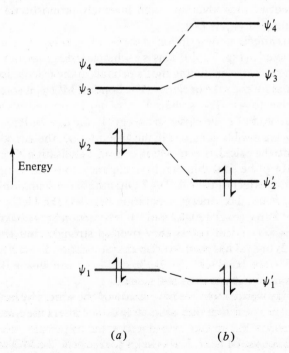

Fig. 7.2. MO energy level diagram for He_2, (a) with 1s–2s interaction neglected and (b) with 1s–2s interaction included.

constituent AOs. Similarly, if two MOs have to be combined to ensure orthogonality the two energy levels are driven apart, e.g.:

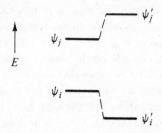

The lower energy MO ψ_i' will be predominantly constructed from ψ_i, with a smaller proportion of ψ_j in its make-up. Conversely, the higher energy MO ψ_j' will be predominantly ψ_j. In general, an MO constructed from two or more simpler functions (which may be pure AOs or linear combinations of AOs) will most closely resemble the AO(s) which lie closest to it in energy. If ψ_j and ψ_i are well separated in energy, the extent of their interaction will be proportional to the extent of their non-orthogonality – i.e. to the extent of their overlap – and inversely proportional to the difference in energy $E(\psi_j) - E(\psi_i)$.

How does this affect our description of the bonding in H_2 and He_2? The situation is depicted in Fig. 7.2(b). This is a schematic diagram, not drawn to scale, and the effect of including the 2s orbitals in the calculations has been exaggerated for the sake of clarity. The bonding MO ψ_1' is somewhat stabilised relative to ψ_1. The stabilisation of ψ_2' compared with ψ_2 is greater, because ψ_2 and ψ_4 lie closer in energy. In the case of He_2, where both ψ_1' and ψ_2' are doubly occupied in the ground state, the introduction of 2s orbitals into the calculation results in overall stabilisation. Could the antibonding MO ψ_2' be so stable that its occupancy does not completely offset the bonding effect of the filled ψ_1'? Quantitative calculations show that the answer is no. The energy separation between the He 1s and 2s orbitals is about 20 eV (over 1900 kJ mol^{-1}). Interaction between orbitals this far apart is very small unless they overlap strongly; but the very diffuse helium 2s orbital has poor overlap characteristics. Even if further AOs are added to the basis set – 2p, 3s, 3p etc. – our conclusion that the He_2 molecule has no right to exist still stands.

The reader may wonder why we have muddied the waters by including 2s orbitals, only to assert that they ultimately do not affect the issue. The foregoing discussion has in fact introduced some important principles which we will make use of later. It also helps to reconcile the MO and VB approaches to the question of the He_2 molecule. In VB language, the formation of a stable He_2 molecule would require promotion of a 1s electron on each atom to the 2s orbital to give two singly-occupied orbitals on each atom which can form a double bond. But the $1s \rightarrow 2s$ promotion energy is too great to offer any hope of compensation by bond formation. The analogous interpretation in MO language similarly attributes the instability of He_2 ultimately to the large 1s–2s separation; if the He 2s orbital were only about 4–5 eV higher in energy than the 1s orbital, it is possible that the filling of ψ'_1 and ψ'_2 would lead to stability with respect to the separated atoms. In the case of H_2, the effect of including the 2s orbitals in the calculation is greater than for He_2 in the sense that the 1s–2s interaction is stronger – the energy separation in the H atom is only 10.2 eV, about half as great as that in the He atom. But the effect on the bonding is slight and can be completely disregarded in any but the most accurate calculations.

The reader may view with some concern our manipulation of MOs, and our alterations to the basis set as if the effect can be switched on or off at will like an electrical appliance. You may ask what the 'real' MOs and their energies are. The answer is that all numerical MO energies are calculated values obtained from one or other of the many recipes at various levels of approximation, and are dependent on the chosen basis set. There is no 'experimental' and hence 'true' value for an MO energy (*pace* some enthusiasts for photoelectron spectroscopy; see Section 2.7). It is true that experimental values for the AO mixing coefficients in MOs can be obtained for some paramagnetic species, but even these depend on approximations and assumptions. For this reason, I have refrained from giving explicit calculated values for MO energies and coefficients because these are meaningless unless the precise method of calculation is fully explained and understood. The games played with MOs in this chapter are perfectly legitimate and often yield more insight into molecular electronic structure than the numerical results of a full calculation.

Hydrogen fluoride HF

This is a simple example of a heteronuclear diatomic molecule which is found in a stable molecular substance. We must first choose the basis set. The only AOs that need to be seriously considered are the hydrogen 1s, fluorine 2s and fluorine 2p, written for brevity as 1s(H), 2s(F) and 2p(F). The fluorine 1s orbital lies very low in energy (700 eV lower than 2p) and is so compact that its overlap with orbitals on other atoms is quite negligible. The fluorine 2p level lies somewhat lower than 1s(H), as indicated by the higher ionisation potential and electronegativity of F. Interaction between 2p(F) and 2s(H) is very small and can be neglected for all practical purposes. One is tempted to discard 2s(F), which lies more than 20 eV below 2p(F); the 2s–2p separation increases

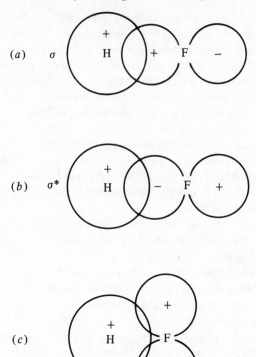

Fig. 7.3. MOs of HF, showing (*a*) bonding combination of 1s(H) and $2p_z$(F), (*b*) antibonding combination and (*c*) lack of overlap between (i.e. orthogonality of) 1s(H) and $2p_{x,y}$(F).

steadily along the Period from He to Ne. Let us first perform a qualitative treatment of HF without 2s(F), and then speculate upon the consequences of including it in the basis set.

It is conventional to label the internuclear axis in a diatomic molecule as z. Thus the three 2p(F) orbitals can be labelled $2p_x$, $2p_y$ and $2p_z$; $2p_x$ and $2p_y$ have their lobes directed perpendicular to the internuclear axis, and have nodal surfaces containing that axis, while $2p_z$ clearly overlaps in σ fashion with 1s(H). (The reader may wonder whether this orientation of $2p_x$, $2p_y$ and $2p_z$ is obligatory, or whether it is chosen for convenience. For a spherically-symmetric atom, there are no constraints in choosing a set of three Cartesian axes. Any set of orthogonal p orbitals can be transformed into another equally acceptable set, by a simple rotation which does not change the electron density distribution of the atom. The overlap integral between a hydrogen 1s orbital and the set of three 2p(F) orbitals is the same regardless of the orientation of the latter with respect to the internuclear axis.)

The overlap between 1s(H) and $2p_z$(F) leads to a bonding MO, which we will call σ, and an antibonding MO σ*, as shown in Fig. 7.3. The $2p_x$ and $2p_y$(F) orbitals do not overlap with 1s(H). This is shown in Fig. 7.3, where we see that the positive overlap with the positive lobe of $2p_x$(F) is exactly cancelled out by the negative overlap with the negative lobe. Thus $2p_{x,y}$(F) constitute a twofold-degenerate pair of *nonbonding* MOs; their occupancy contributes nothing – whether in a positive or negative sense – to the bonding in the molecule. As we shall see, we can have other types of nonbonding MO where we have about equal amounts of bonding and antibonding characteristics. We have a total of six electrons to feed into the MOs in Fig. 7.4(a), one contributed by the H atom and five by the F atom (remembering that the fluorine 1s and 2s orbitals are deemed to be part of the inner core, the configuration $1s^2 2s^2 2p^5$ contributes only the five

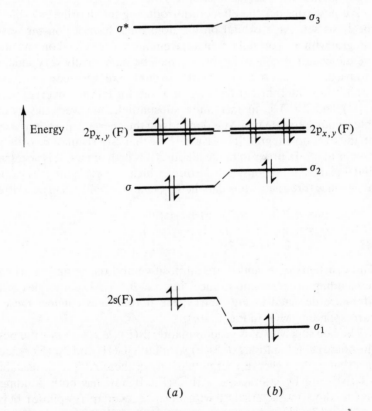

(a)   (b)

Fig. 7.4. MO energy diagram for HF, (a) neglecting the 2s(F) AO and (b) including 2s(F) in the basis set.

2p electrons to the bonding). The ground state of the HF molecule is obviously defined by the configuration $(\sigma)^2(2p_{x,y})^4$ and the reader should recognise immediately the molecule's 'right to exist'. We have one filled bonding MO, no occupancy of the corresponding antibonding MO and all other electrons in nonbonding orbitals.

The bonding MO σ will be predominantly $2p_z(F)$ in character, because it lies closer in energy to the fluorine 2p orbitals than to the hydrogen 1s; conversely, the unoccupied antibonding MO σ^* is largely 1s(H) in character. If, for accounting purposes only, we suppose that the H and F atoms each contribute one electron to the filling of σ, it will be apparent from the composition of this bonding MO that the formation of the molecule leads to a drift of electron density from H to F; if we were to partition the bonding pair between 1s(H) and 2p(F), the latter would have the lion's share. Thus the F atom bears a fractional negative charge while the H atom is positively charged. This, of course, is consistent with the electronegativities of the atoms, which are related to the energies of their valence orbitals (see Section 4.4.).

We now consider the effect of introducing the fluorine 2s orbital into the basis set. An s orbital on an atom must have a non-zero overlap integral with an s orbital on an adjacent atom (or indeed on any atom in the molecule) although the overlap may be numerically very small if the atoms are far apart and/or if the orbitals are very compact. At the equilibrium internuclear distance of 91 pm for HF, the overlap between 1s(H) and 2s(F) is in fact quite substantial; however, the extent of interaction between two AOs depends on the energy separation as well as on the overlap integral. It is easy to show that 2s(F) cannot be orthogonal to σ or to σ^*. In order to make all our MOs orthogonal, it is necessary to find linear combinations of these three which are orthogonal. The effect is to produce three new MOs, labelled σ_1, σ_2 and σ_3, which can be written as:

$$\sigma_1 = a_1 2s(F) + b_1 2p_z(F) + c_1 1s(H)$$
$$\sigma_2 = a_2 2p_z(F) + b_2 1s(H) - c_2 2s(F)$$
$$\sigma_3 = a_3 1s(H) - b_3 2p_z(F) - c_3 2s(F)$$

The constants a_i, b_i and c_i are all positive and the terms are written in descending order of importance, i.e. $a_i > b_i > c_i$. The energies of these MOs are depicted in Fig. 7.4(b); the dotted lines connect these with corresponding MOs in Fig. 7.4(a).

The bonding MO σ_1 is predominantly 2s(F) $(a_1 > b_1 > c_1)$; the positive (in-phase) combinations of 2s(F) with both 1s(H) and $2p_z(F)$ reflect the fact that σ_1 lies lower in energy than any of these AOs. σ_2 is destabilised relative to σ by interaction with 2s(F). It now has both bonding and antibonding characteristics, with positive overlap (in-phase) between 1s(H) and $2p_z(F)$ and negative overlap (out-of-phase) between 1s(H) and 2s(F). The former is the more important interaction – given the greater proximity in energy of 2p(F) to 1s(H) – so that σ_2 is overall still a bonding MO. The overall contribution made to the bonding by including 2s(F) is quite small; the stabilisation of σ_1 relative to 2s(F) is partly (but not wholly) offset by the destabilisation of σ_2 relative to σ. For this reason, we would be justified in leaving 2s(F) out of the picture, especially when it becomes apparent that its inclusion tends to complicate the description.

The MO description of the bonding in HCl, HBr and HI will proceed along very similar lines, except that: (i) as the electronegativity of the halogen atom decreases with increasing atomic number, the energy of the np orbitals relative to 1s(H) will increase, and (ii) the ns–np separation decreases as we go down the Group, so that participation of the halogen ns orbital becomes more important; the effect of this will be to make σ_1 more bonding and σ_2 less bonding.

Let us compare the MO and VB descriptions of HX molecules. The simplest VB treatment would consider simply the overlap between the singly-occupied AOs 1s(H) and $2p_z$(F), leading to the formation of a σ bond. The filled 2s(F) orbital together with the filled $2p_{x,y}$(F) orbitals would constitute the three lone pairs; the valence state corresponds to the configuration $s^2p^2p^2p^1$. If we admit that 2s(F) should be considered, the VB procedure would be to construct a pair of orthogonal s–p hybrid orbitals:

$$h_1 = N_1(\lambda s + p)$$
$$h_2 = N_2(s - \lambda p)$$

The hybrid h_1 has its electron density concentrated towards the H atom, and will give better overlap than either pure ns or np. The hybrid h_2, with its electron density concentrated away from the H atom, will overlap rather less; the different signs of the coefficients of ns and np imply some reduction in the amplitude of the wave function (destructive interference) in the internuclear region. The valence state can be described as $h_1^1 h_2^2 p^2 p^2$, i.e. h_1 forms the bond while h_2 is one of the lone pairs. The price to be paid for such s–p hybridisation and the better overlap it allows is the higher promotion energy required to attain the valence state, now that we have a doubly-occupied lone pair orbital having some np character. We may expect that the mixing coefficient λ will be fairly small – certainly less than unity – in the case of HF, given the large 2s–2p energy separation. The reader should be able to reconcile these two approaches – MO and VB – without difficulty. In either case, the introduction of the ns orbital has its pros and cons, and the net effect is probably quite small.

Homonuclear diatomics of the first short period (Li_2–Ne_2)

Most readers will have made an elementary acquaintance with this topic, and the MO diagram shown in Fig. 7.5(a) will be familiar. By

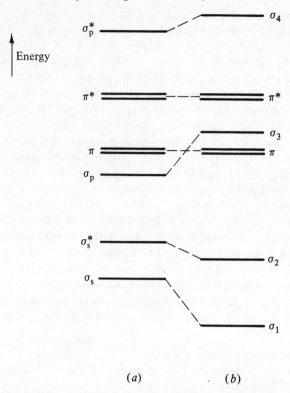

Fig. 7.5. MO energy diagram for homonuclear diatomic molecule X_2 (where X is an atom in the Period Li→Ne), (a) with neglect of 2s–2p overlap and (b) with 2s–2p overlap included.

feeding in the appropriate number of electrons, bond orders consistent with the observed bond dissociation energies and bond lengths can be obtained. For example, the ground state of the O_2 molecule arises from the configuration $(\sigma_s)^2(\sigma_s^*)^2(\sigma_p)^2(\pi)^4(\pi^*)^2$, with two π^* antibonding MOs singly occupied by electrons of the same spin to minimise interelectron repulsion. The bond order B is given by:

$$B = \tfrac{1}{2}(N_b - N_a)$$

where N_b and N_a are respectively the numbers of bonding and antibonding electrons, so that B for the O_2 molecule is 2. In other words, the bonding effect of the filled σ_s^* MO is countered by the filling of its antibonding counterpart σ_s^*; the filling of σ_p leads to a σ bond, because o_p^* is left empty, and one of the two π bonds formed by filling the two degenerate π MOs is cancelled out by half-filling the antibonding π^* level. Long before the advent of quantum mechanics, chemists were accustomed to describe the O_2 molecule as O=O; the bond length, dissociation energy and force constant are all consistent with a double bond. Moreover – and this is cited as one of the great triumphs of MO theory over VB theory – the observed paramagnetism of O_2 is accounted for; a simple VB treatment of O_2 leads to the formation of a σ and a π bond but does not predict the presence of unpaired electrons (a more refined VB treatment, in which 'three-electron' bonds are invoked, has been developed).

But the simple MO diagram in Fig. 7.5(a) is not appropriate for all the homonuclear diatomics of the Period. Thus the B_2 molecule is found to be paramagnetic in its ground state, suggesting the configuration $(\sigma_s)^2(\sigma_s^*)^2(\pi)^2$ instead of $(\sigma_s)^2(\sigma_s^*)^2(\sigma_p)^2$; the π level must therefore lie below σ_p. This ordering is also consistent with the observed diamagnetism of C_2, where the π level is filled and σ_p is empty. The N_2^+ ion has one electron more than C_2 and it is apparent from the photoelectron spectrum of N_2 (see Fig. 2.4 in Section 2.7) that this odd electron is in a σ MO. Thus it appears that – for the aforementioned molecules at least – Fig. 7.5(a) gives the wrong ordering of σ_p and π. Before we attempt a more refined MO description, we have to look in more detail at the MOs of Fig. 7.5(a). These are shown pictorially, together with the algebraic wave functions, in Fig. 7.6.

Why, in Fig. 7.5(a), have we placed π higher in energy than σ_p? It is generally assumed that – other things being equal – π overlap is always weaker than σ overlap. This is not always true; the π overlap between $2p_x$ orbitals at a very short internuclear distance can be greater than the σ overlap between the $2p_z$ orbitals, if the 2p orbitals are relatively diffuse, as is the case towards the left-hand side of the Period. In the case of B_2, π overlap is significantly stronger than σ, but in C_2 and N_2 the situation is reversed; for O_2 and F_2, σ overlap is decisively stronger. This explains why B_2 is paramagnetic, with π more stable than σ_p. But it does not explain the ground states of C_2 and N_2^+. If π is always lower than σ_p, it would be logical to place π^* higher than σ_p^*, in which case O_2 should be diamagnetic. The problem is best resolved by acknowledging the fact that the 2s orbital on one atom and the $2p_z$ orbital on the other in X_2 overlap (cf. the overlap between 1s(H) and $2p_z$(F) in Fig. 7.3). This being so, the MOs in Fig. 7.6 cannot be orthogonal. Our mathematically-competent reader will be able to show that the integrals $\int\sigma_s\sigma_p d\tau$ and $\int\sigma_2^*\sigma_p^* d\tau$ are non-zero because of 2s–2p overlap. All other integrals between the MOs of Fig. 7.6 are properly zero.

The usual remedy for non-orthogonality must be applied. We need to take linear combinations of σ_s and σ_p, and of σ_s^* with σ_p^*, which are orthogonal to each other. The consequences of such mixing upon the MO energy diagram are shown in Fig. 7.5(b). We now have four MOs where the overlap is of σ type, designated σ_1, σ_2, σ_3 and σ_4; the s and p subscripts are no longer appropriate. The new MOs are depicted in Fig. 7.7. The

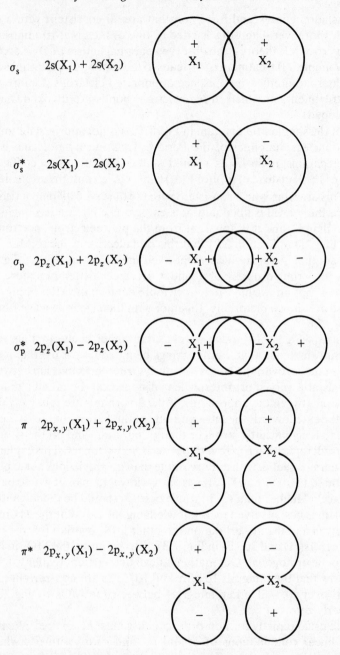

σ_s $2s(X_1) + 2s(X_2)$

σ_s^* $2s(X_1) - 2s(X_2)$

σ_p $2p_z(X_1) + 2p_z(X_2)$

σ_p^* $2p_z(X_1) - 2p_z(X_2)$

π $2p_{x,y}(X_1) + 2p_{x,y}(X_2)$

π^* $2p_{x,y}(X_1) - 2p_{x,y}(X_2)$

Fig. 7.6. MOs of homonuclear diatomic molecule X_2 (normalisation constants ignored), without allowance for 2s–2p mixing.

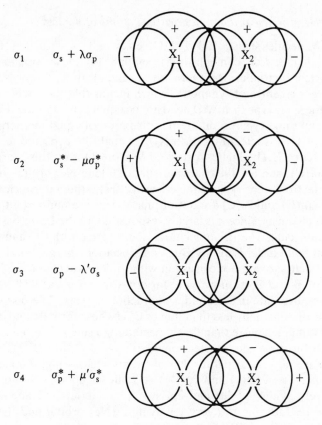

Fig. 7.7. σ MOs of X_2, after inclusion of 2s–2p overlap.

most stable MO σ_1 is bonding with respect to s–s, p–p and s–p overlaps. σ_2, which resembles σ_s^* most closely among the 'original' MOs, is antibonding as far as s–s and p–p overlaps are concerned; but the s–p overlaps are positive (in-phase). This MO is overall antibonding but it has some bonding character, and is more stable than would be calculated for σ_s^*. σ_3 is bonding as far as s–s and p–p interactions are concerned, but the s–p interaction is antibonding; σ_3 is destabilised relative to σ_p, pushing it higher than π where the s–p interaction is sufficiently great. σ_4 is a strongly-antibonding MO, with all overlaps negative (out-of-phase). The extent of s–p interaction – leading *inter alia* to destabilisation of σ_3 compared with σ_p – will fall off as the s–p energy separation increases towards the right-hand side of the Period. Thus it is apparent that for B_2, C_2 and N_2 the energy ordering is $\pi < \sigma_3 < \pi^* < \sigma_4$, but for O_2 and F_2 we have $\sigma_3 < \pi < \pi^* < \sigma_4$.

Heteronuclear diatomics between atoms of the first short period

Molecules such as CO, CN, OF etc. can be described in terms of MOs and electron configurations similar to those obtained for homonuclear diatomics. But there are two important differences. There is no longer a centre of symmetry, which means that the coefficients of AOs of the same type in an MO need not be equal in magnitude. Further, AOs of the same type on different atoms are not equal in energy. For example, CO is isoelectronic with N_2 and might be expected to have a bond order of 3. The highest occupied MO in CO is σ_3; this is predominantly $2p_z(C)$, and has the characteristics of a lone-pair orbital, most of whose electron density is directed away from the internuclear axis because $2p_z(C)$ and $2s(C)$ are out-of-phase. This is an important MO to inorganic chemists, since it is largely responsible for the Lewis basicity of the CO molecule. The BF molecule is isoelectronic with CO and N_2 and its formal bond order is also 3. But its dissociation energy, bond length, force constant etc. are not consistent with triple-bond character. The 2p orbitals of B and F are now quite far apart in energy as well as being somewhat disparate in size, and π interaction is weak. The dissociation energy of BF is actually less than that of BO (whose formal bond order is 2.5) and not much more than that expected for a single B–F bond. What of the BeNe molecule, isoelectronic with N_2, CO and BF? The AO energies are now so disparate that diagrams such as Fig. 7.6 are quite misleading. The 2s(Be) and 2p(Be) AOs lie at about -9 and $-6\,eV$ respectively, while 2s(Ne) and 2p(Ne) are at about -52 and $-23\,eV$ respectively. Ignoring the very low-lying 2s(Ne) orbital and the high-lying, rather diffuse 2p(Be) orbitals, we obtain an MO diagram as shown in Fig. 7.8. The 'antibonding effect' mentioned in the previous section will ensure that the highest occupied MO is sufficiently antibonding to destabilise the molecule. Inclusion of the 2p(Be) AOs will have a moderating influence; there will be weak π overlap and the $2p_{x,y}(Ne)$ orbitals will become somewhat bonding, while the inclusion of $2p_z(Be)$ will 'buffer' the antibonding effect to some extent. Even so, the prospects for the BeNe molecule are not particularly bright, and little is known about it experimentally.

This brings us to the question of how far MO theory accounts for the existence/nonexistence of diatomic molecules. The likes of He_2 and BeNe are expected to be unstable in the sense that the molecules are likely to fly apart; they are thermodynamically unstable with respect to the free atoms. Other molecules whose MO energy diagrams indicate stability may be collectively unstable as molecular substances, because the molecules react with each other to give more stable products, usually via dimerisation, polymerisation or disproportionation. It is not immediately obvious from a qualitative MO treatment that both NO(g) and

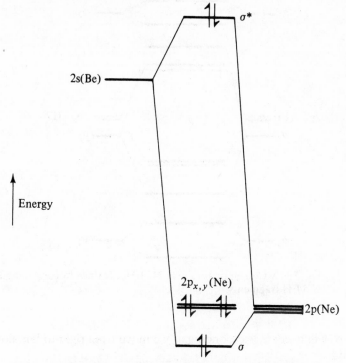

Fig. 7.8. MO diagram for BeNe (ignoring 2p(Be) and 2s(Ne)).

OF(g) are unstable as substances with respect to disproportionation to the respective elemental gaseous substances. The fact that NO(g) can nevertheless be isolated while OF(g) cannot is a kinetic phenomenon, no doubt related to the much greater dissociation energy of NO ($630\,kJ\,mol^{-1}$, compared with $222\,kJ\,mol^{-1}$). It has already been noted (Section 6.4) that NO is stable to dimerisation; OF dimerises to FOOF.

The tendency towards dimerisation – and perhaps ultimately to polymerisation – is more easily predictable in many cases. For example, let us consider the possible dimerisation of HE to give linear HEEH, where E is an element in the Period Li–Ne. The simplest approach here is to start with the MOs of HE and to see how two molecules can be fitted together. In Fig. 7.9 we show the result of combining the corresponding MOs of the two EH fragments, to give bonding and antibonding combinations. The MOs of EH are essentially those already obtained for HF; note, however, that the extent of 2s participation in the bonding is likely to become more important along the series HF < HO < HN < HC < HB < HBe < HLi, as the 2s orbital rises in energy and the 2s–2p separation diminishes. This will mean that the

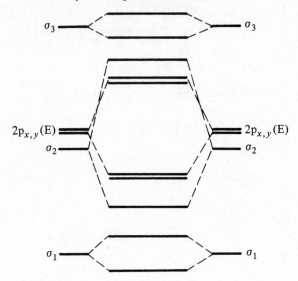

Fig. 7.9. MO diagram for linear HEEH, obtained by combining MOs of two EH fragments.

MO σ_2 will become less bonding as we move from right to left along the Period.

The σ_1 MOs interact rather weakly; being composed mainly of 2s(E), they are relatively compact and low-lying. The σ_2 MOs of HE overlap strongly, because the electron density is concentrated along the molecular axis away from the H atom. The $2p_{x,y}(E)$ orbitals will give rise to a π pair of bonding MOs and a π^* pair of antibonding MOs. The diffuse, high-energy σ_3 MOs have little interaction.

The most stable linear HEEH molecule will be formed by dimerisation of HC, where we have a total of 10 electrons (four from each C and one from each H) to insert into the MOs. These will lead to a triple bond, i.e. we have one filled σ bonding MO and two filled π bonding MOs after disregarding the MOs derived from σ_1. This is consistent with the properties of acetylene HC≡CH. Linear HFFH offers no advantages compared with HF: all F–F bonding cancels out if we have 16 electrons, filling all eight MOs in Fig. 7.9.

The above is an example of the *frontier orbital* approach. In constructing the MOs of a complex molecule XY by combining the X and Y AOs, the MOs of the fragments that we need to look at are the highest occupied MO (HOMO), which is usually nonbonding or weakly bonding and is not absolutely essential to the integrity of the fragment itself, and often the lowest unoccupied MO (LUMO) as well, unless this is strongly antibonding and very high in energy. These are the MOs that will be directly involved in redox or in donor/acceptor reactions of X or Y, and are called the frontier orbitals.

7.3 Polyatomic molecules AB_n

Explicit consideration of molecular symmetry is virtually mandatory in tackling an MO treatment of a polyatomic molecule AB_n. This requires the application of group theory, a branch of abstract algebra which transforms our subjective perceptions of symmetry into rigorous mathematical terminology and notation. A detailed account of group theory and molecular symmetry is outwith the scope of this book; many excellent accounts are available, some in comprehensive inorganic texts, others in monographs, such as the one cited at the end of this chapter. For the benefit of those readers who have some acquaintance with the subject, occasional use will be made of group-theoretical terminology; readers who have yet to be exposed to this fascinating and elegant subject may have their appetites whetted and should be in a position to identify those aspects that they will really need to master in order to study inorganic chemistry. As with other topics in which inorganic chemists claim a proprietary interest, group theory/symmetry is often treated in allegedly 'inorganic' texts and courses at a level far beyond that needed for the study of descriptive inorganic chemistry; the subject has its intrinsic delights and, like thermodynamics, is a worthy intellectual inquiry for its own sake. We will try to introduce the principles of MO theory as applied to polyatomic molecules with the barest minimum of group-theoretical terminology, but the reader must understand that a course on group theory is an essential part of modern chemical education.

Symmetry and MO theory

In a molecule AB_n, a conventional Cartesian axis system has the A nucleus at the origin. If in such a system we can define a coordinate transformation (represented mathematically by a matrix) such that every point (x_i, y_i, z_i) is either left unchanged or changes places with a chemically-equivalent point (x_i', y_i', z_i'), we have defined a *symmetry operation* of the molecule. These operations can be visualised as rotations about axes, reflections in planes or as inversions about centres of symmetry; these geometrical features constitute *symmetry elements*: a given symmetry element may generate more than one symmetry operation. Not all possible combinations of symmetry operations can be realised: those combinations which are acceptable are called *point groups*, because they fulfil the algebraic requirements for a group and all the symmetry elements intersect at a point, the origin of the Cartesian axis system.

If a molecule has any symmetry at all, its electron density distribution must conform to its symmetry. Obviously, the electron density must be equal at all equivalent points: in other words, the performance of a symmetry operation cannot affect the electron distribution, and the total molecular wave function must lead to an electron density distribution having the full symmetry of the molecule. In an MO treatment, we work out the MOs for a particular nuclear framework and then feed in the appropriate number of electrons. This means that – qualitatively, disregarding the exact values obtained for the energies and mixing coefficients – the MOs are independent of the number of electrons in the molecule; thus, *mutatis mutandis*, the MOs of N_2^+ and N_2^- are the same as those for N_2; a quantitative calculation would lead to different energies, and different relative amounts of 2s and 2p character in the σ MOs. The same applies to polyatomic molecules, assuming that the addition or subtraction of electrons does not alter the gross geometry. Thus – assuming that all the species have the same shape and symmetry in the ground state – a rough MO diagram for H_2O^+ or H_2O^- should be the same as that we obtain for H_2O. Now if the addition or subtraction of electrons does not affect the symmetry of the molecule/ion, if indeed the MOs are (qualitatively) independent of the number of electrons ultimately inserted, then it follows – given that the total wave function must be consistent with the full molecular symmetry – that *each and every MO must have symmetry such that its occupancy leads to an electron distribution having the full symmetry of the molecule.* That is, the electron distribution corresponding to occupancy of any MO must be such that equivalent points in space around the molecule have the same electron density. This does *not* mean that every MO must have the full molecular symmetry; an MO whose sign is reversed at all points in space by the application of a symmetry operation will have a totally-symmetric electron distribution if we remember that the electron density is proportional to the *square* of the wave function.

A group-theoretical treatment of this symmetry contraint leads to the requirement that an MO must belong to an *irreducible representation* of the point group. A representation is a set of matrices – one for each symmetry operation – which constitutes a group isomorphous with the group of symmetry operations and can be used to 'represent' the symmetry group. When we say that a function 'belongs to' (or 'transforms as', or 'forms a basis for') a particular representation, we mean that the matrices which constitute the representation act as operators which transform the function in the same way as the symmetry operations of the molecule. (The reader who knows little about matrices and their application as transformation operators can skip over such remarks.) An irreducible representation is one whose matrices cannot be simplified to sets of lower order.

The number of possible irreducible representations of a point group is equal to the number of classes of symmetry operations (two or more operations that are equivalent in the sense that they are related via symmetry operations of the molecule are grouped together as one class). Where the symmetry is too low to permit orbital degeneracy, all the irreducible representations consist of 1×1 matrices which are just simple numbers, $+1$ or -1. This means that the effect of any symmetry operation upon an MO can only be to leave it unchanged, or to reverse its sign. Where the symmetry is higher – where twofold or threefold orbital degeneracies are possible because two or three Cartesian axes are equivalent – some irreducible representations consist of 2×2 or 3×3 matrices. A pair of degenerate MOs – like the π and π^* MOs encountered in the homonuclear diatomic molecules discussed in the previous section – will belong to a twofold-degenerate irreducible representation consisting of 2×2 matrices. The effect of a symmetry operation on one member of the degenerate pair may be: (i) to leave it unchanged, (ii) to reverse its sign, (iii) to transform it into its degenerate partner or (iv) to transform it into a linear combination of itself and its degenerate partner. A threefold-degenerate set of MOs will belong to an irreducible representation of 3×3 matrices, and the effect of a symmetry operation on any member of the set is similar to that for members of twofold-degenerate sets.

Two functions that belong to different irreducible representations are necessarily orthogonal to each other. If they belong to the same irreducible representation, they may not be orthogonal, and must be combined to produce a pair of orthogonal linear combinations. Thus the application of group-theoretical principles and the exploitation of molecular symmetry help to fulfil essential quantum-mechanical requirements in the construction of MOs. We now illustrate these principles by looking at the MOs of the H_2O molecule.

MO treatment of H_2O molecule

The first step is to specify a coordinate system. This is shown below:

$$z$$

$$0 \longrightarrow x$$

$$H_1 \qquad\qquad H_2$$

The z axis is a twofold proper axis of symmetry, designated C_2. A molecule possesses a C_n axis of symmetry if the operation of rotation by an angle $(360/n)$ is a symmetry operation, i.e. it transforms the molecule into an indistinguishable configuration. This is the only symmetry axis present and it is conventional to label the principal symmetry axis as z. The x and y axes are chosen so as to coincide with, or to be contained by, other symmetry elements (see below). Having established the coordinate system, the presence of a C_2 axis can be alternatively described by saying that the operation C_2 transforms a point (x_1, y_1, z_1) to $(-x_1, -y_1, z_1)$ and all pairs of points having this relationship are chemically equivalent. The other symmetry elements present are two planes: the molecular plane xz is a plane of symmetry (as it must be for all planar molecules) and is designated $\sigma_v(xz)$. The yz plane, designated $\sigma_v(yz)$, is likewise a plane of symmetry. Each of these planes slices the molecule into two equal halves, which are mirror images of each other in the plane. The effect of the operation $\sigma_v(xz)$ – reflection in the xz plane – is to transform the point (x_1, y_1, z_1) into $(x_1, -y_1, z_1)$ while $\sigma_v(yz)$ transforms (x_1, y_1, z_1) into $(-x_1, y_1, z_1)$. Apart from $\sigma_v(xz)$, these equivalences stem from the indistinguishability of the two H atoms. These coordinates represent chemically-equivalent points in space about the molecule.

Note that the effect of applying $\sigma_v(xz)$ and then applying $\sigma_v(yz)$ is to transform (x_1, y_1, z_1) into $(-x_1, -y_1, z_1)$, which is the same as C_2. Relationships like this endow the set of symmetry operations with group properties, and restrict the number of possible combinations of symmetry operations to those specified by the point groups. The point group to which the H_2O molecule belongs is designated C_{2v}.

We now choose the basis set. The H atoms contribute only the AOs $1s(H_1)$ and $1s(H_2)$. For the O atom, we consider for the meanwhile only the 2p orbitals; since O lies to the right of the Period, its 2s orbital is very low in energy and is expected to contribute less to the bonding than the 2p orbitals. Later, we will consider the effect on the MO description of including $2s(O)$. In constructing the MOs we must take care that they conform to the symmetry of the molecule. Consider, for example, the following combinations of $1s(H_1)$, $1s(H_2)$ and $2p_x(O)$ that might be considered in constructing the MOs of H_2O by a trial-and-error procedure:

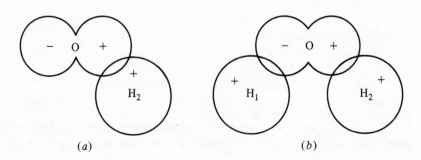

(a)   (b)

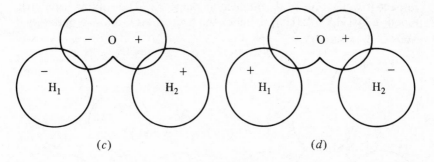

(c)   (d)

The function (a) above is unacceptable because the electron density arising from the occupancy of any MO must have the full molecular symmetry. Thus the electron density at a point in the region between the O nucleus and the H_1 nucleus must be equal to the electron density at an equivalent point between O and H_2, obtained either by C_2 or $\sigma_v(yz)$. Another way of specifying the constraints is to say that because the two H atoms and the AOs $1s(H_1)$ and $1s(H_2)$ are equivalent, these two AOs must contribute equally in any MO. Recognition of this principle might suggest the function (b) as a candidate for MO status. A closer inspection shows that (b) is unacceptable too; we have positive overlap, and a build-up of electron density between O and H_2, but negative overlap and a nodal surface between O and H_1. The contribution to the overall electron density made by occupancy of (b) will not have the full molecular symmetry. The functions (c) and (d) are acceptable, however.

The reader should perceive that the problem is one of matching the symmetry of the $2p_x(O)$ orbital with the $1s(H)$ orbitals. The best way to find the acceptable MOs for AB_n, having regard for symmetry constraints and the Law of Conservation of Orbitals, is to start by combining sets of equivalent AOs – e.g. the $1s(H)$ orbitals in H_2O – to give new orthogonal sets of *group orbitals* (or *symmetry orbitals*) each of which belongs to a specific irreducible representation of the point group. If a group orbital and an AO on the central atom A both belong to the same irreducible representation – i.e. if they have the same symmetry properties – they are not orthogonal and linear combinations must be taken to give an antibonding and a bonding MO. The two functions labelled (c) and (d) above for the H_2O molecule are bonding and antibonding MOs obtained by combining $2p_x(O)$ with the out-of-phase combination of hydrogen 1s orbitals. They may be labelled b_1 and b_1^* respectively, since the functions $2p_x(O)$ and the group orbital $[1s(H_1) - 1s(H_2)]$ both belong to the irreducible representation b_1 in the C_{2v} point group. In C_{2v}, the labels a and b mean that the functions are respectively symmetric or antisymmetric with respect to rotation by $180°$ about the z axis, while the subscripts 1 and 2 denote respectively symmetry or antisymmetry with respect to reflection in the molecular plane xz. The totally-symmetric function $[1s(H_1) + 1s(H_2)]$ matches the $2p_z(O)$ AO, and we obtain two MOs of a_1 symmetry.

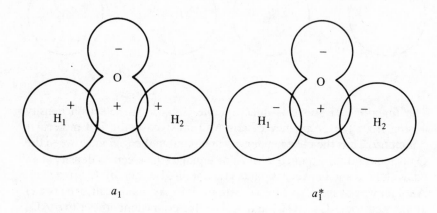

The relative energies are shown in Fig. 7.10(a). $2p_y(O)$ (not shown above) has its lobes directed perpendicular to the molecular plane and does not overlap with either of the hydrogen 1s orbitals. It is a purely nonbonding orbital, with the symmetry label b_2. The oxygen 2s orbital is considered for the meanwhile to be part of the inner core. The bonding MO b_1 is placed lower in energy than a_1 because the H–O–H angle is known experimentally to be 104.5° and simple trigonometric considerations show that for an angle greater than 90°, $2p_x(O)$ will overlap more than $2p_z(O)$ with hydrogen 1s orbitals; if the angle were 180°, $2p_z$ would have no overlap at all. Now the ground state configuration of the O atom is $[He]2s^22p^4$; including the two 2s electrons and one electron from each H atom, we have eight electrons to feed into the energy levels in Fig. 7.10(a). The ground state configuration of the H_2O molecule can then be described as $(2s)^2(b_1)^2(a_1)^2(b_2)^2$, with two filled bonding MOs and no occupied antibonding MOs. This MO description can be reconciled with the corresponding VB treatment quite easily. Taking the O atom in its ground state $2s^22p_y^22p_x^12p_z^1$, we can form two O–H bonds by overlap of the singly-occupied $2p(O)$ orbitals with hydrogen 1s orbitals. The filled 2s and $2p_y$ orbitals constitute two lone pairs; these may be combined to give two equivalent sp hybrids without affecting the electron density distribution. An MO treatment which leads to two filled bonding MOs, with their antibonding counterparts left empty, corresponds to a VB description with two single bonds. By taking linear combinations of the two bonding MOs a_1 and b_1, so that in each case the contribution of one hydrogen 1s

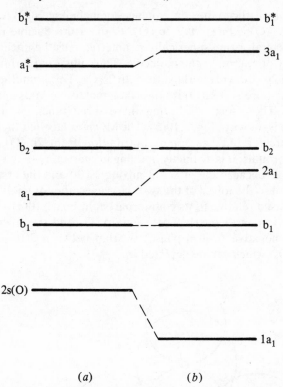

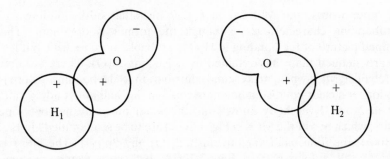

(a)   (b)

Fig. 7.10. MO diagram for H_2O: (a) with 2s(O) excluded from the basis set and (b) with 2s(O) included.

orbital is eliminated, it is possible to devise two equivalent, localised bond orbitals thus:

The MO and VB descriptions for a molecule are related by such transformations.

Let us now investigate the consequences of including 2s(O) in the MO scheme. This AO belongs, like $2p_z(O)$, to the a_1 irreducible representation; an s orbital, being spherically symmetric, must be unchanged by any symmetry operation of the molecule. Thus any group orbital which overlaps with $2p_z(O)$ will overlap also with 2s(O). This is another way of saying that the pure AO 2s(O) is not orthogonal to the MOs a_1 and a_1^* in Fig. 7.10(a). The effect of altering these three functions to ensure orthogonality is shown in Fig. 7.10(b). The MO now labelled $1a_1$ is mostly 2s(O) in character but has some contribution from $2p_z(O)$ and the hydrogen 1s orbitals; it is distinctly bonding in character, all the overlaps being in-phase, because the effect of mixing 2s(O) and the erstwhile a_1 MO is to stabilise the lower of the two. However, the MO now labelled $2a_1$ is destabilised relative to its counterpart a_1 in Fig. 7.10(a). This MO has positive (in-phase) overlap between $2p_z(O)$ and the hydrogen 1s orbitals, but negative (out-of-phase) overlap between 2s(O) and the hydrogen AOs, which can be depicted as:

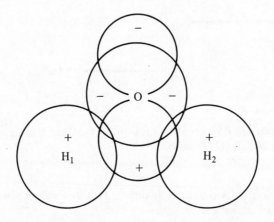

In other words, the MO $2a_1$ in Fig. 7.10(b) has both bonding and antibonding characteristics, although the former is dominant. This refined picture of the bonding in H_2O is probably more realistic but less clearly defined than that obtained by neglecting 2s(O). There are now three filled MOs which make some contribution to the bonding. There is no longer a clear-cut distinction between bonding and nonbonding pairs of electrons. Everybody knows that the water molecule has two lone pairs; which two of the MOs in Fig. 7.10(b) are to be so identified? In Fig. 7.10(a), the lone pairs are obviously 2s(O) and $2p_y(O)$. The latter is clearly still a lone pair in Fig. 7.10(b); but there would be some justification for identifying $2a_1$ as the second lone pair, since it is only weakly bonding and, being relatively high in energy, might be expected to come into play when the molecule functions as a ligand. In other words, $2a_1$ is surely a frontier orbital.

The most direct translation of the MO description into VB language is achieved by combining the MOs to give localised bond orbitals and lone-pair orbitals. In the case where $2s(O)$ was neglected, this was a straightforward procedure; we had two bonding and two nonbonding MOs to combine into two bonds and two lone pairs. Now, with the MOs of Fig. 7.10(b), we have to take linear combinations of three MOs $1a_1$, $2a_1$ and $1b_1$ to obtain two localised bond orbitals and one lone pair; the latter can be combined with the 'pure' nonbonding orbital $2p_y(O)$ to give two equivalent lone pairs. The discrimination between bonding and nonbonding MOs, and the identification of lone-pair orbitals, is no longer straightforward.

In VB theory, the participation of the $2s(O)$ orbital requires the use of sp^n hybrid orbitals for bonding. The injection of some 20% s character into the bonding hybrids is necessary in order that they point directly at the H nuclei. This means that the lone pair hybrids have more p and less s character: the compositions of the four hybrids must add up to one 2s and three 2p. More promotion energy is needed if hybrids which are doubly (rather than singly) occupied in the valence state are given less s character, but this is recompensed by the better overlap afforded by hybrids. The reader should be able to see the correspondences between the MO and VB accounts of the consequences of $2s(O)$ involvement.

There is, of course, an important difference between the MO and VB descriptions of the H_2O molecule. In the latter case, the two pairs of bonding electrons are in equivalent bond orbitals; the two lone pairs can be allocated to equivalent or non-equivalent hybrids according to whim, but they are usually depicted as 'rabbit's ears'. In the MO description, we have four filled MOs, all of different energy. The photoelectron spectrum of H_2O shows four peaks corresponding to binding energies between 12 and 36 eV, and even a fairly crude numerical MO calculation produces MO energies in good agreement with the experimental binding energies. It is sometimes cited as evidence of VB theory's inadequacies that it wrongly predicts the number of peaks in the photoelectron spectrum of H_2O. However, as pointed out in Section 2.7, VB theory does not purport to provide a theoretical framework for a discussion of the electron states of the radical cation H_2O^+.

More complex polyatomic molecules AB$_n$ are tackled in essentially the same way as H_2O. In a qualitative treatment, the most difficult part is working out the group orbitals. In the simple case of H_2O, this was done by inspection. Group-theoretical techniques are available in cases where this would not be practicable. The resulting MO energy diagram, after feeding in the appropriate number of electrons, will usually confirm or deny a molecule's 'right to exist'. A molecule in which the highest occupied MO (HOMO) is strongly antibonding is likely to undergo ionisation. As an example, consider NH_4. Assuming a tetrahedral configuration, the extra electron compared with methane CH_4 must go into an antibonding MO; the eight MOs constructed for tetrahedral EH_4 comprise four bonding and four antibonding orbitals. Thus NH_4^+ is readily formed. NH_4 might conceivably be stable with respect to the decomposition:

$$NH_4(g) \rightarrow NH_3(g) + H(g)$$

but it is certainly unstable with respect to:

$$NH_4(g) \rightarrow NH_3(g) + \tfrac{1}{2}H_2(g)$$

A molecule with vacancies in strongly-bonding MOs will be unstable with respect to electron capture (i.e. reduction), e.g. BH_4. One with partly-filled HOMOs is likely to be unstable with respect to dimerisation. An example is CH_2. This is predicted (by VSEPR theory) to be bent; the H–C–H angle is found to be 130°, rather greater than for H_2O. At this angle, the overlap between $2p_z(C)$ and the hydrogen 1s orbitals is less than the corresponding overlap for H_2O; moreover, the smaller 2s–2p separation means that the 2s(C) orbital is much more important and the $2a_1$ MO is now virtually nonbonding, about equal in energy with b_2, so that the ground state of CH_2 is a triplet, corresponding to the configuration $(1a_1)^2(1b_1)^2(2a_1)^1(b_2)^1$. The molecule is stable in the sense that it does not decompose spontaneously to $C(g) + 2H(g)$ under ordinary conditions; there are two genuine C–H 'bonds'. It will, however, be unstable as a substance because of a tendency to dimerise to give ethylene H_2CCH_2. The reader should be able to construct an MO energy diagram for ethylene by taking the frontier orbitals of two CH_2 radicals and combining them to give a bonding/antibonding pair of σ MOs and a corresponding pair of π MOs. There are enough electrons to fill the two new bonding MOs, giving a σ and a π bond between the C atoms (cf. the treatment of acetylene HCCH in Section 7.2). Many complex molecules are conveniently treated by breaking them down into AB_n fragments and putting these together via their frontier orbitals.

MO theory for complexes of the d block elements is discussed in Section 8.2.

7.4   Three-centre bonds

The need to speak of 'three-centre' bonds in the context of MO theory may surprise the reader. The MO equivalent of a bond as understood in the language of Lewis or VB theory is a filled bonding MO whose antibonding counterpart is unoccupied. In a molecule AB_n (or in a more complex molecule constructed from AB_n fragments) such an MO will necessarily be delocalised over three or more atomic centres. Three-centre bonds are invoked in cases where an MO description cannot readily be translated into Lewis/VB language with two-centre electron-pair bonds.

For any plausible molecule AB_n, it is always possible to assign to each atom B an AO which has electron density along the B–A axis and which can engage in σ overlap if a suitable AO on the central atom A is available. The orthogonality of all AOs belonging to a given atom ensures that there can be only one such AO per B atom. Hence only n group orbitals of σ type can be constructed. It then follows that the maximum possible number of σ bonding MOs ie equal to n, and this is attainable only if we have n AOs on the central atom having the correct symmetries for overlap with the group orbitals. The most stable AB_n molecules – including most which form stable molecular substances under ordinary conditions – are those where we have indeed n filled σ bonding MOs, with other valence electrons (if any) accommodated in nonbonding MOs. In such cases, an MO description is easily translated into Lewis/VB terminology, where AB_n has n σ A–B bonds.

The number of π bonding MOs that can be formed is usually restricted by the number of suitable AOs on the central atom A left over after taking care of the σ bonding, which takes precedence. For example, in CO_3^{2-} (and in the isoelectronic NO_3^-) we have three filled σ bonding MOs (cf. three sp^2 hybrid bonds in VB language) with one central atom $2p$ orbital, having lobes directed perpendicular to the molecular place, left over. It is possible to construct a matching group orbital and hence a bonding and antibonding pair of MOs: the bonding combination is illustrated below.

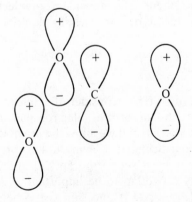

After adding the requisite number of electrons to the MOs, we obtain a ground state in which we have three filled σ bonding MOs, and one filled π bonding MO, all other valence electrons being accommodated in nonbonding MOs based on the O atoms. This is consistent with the Lewis/ VB description which proposes two single bonds and one double bond, or three σ bonds and one π bond. But the π bond in the MO description is shared equally among the three C–O bonds, i.e. it is delocalised over the whole ion. This, of course, is consistent with the experimental observation of three equal C–O distances: the ion describes an equilateral triangle with the C atom at its centroid. This is accommodated in VB theory by the concept of resonance:

Resonance is usually invoked in order to deal with delocalised π bonding and – if clearly understood – is a perfectly acceptable device which should cause no difficulty.

Three-centre bonding is invoked in situations where the σ framework cannot be described in terms of two-centre, electron-pair bonds, although it can often be accommodated by postulating resonance of a different type from that usually encountered. Two types of three-centre bond can be distinguished. The first is often postulated in hypervalent molecules/polyatomic ions AB_n where the central atom exceeds the octet in its Lewis formulation, as an alternative to the use of d orbitals which many chemists find objectionable. The second type occurs where there appear to be insufficient electrons – regardless of the supply of orbitals – to form the requisite number of bonds in a Lewis/VB description. In other words, the first type is postulated where we have an insufficiency of orbitals, and the second where there is a deficiency of electrons; compounds containing the latter type are often described as 'electron-deficient'.

Three-centre, four-electron bonds

As an example of the first type, consider the linear XeF_2 molecule. A Lewis structure with two-centre, electron-pair bonds offends the octet rule. This is not a problem if we allow the use of the Xe 5d orbitals. However, as discussed in Section 6.1, many authors prefer to avoid such use. If we restrict our basis set to the 5p(Xe) and 2p(F) orbitals, and if we consider only σ overlap to be important, we need consider only three AOs: $2p_z(F_1)$, $2p_z(F_2)$ and $5p_z(Xe)$, where the molecular axis is labelled z. Thus we will obtain three MOs ψ_1, ψ_2 and ψ_3 as shown below:

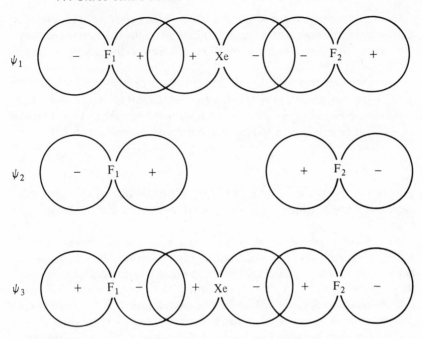

These can be derived by trial-and-error, having regard for the rules of the game outlined in earlier sections of this chapter. More neatly, we construct two group orbitals by in-phase and out-of-phase combination of the $2p_z(F)$ orbitals. The latter combination overlaps with $5p_z(Xe)$ so that a bonding and an antibonding MO can be constructed; these are ψ_1 and ψ_3 respectively. The other group orbital has nothing to overlap with and remains a nonbonding MO, ψ_2. From the ground state configurations $F = [He]2s^2 2p^5$ and $Xe = [Kr]4d^{10}5s^2 5p^6$, we conclude that four electrons have to be inserted into our three MOs; for book-keeping purposes, we might like to think of a filled $5p(Xe)$ orbital interacting with two singly-occupied $2p(F)$ orbitals, although such thinking is not altogether in keeping with the spirit of MO theory. Thus the ground state of XeF_2 can be described as $(\psi_1)^2 (\psi_2)^2$.

The electron pair in ψ_1 will tend towards the F atoms, since the $2p(F)$ orbitals are lower in energy than $5p(Xe)$. The electron pair in ψ_2 'belongs' wholly to the F atoms. Remembering that – formally – the Xe atom contributed two of the four electrons being considered, there must have been a considerable movement of electron density from Xe to F upon molecule formation. This, of course, is consistent with the higher electro-negativity of F. The MO ψ_1 is strongly bonding. But ψ_2 is a very stable nonbonding MO. The fact that there is some positive overlap between $2p_z(F_1)$ and $2p_z(F_2)$ is not very significant because the two F atoms are too far away. More important will be the effect of the exposure of the electron pair in ψ_2 to the large effective nuclear charge on the fractionally-positive Xe atom; the electrostatic attraction will lower the energies of the 'nonbonding' electrons.

In translating this MO description into VB language, we speak of a three-centre, four-electron bond, often abbreviated as (3c, 4e); three orbitals rather than four are used to hold the three atoms together. An equivalent description in VB language involves resonance structures:

$$\overset{+}{F-Xe} \quad \overset{-}{F} \quad \leftrightarrow \quad \overset{-}{F} \quad \overset{+}{Xe-F}$$

The term 'three-centre' bond in XeF_2 is analogous to the term 'delocalised π bond' which might be applied in describing the π bonding in CO_3^{2-} or NO_3^-.

Nearly all hypervalent molecules and ions contain linear X–E–X moieties where X = halogen or some highly-electronegative group; molecules such as tetrahedral $POCl_3$ do not qualify as hypervalent because they can be described by Lewis structures that do not violate the octet rule, by use of E $\rightarrow$ O dative bonds. The description of X–E–X as a three-centre, four-electron bond helps to account for their existence, and indeed for their geometries. Trigonal bipyramidal PX_5, and structures derived from it in VSEPR terminology with lone pairs replacing bond pairs as in SF_4, ClF_3 and XeF_2, all contain one three-centre bond. The central atom furnishes one pure np orbital for this purpose; the other bonds, and the lone pairs, can be accounted for using the ns orbital and the remaining np orbitals. Molecules such as XeF_4 (square planar) and BrF_5 (square pyramidal) have two three-centre bonds, while octahedral SF_6 requires three. As might be expected, the E–X distances in such three-centre bonds are rather longer than the other E–X bonds in hypervalent molecules. Thus the axial bonds in PF_5 are longer than the equatorial bonds, while the axial bond in BrF_5 is shorter than the four equatorial bonds:

The fact that most hypervalent molecules are fluorides or oxofluorides can be easily explained in terms of the three-centre bond approach. Crucial to the viability of the linear X–E–X moiety is the stabilisation of the 'nonbonding' MO ψ_2. This is optimised if X is relatively small in size and high in electronegativity, so that X bears a large fractional negative charge; the central atom E should be of relatively low electronegativity in order to maximise the latter. This helps to explain why molecules like NF_5 are unstable. It helps also to explain 'anomalies' such as the low stability of $AsCl_5$ compared with PCl_5 and $SbCl_5$; according to the Allred–Rochow values (Table 4.5), the electronegativities follow the sequence $P < As > Sb$. The model is also consistent with the relative stabilities of geometrical isomers in systems such as PF_nCl_{5-n}. The importance of the role played by the three-centre bond in stabilising the molecule is emphasised by the preference for F atoms in axial positions in the trigonal bipyramid; the F–E–F (3c, 4e) bond is much stronger than F–E–Cl or Cl–E–Cl (cf. the discussion of PCl_nF_{5-n} in Section 2.3).

Before leaving this type of three-centre bond, we must mention the linear, symmetric [F—H—F]$^-$ ion which occurs in a number of ionic solids, such as KHF_2. For book-keeping purposes, we might render this as $(H^-)F_2$, isoelectronic with HeF_2 and hence an analogue of XeF_2. However, the three-centre bonding must involve the hydrogen 1s orbital, rather than an np orbital. The relevant MOs are therefore:

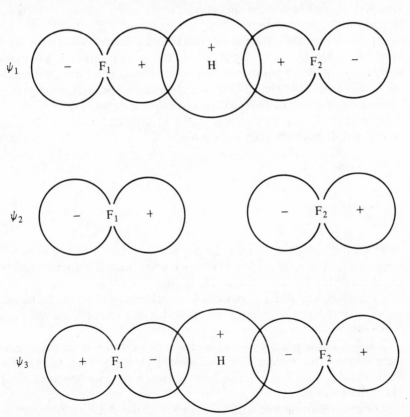

With each F atom contributing one electron, plus the hydrogen atom's single electron and the negative charge, we have four electrons so that ψ_1 and ψ_2 are filled. The difference between this situation and that described above for XeF_2 is that the nonbonding MO ψ_2 appears to be somewhat antibonding, with negative or out-of-phase overlap between the $2p_z(F)$ orbitals. But this MO is sufficiently stable for its occupancy to be favoured, especially when we realise that the nakedness of the proton will be most indecently exposed by the electronegative F atoms to the electrons in ψ_2. An alternative description of HF_2^- is to consider the interaction of F^- with HF, via overlap of a filled $2p_z$ orbital on the former with the antibonding MO of HF (see Fig. 7.4). This may be more appropriate where the anion is distinctly unsymmetrical, as is sometimes the case. The HCl_2^-, HBr_2^- and HI_2^- ions are all known in crystalline salts MHX_2 but these are less stable than MHF_2, readily losing HX.

Three-centre, two-electron bonds

This term is used to describe a three-centre bond – usually of the type E–H–E and non-linear, in contrast to the other variety – which we imagine to be formed by three orbitals (one on each atom) and two electrons (one furnished by the central H atom). Such bonding is invoked in 'electron-deficient' compounds where there are insufficient electrons to form the requisite number of two-centre bonds in a Lewis/VB treatment. Electron-deficient compounds abound in boron chemistry; the classic case of diborane B_2H_6 will be discussed in detail.

Diborane is a gaseous substance, the simplest hydride of boron that can be isolated. It has the dimeric structure:

in which the subscripts t and b denote terminal and bridging H atoms respectively. The arrangement of H atoms about each B is approximately tetrahedral; the H_t–B–H_t and B–H_b–B angles are respectively 122° and 97°. The B–H_t and B–H_b distances are respectively 119 and 133 pm, and the B–B distance is 177 pm (cf. the B–B distance of *c*. 170 pm found in X_2B—BX_2).

The simplest and most informative way of performing an MO treatment is to view the molecule as two angular BH_2 fragments held together by two bridging hydrogens. The MOs of BH_2 will be similar to those of H_2O; the main differences arise from the much smaller 2s–2p energy separation in B compared with O. In fact, because the 2p(B) orbitals lie higher in energy than 1s(H), it would appear that 2s(B) actually interacts more strongly with 1s(H) than 2p(B) does. This means that (referring to the MO diagram for H_2O in Fig. 7.10(b)) $1a_1$ and $1b_1$ are bonding MOs while $2a_1$ is now essentially nonbonding alongside b_2. These nonbonding MOs will have to bear the main responsibility for the interactions that hold the two BH_2 fragments together (with the help of the bridging hydrogens) since the filled bonding MOs $1a_1$ and $1b_1$ are preoccupied with binding of the terminal H atoms. Thus the bridge-bonding requires the consideration of six orbitals: two nonbonding MOs on each B atom and the two 1s(H) orbitals. These lead to the construction of six MOs; from the symmetry considerations set out in this chapter, these can readily be obtained and are depicted in Fig. 7.11. The four relevant nuclei lie in a plane which is perpendicular to the planes of the BH_2 fragments. The circles represent the $1s(H_b)$ orbitals; the b_2 nonbonding MOs of the BH_2 fragments are shown as the pure 2p orbitals they are, while the more complicated a_1 MOs are shown as lobes pointing along the B–B axis. You can easily construct the MOs by taking in-phase and out-of-phase combinations of equivalent BH_2 nonbonding MOs, and matching each (where possible) with a combination of $1s(H_b)$ orbitals.

The MO energy level diagram is shown in Fig. 7.12. ψ_1 and ψ_2 are strongly bonding, having positive B–H and B–B overlap. ψ_3 and ψ_4 are nonbonding as far as B–H interaction is concerned but – bearing in mind the relatively short B–B distance – they are distinctly antibonding with respect to B–B interaction, especially ψ_4 which involves out-of-phase σ overlap. The remaining MOs ψ_5 and ψ_6 are both strongly antibonding. How many electrons are available to be inserted into the diagram? In each BH_2 fragment, we have five valence electrons but four of these are in bonding MOs which hold the H_t atoms. Thus each fragment has just one electron to contribute to the bridge-bonding, making a total of four after counting the two H_b electrons. Thus ψ_1 and ψ_2 are filled and all other AOs are empty. Two pairs of bonding electrons, and no occupied antibonding MOs, translates into two bonds in Lewis/VB language. By taking linear combinations of ψ_1 and ψ_2, we can obtain two equivalent three-centre bond orbitals, thus:

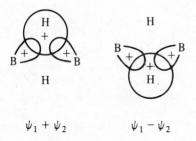

$$\psi_1 + \psi_2 \qquad\qquad \psi_1 - \psi_2$$

Hence the description 'three-centre, two-electron' or (3c, 2e) bonds. It is clear that additional electrons would be no help at all, since they would have to go into distinctly antibonding MOs.

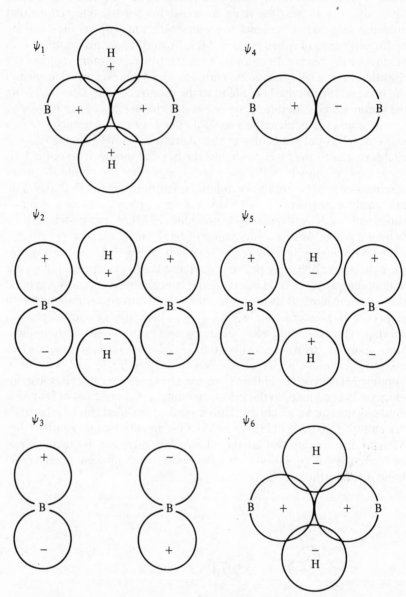

Fig. 7.11. MOs of B_2H_6, excluding those primarily engaged in B–H$_t$ bonding.

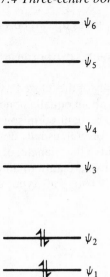

Fig. 7.12. MO energy diagram for bridge-bonding MOs in B_2H_6.

The 'right to exist' of the B_2H_6 molecule seems well established from the MO treatment. What is less clear is why B_2H_6 is stable with respect to decomposition to the monomer BH_3. An MO or VB treatment of the latter is straightforward; in either case, one boron 2p orbital (with lobes directed perpendicular to the molecular plane) is left empty and nonbonding. There are as many filled bonding MOs in two BH_3 molecules as there are in B_2H_6 (six). There is no simple explanation for the tendency of BH_3 to dimerise; a useful generalisation is the principle that atoms like to use all their primary valence orbitals for bonding, or for the accommodation of nonbonding electrons. Molecules in which a valence AO of any atom is left empty tend to be unstable with respect to chemical changes which enable this orbital to find gainful employment. To say that the B atoms in B_2H_6 are using all their valence orbitals for bonding is tantamount to the conclusion that they have attained their Lewis octet configurations, albeit in a devious manner.

Diborane is the simplest of more than 20 molecular hydrides B_mH_n that have been characterised; these, and many thousands of their derivatives, are classed as electron-deficient. Their structures can often be rationalised by a simple procedure which involves counting boron orbitals and electrons. The types of bond encountered can be classified as:

(a) (2c, 2e) B–H_t bonds. Each requires a contribution of one orbital and one electron from the B atom.

(b)   (3c, 2e) B–H–B bridge bonds. Each takes up two boron orbitals and one boron electron.

(c)   (2c, 2e) B–B bonds. One of these requires two boron orbitals and two boron electrons.

(d)   (3c, 2e) 3B bonds. These are normally of the 'closed' type, i.e. a triangle of B atoms each of which is about equally bonded to the other two (an 'open' variety is also recognised in some derivatives). Each B atom contributes one orbital, usually described as an sp^n hybrid pointing to the centroid of the triangle; of the three resulting MOs, one is bonding and two are antibonding, so that the optimum number of electrons is two. This type of bond is usually depicted as:

   In the simple case of B_2H_6, the two B atoms contribute a total of eight valence orbitals and six valence electrons to the bonding. The four B–H_t bonds take up four orbitals and four electrons, leaving four orbitals and two electrons for bridge-bonding. This enables two (3c, 2e) B–H–B bonds to be formed. Consider now the more difficult case of B_5H_9. The five B atoms describe a square pyramid; each B atom forms one B–H_t (2c, 2e) bond and there are four B–H–B bridge bonds in the basal plane:

Starting with four orbitals and three electrons per B atom, we have 20 orbitals and 15 electrons. The five B–H_t bonds take up five orbitals and five electrons, and the four B–H–B bridge bonds require eight orbitals and four electrons. Thus we are left with seven orbitals and six electrons to hold the apical B atom (i.e. the B atom at the apex of the pyramid) to the four in the basal plane. These can form two (2c, 2e) B–B bonds and one (3c, 2e) closed 3B bond. Thus the bonding can be represented by the schematic diagram:

Since the four B–B distances between the apex and the basal plane are equal, we have to invoke resonance among structures in which the (2c, 2e) and (3c, 2e) bonds interchange.

The reader may wonder what place the above description has in a chapter supposedly devoted to MO theory. What we have performed is basically a VB treatment onto which we have grafted the idea of three-centre bonds, which are best understood within the framework of MO theory, and are therefore best introduced at this stage. Why not carry out a full-blown MO treatment of a molecule like B_5H_9, rather than the hybrid treatment offered above? Chemists are still reluctant, with good reason, to discard the (2c, 2e) description which is successful for such a large proportion of molecules and polyatomic ions. From the viewpoint of rationalising the structure of B_5H_9, and its very existence, a description which preserves as much as possible of Lewis/VB theory is more edifying than one which rejects completely the concept of the localised chemical bond. As we shall see later in this section, there are many cases which are beyond redemption by such measures, and only a full MO treatment, with multicentre bonding, will do.

Apart from the vast number of compounds containing B–H–B bridge bonds, a good many other three-centre E–H–E' links are found, especially where E or E' is B, Be or Li. Beryllium hydride BeH_2, is a one-dimensional polymeric solid (isostructural with $BeCl_2$ and SiS_2: see Section 3.3), whose structure can be rationalised in terms of Be–H–Be (3c, 2e) bridge bonds:

The polymerisation of the linear BeH_2 molecule is exactly analogous to the dimerisation of BH_3. The Be–H–Be bridge is found in a number of molecular substances, e.g. in complexes having the empirical formula BeHRL, where L is a neutral ligand and R is an alkyl group:

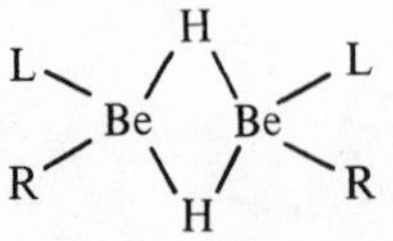

E–H–B bridge-bonding is found in borohydrides $E(BH_4)_n$, except for MBH_4 where M is one of the heavier Group 1 elements; thus KBH_4 contains K^+ ions and BH_4^- ions but $LiBH_4$ appears to involve Li–H–B bridge-bonding. Complexes of the type L_2CuBH_4, where L is a neutral ligand, contain Cu–H–B three-centre bonds.

Hydride derivatives of the d block carbonyl complexes often contain H atoms in environments which demand three- (or more) centre bonding. The case of $[Cr_2(CO)_{10}H]^-$ is mentioned in Section 8.6; here we have a Cr–H–Cr bridge with significant Cr–Cr bonding as well. In more complex polyhedral carbonyl systems (see Section 8.5), a hydrogen atom may be triply-bridging, sitting atop a triangle of three M atoms. It may even sit at the centre of an M_6 octahedron.

Hydrogen is not unique in forming the middle partner in (3c, 2e) bridge bonds. A number of aluminium alkyls and aryls AlR_3 have dimeric structures similar to diborane. The CH_3 group behaves very much like an H atom in many circumstances, and furnishes a singly-occupied orbital for the formation of (3c, 2e) bonds.

Finally in this section, we note that multicentre bonding must sometimes be admitted and cannot be simplified into a VB (or pseudo-VB) description. An example is methyllithium, which consists in the solid state of $Li_4(CH_3)_4$ molecules. The structure is a distorted cube, which may be seen as interpenetrating Li_4 and $(CH_3)_4$ tetrahedra:

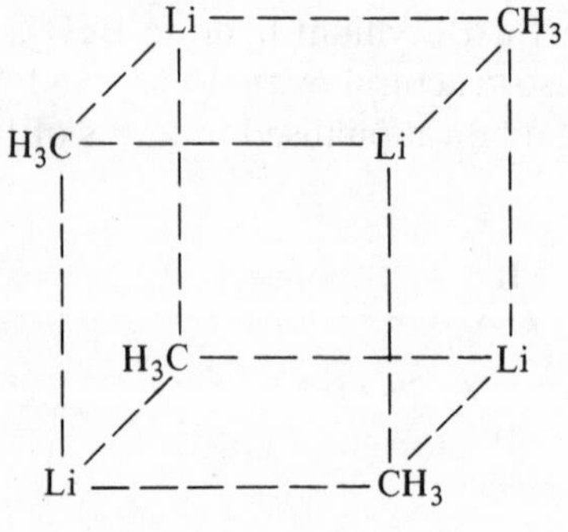

Unless 'bond/no bond' resonance among a number of canonical structures is postulated, it is impossible to devise a satisfactory VB description
for this molecule. An MO treatment begins by looking at the Li_4 and C_4
tetrahedra. Each atom contributes one orbital to the bonding, the 2s
orbital in the case of Li and a nonbonding MO which might be described
as a carbon sp^3 hybrid in the case of CH_3. Each tetrahedron thus gives rise
to four MOs; one of these – a totally-symmetric combination – is strongly
bonding, while the others form a threefold-degenerate set of antibonding
MOs. Fig. 7.13 shows what happens when the two tetrahedra are allowed
to interact with each other. We finish up with four bonding MOs. The
lowest, nondegenerate level has in-phase overlap of all eight AOs. The
threefold-degenerate bonding MOs are antibonding with respect to Li–Li
and CH_3–CH_3 interaction, but are bonding with respect to Li–CH_3. Since
each Li atom and each CH_3 radical has one electron to contribute to the
bonding, we have four filled bonding MOs giving a stable, closed-shell
configuration. If we include the lithium 2p orbitals in the basis set, the
combinations of these in the Li_4 tetrahedron include one having the same
symmetry (a_1) as the nondegenerate bonding MO and two threefold-
degenerate sets having the same t_2 symmetry as the other bonding MOs.
When functions having the same symmetry are allowed to interact, the

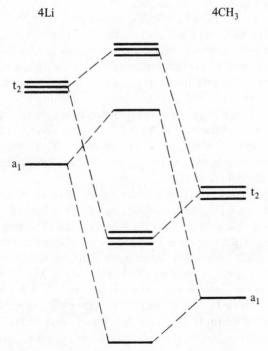

Fig. 7.13. MO energy diagram for $Li_4(CH_3)_4$.

lower ones are depressed in energy. Thus the effect (as usual) of including higher orbitals in the basis set is to stabilise the lowest MOs and strengthen the bonding.

The MO description of bonding in terms of filled MOs delocalised over all the atoms in the molecule reaches its fullest expression in the theory of metallic bonding, to which the rest of this chapter is devoted.

7.5   MO theory of bonding in metallic substances

Metallic elemental substances

The majority of elemental substances, and a large number of compounds, have metallic properties (see Section 3.3). Metallic elemental substances are characterised by three-dimensional structures with high coordination numbers. For example, Na(s) has a body-centred cubic (bcc) structure in which each atom is surrounded by eight others at the corners of a cube, each at a distance of 371.6 pm from the atom at the centre. The Na atom also has six next-nearest neighbours in the form of an octahedron, with Na–Na distances of 429.1 pm. A fragment of this lattice is shown in Fig. 7.14. These distances may be compared with the Na–Na bond length of 307.6 pm in the Na_2 molecule, which can be studied in the gas phase by vaporisation of sodium metal.

A satisfactory theory of metallic bonding must account for the characteristic properties of high electrical and thermal conductivity, metallic lustre, ductility and the complex magnetic properties of metals which imply the presence of unpaired electrons. The theory should also rationalise the enthalpies of atomisation ΔH^0_{atom} of metallic elemental substances. ΔH^0_{atom} is a measure of the cohesive energy within the solid, and we saw in Chapter 5 how it plays an important part in the thermochemistry of ions in solids and solutions. The atomisation enthalpies of elemental substances (metallic and nonmetallic) are collected in Table 7.1. There is a fair correlation between ΔH^0_{atom} and physical properties such as hardness and melting/boiling points.

The reader is probably familiar with a simple picture of metallic bonding in which we imagine a lattice of cations M^{n+} studded in a 'sea' of delocalised electrons, smeared out over the whole crystal. This model can rationalise such properties as malleability and ductility; these require that layers of atoms can slide over one another without undue repulsion. The 'sea' of electrons acts like a lubricating fluid to shield the M^{n+} ions from each other. In contrast, distortion of an ionic structure will necessarily lead to increased repulsion between ions of like charge while deformation of a molecular crystal disrupts the Van der Waals forces that hold it together. It is also easy to visualise the electrical properties of metals in

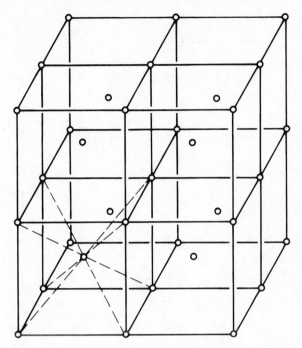

Fig. 7.14. Part of body-centred cubic (bcc) metallic lattice, showing its relationship to the CsCl ionic lattice with cubic eight-coordination of each atom.

terms of this model. For a more detailed treatment of metallic bonding, MO theory will be the obvious choice, given the lack of directional bonds and the apparent delocalisation of electrons in a metallic crystal.

We first consider elemental lithium, which has a bcc structure similar to that described above for sodium. Upon vaporisation, Li(s) yields Li_2(g). The bonding in this diatomic molecule can be readily described along the lines discussed in Section 7.2 and the MO diagrams shown in Fig. 7.5(a) and (b). In the simplest model, we consider only the 2s orbitals, in which case Figs. 7.5(a) and 7.6 are appropriate. With just one valence electron per Li atom available, the ground state is associated with the configuration $(\sigma_s)^2$. A more refined picture allows for the inclusion of the 2p orbitals in the basis set; the effect is to stabilise both σ_s and σ_s^*, which become σ_1 and σ_2 in Fig. 7.5(b). One filled bonding MO corresponds to a single bond.

Clearly, s–p mixing will have a stabilising effect upon the Li_2 molecule, although the bonding can be qualitatively described in terms of s–s overlap alone. The dissociation energy of Li_2 is only $105\,kJ\,mol^{-1}$; compared with the covalent bond energies in Table 6.2, this is rather

Table 7.1 Atomisation enthalpies of elemental substances (kJ mol^{-1}, 25°C), or enthalpies of formation of gaseous monatomic substances E(g)

H 218																	
Li 161	Be 324											B 573	C 717	N 473	O 249	F 79	
Na 109	Mg 148											Al 326	Si 452	P 315	S 279	Cl 121	
K 90	Ca 177	Sc 378	Ti 470	V 514	Cr 397	Mn 281	Fe 418	Co 425	Ni 430	Cu 339	Zn 129	Ga 286	Ge 377	As 302	Se 227	Br 112	
Rb 86	Sr 164	Y 423	Zr 609	Nb 726	Mo 658	Tc 677	Ru 643	Rh 556	Pd 378	Ag 284	Cd 112	In 243	Sn 302	Sb 262	Te 197	I 107	
Cs 79	Ba 180	La 431	Hf 619	Ta 782	W 849	Re 770	Os 791	Ir 665	Pt 565	Au 366	Hg 61	Tl 182	Pb 196	Bi 207	Po 144	At 92	
	Ra 159	Ac 406															

Ce 423	Pr 356	Nd 328	Pm (300)	Sm 207	Eu 175	Gd 398	Tb 389	Dy 290	Ho 301	Er 317	Tm 232	Yb 152	Lu 428
Th 598	Pa 607	U 536											

feeble. The 2s and 2p orbitals of lithium are rather diffuse and their mutual interaction relatively weak.

Consider now a macroscopic crystal assembled from n Li atoms, where n may be, say, 10^{20}. Considering for the moment only the Li 2s orbitals, these can be combined to give n MOs. When plotted on an energy scale, a band of very closely-spaced levels results, ranging from strongly-bonding to strongly-antibonding MOs. The density of states within this s-band – i.e. the number of MOs per unit of energy – is not uniform, and reaches a maximum around the middle of the band. Thus only a small proportion of the levels can be regarded as strongly bonding or antibonding, and most are relatively nonbonding. If we insert n electrons (one per atom) into this band, the lowest $n/2$ levels will be doubly occupied. At room temperature, however, there is enough thermal energy to allow electrons to be promoted to levels in the upper half of the band; the spacing between levels is several orders of magnitude less than kT, where k is Boltzmann's constant. Thus we have a region of singly-occupied MOs around the middle of the band; these account for the complex magnetic properties of metallic substances. The easy promotion of electrons to higher levels within the band leads to the characteristic metallic properties of high thermal and electrical conductivity (see textbooks of physics for further details). Metallic lustre (or specular reflectance) occurs where there exists a band of excited electronic states covering the whole of the visible region of the spectrum; this happens where we have a partly-filled band of closely-spaced MOs.

It might be thought that, even at absolute zero, there should be some promotion of electrons in the band to give a region of partly-filled MOs, since spin-pairing energy should dominate over the very tiny increments in orbital energy needed to achieve such promotion (cf. the occurrence of high-spin ground states in transition metal complexes, as interpreted by CF theory). However, the electrons in a metallic crystal are so delocalised that spin-pairing energies are unimportant.

The double-occupancy of the strongly-bonding MOs at the bottom of the band in metallic Li(s) is apparently enough to stabilise the crystal compared with molecular Li_2:

$$Li(s) \rightarrow Li(g) \qquad \Delta H^0 = 161 \text{ kJ mol}^{-1}$$
$$\tfrac{1}{2}Li_2(g) \rightarrow Li(g) \qquad \Delta H^0 = 53 \text{ kJ mol}^{-1}$$

Consider now a metallic assembly of n Be atoms, using for the meanwhile only the 2s orbitals. As for Li(s), we obtain a band of n MOs; but we now have $2n$ electrons, so that the band should be filled. Not only should this crystal have no metallic properties but it should also have zero cohesive energy since the occupancy of antibonding MOs should cancel out the effect of the occupied bonding MOs. But Be(s) does crystallise as a metallic solid, whose enthalpy of atomisation is twice as great as that of

Li(s). Our error has been to ignore the 2p orbitals. These interact sufficiently to form a band of $3n$ MOs – which we may call the p-band – which overlaps with the s-band, as shown below:

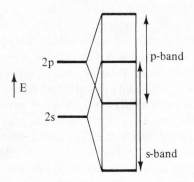

Thus we have a continuous band of $4n$ MOs, into which $2n$ electrons have to be inserted; this should give rise to metallic properties. The strict partitioning of the MOs between the s-band and the p-band is invalidated by s–p mixing, i.e. the significant overlap between the 2s orbital on one atom with the 2p orbitals of its neighbours. As in the case of Li_2 discussed previously, this will tend to stabilise the lowest levels in the band and thus enhance the cohesive energy of the metallic crystal. Since the 2s and 2p orbitals of Be are only about $300 \, kJ \, mol^{-1}$ apart in the free atom, s–p mixing is expected to be quite strong, and the n MOs which are occupied at absolute zero should all be distinctly bonding. This accounts for the high atomisation enthalpy of Be compared with Li, as well as its much higher melting/boiling points and its greater hardness.

Going down Groups 1 and 2, we might expect steady decreases in the atomisation enthalpies; the atoms are increasing in size, the valence orbitals are becoming more diffuse and their interaction correspondingly weaker. This expectation is fulfilled in the case of Group 1, where the steadily-decreasing elemental atomisation enthalpies down the Group parallel the decreasing dissociation energies of the gaseous diatomic molecules. But in Group 2, we find that Ca(s) has a higher atomisation enthalpy than Mg(s), and a similar 'anomaly' is observed in going from Sr to Ba. As with the Periodic 'anomalies' we encountered in Chapter 4, we have to be careful to compare 'like with like'. Calcium differs from magnesium in having five 3d orbitals at about the same energy as its 4p orbitals in the free atom. These form a narrow, dense band in a metallic crystal composed of Ca atoms and are unoccupied. However, s–d interaction stabilises the bonding MOs at the bottom of the s-band. The intervention of d orbitals leads to a discontinuity in the atomisation enthalpies of the Group 2 elemental substances. In Ba, the 5d orbitals are distinctly lower in energy than the 6p orbitals and play a more important role in the band structure than the 3d and 4d orbitals do for Ca and Sr.

Let us now look at Zn, Cd and Hg. These as elemental metallic substances have very low atomisation enthalpies, within the range found for the Group 1 metals and much lower than those of the Group 2 elemental substances. The metallic bonding in the Group 12d elemental substances involves mainly the ns and np orbitals, the filled $(n - 1)$d subshell being effectively buried in the atomic core. However, the np orbitals are relatively higher in energy than the ns orbitals compared with the Group 2 atoms. The poor shielding from the nucleus afforded by the filled $(n - 1)$d subshell (and, in the case of Hg, the filled 4f subshell) affects the penetrating ns orbitals more than the np. The $ns \rightarrow np$ promotion energies for the Group 12d atoms are in the range 400–500 kJ mol^{-1}, much greater than for their Group 2 cousins. This would suggest that s–p mixing will be relatively small for Group 12d elemental substances; remembering that such mixing is responsible for the stability of the Group 2 elemental substances, we can understand why the Group 12d metals have such low cohesive energies; so low that mercury is a liquid at room temperature.

The d block elemental substances have relatively high atomisation enthalpies. It is a fair approximation to regard these as having a narrow but dense d-band in the middle of a much broader, but less dense, s-band, as shown schematically below:

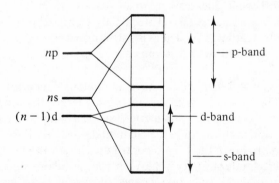

The occupancy of the s-band amounts to approximately 0.6 of an electron per atom, the remaining electrons being accommodated in the d-band. The density of states in the d-band is rather complex and accounts for the rather irregular variation in atomisation enthalpy across each transition series. As a broad generalisation, the highest atomisation enthalpies are found for the early members of each series, with all valence electrons accommodated in the bonding MOs of the s-band and the most bonding levels in the d-band. Clearly, s–d mixing will stabilise the bonding levels in the band and will be most important towards the left-hand side of each series, where the ns and $(n-1)$d orbitals are closest in energy; it will be less important towards the end of a series, as the $(n-1)$d orbitals tend to drop into the core.

The atomisation enthalpies of the lanthanides as metallic elemental substances exhibit very different trends. From La to Eu, we see a steady decrease, followed by an abrupt increase at Gd. The atomisation enthalpies then decrease (not quite monotonically) to Yb, then increase at Lu. These trends may be rationalised as follows. According to magnetic studies, the lanthanide atoms in the elemental substances have the electronic configurations $6s^2 5d^1 4f^n$; Eu and Yb are exceptions, discussed further below. The band structure is evidently complex and will not be described in detail. The atomisation enthalpy can be broken down for thermochemical purposes into two steps:

$$M(s) \rightarrow M(g)^* \quad \Delta H^0(1)$$
$$M(g)^* \rightarrow M(g) \quad \Delta H^0(2)$$

In $M(g)^*$ the atoms are their 'valence states', having in most cases the configuration $6s^2 5d^1 4f^n$, while in $M(g)$ the atom is in its ground state, which usually arises from the configuration $6s^2 4f^{n+1}$. It appears that $\Delta H^0(1)$ – which must, of course, be positive – is roughly constant at $c.\ 420\ \text{kJ mol}^{-1}$ for all the lanthanides. For La, Ce, Gd and Lu, $\Delta H^0(2)$ is zero, since these have the atomic ground states $6s^2 5d^1 4f^n$ and $M(g)$ is identical to $M(g)^*$. For atoms whose atomic ground states are $6s^2 4f^{n+1}$, $\Delta H^0(2)$ will be exothermic, increasingly so as we move from left to right along each half of the series, i.e. La $\rightarrow$ Eu and Gd $\rightarrow$ Yb, as the 4f orbitals become more stable relative to 5d. There is a striking parallel between the atomisation enthalpies and the third ionisation energies of the gaseous atoms (see Fig. 5.2).

As noted in Section 4.2, Eu and Yb have anomalous structures in the elemental state. Their metallic radii are apparently larger than those of the other lanthanides, and their magnetic properties suggest that the electron configuration of the atom in the elemental substance is $6s^2 4f^7$ for Eu and $6s^2 4f^{14}$ for Yb. Since the f-band contributes little to the cohesive energy – the 4f orbitals interact very weakly with one another – we can say that Eu and Yb contribute only two valence electrons to the metallic bonding, instead of three for the other lanthanides. Thus although $\Delta H^0(2)$ is zero for Eu and Yb, $\Delta H^0(1)$ will be much smaller. It is quite wrong, however, to attribute the low atomisation enthalpies of Eu and Yb to their anomalous structures; evidently the atomisation enthalpies would be lower still if they adopted the same structure as the other lanthanides. Even if we ignore their structural differences, the atomisation enthalpies of Eu and Yb fit into the trends begun earlier in each half of the series. The effect of these trends on the stabilisation of the II oxidation state has been mentioned in Sections 5.2 and 5.4.

Metallic compounds

Five types of compound can be distinguished within this category:

(i) A class of compounds often called *substitutional alloys* are formed by interaction between metallic elemental substances. In many cases, atoms of different elements are randomly distributed over the sites of a typical metallic lattice, forming a substitutional solid solution. Varying degrees of ordering are found; in the ultimate case of a stoichiometric, homogeneous, intermetallic compound, all atoms of a given element have the same environment. For example, the stoichiometric alloy known as β-brass has the composition $CuZn$. The atoms are arranged in a bcc structure, like that described above for metallic lithium (Fig. 7.14). In the high-temperature form, Cu and Zn atoms are randomly distributed over the lattice sites. The low-temperature form still has a bcc lattice but could now be viewed as a CsCl structure; each Cu atom is at the centre of a cube with one Zn atom at each corner, and vice versa. The ordered structure apparently has the stronger bonding, but entropy considerations favour the disordered form at high temperatures ($>470\,^{\circ}C$). One might be tempted to regard the ordered form as an ionic solid Cu^+Zn^- or Zn^+Cu^-; after all, it has the CsCl structure, as exhibited by a number of typical ionic solids. Such a description makes little sense, partly because the species Zn^-, Cu^- and Zn^+ are distinctly unfamiliar in stable compounds and partly because CuZn has none of the characteristics of an ionic solid but does have the hallmarks of a metal. The bonding has to be described in terms of band theory, like elemental copper or zinc.

(ii) *Interstitial alloys* are compounds formed between the d block elements and (principally) H, B, C or N. The d block atoms form a metallic-type lattice with the small interstitial atoms occupying octahedral or tetrahedral holes. Well-known examples of stoichiometric compounds include TiH_2, VN and Fe_3C; numerous non-stoichiometric phases can be obtained. It is too simplistic to view these as essentially unchanged metallic structures into which 'foreign' atoms have been inserted; the array of d block atoms is often different from that in the elemental substance. However, interstitial alloys do behave usually as typical metals. Band theory is again called for, with the interstitial atoms contributing their valence orbitals and electrons to the bonding.

(iii) Many binary compounds whose structures and stoichiometries might, *prima facie*, suggest an ionic description have at least some metallic characteristics. Here we are usually dealing with compounds where the electronegativities of the atoms, and the polarising powers/

polarisabilities of the putative ions, would tend to rule out an ionic description. Suppose we perform an MO calculation on a typical ionic solid, such as NaCl, having regard for extended interactions over a macroscopic crystal. We obtain a band of bonding and nonbonding MOs with just enough electrons to fill them; these consist mainly of AOs based on the more electronegative atom. Well separated in energy from this band is another band of empty antibonding MOs, to which AOs based on the less electronegative atom make the major contribution. The resulting ground state leads to an insulator. It is possible to excite an electron from the filled band to the empty band. This usually requires ultraviolet radiation; for example, in the case of KCl absorption in the UV begins at about 60 000 cm^{-1}, which corresponds to an energy of about 700 kJ mol^{-1}. This is a charge transfer transition (see Section 2.6), the electron jumping from an MO based mainly on the Cl atoms to an MO comprised mainly of potassium AOs. Ionic solids like KCl are transparent to visible light.

But if the atoms in an 'ionic' solid are closer together in electronegativity – i.e. if their valence AOs are comparable in energy – the gap between the filled band of bonding MOs (the *valence band*) and the empty band of antibonding MOs (the *conduction band*) may become quite small. Spectroscopic transitions across the band gap may now appear at lower energy (longer wavelength) and may intrude into the visible region, so that the substance is coloured. For example, CdO is a white solid (NaCl structure) having the usual characteristics of an ionic solid. CdS, CdSe and CdTe all have tetrahedral 4:4 coordinate structures. CdS and CdSe exhibit fluorescence, which is associated with electron transfer across the band gap, while CdTe is a good semiconductor (see Section 7.6). CdS is yellow, while CdSe and CdTe are grey and opaque. There is thus some tendency towards increasing metallic and decreasing ionic character in CdX as X decreases in electronegativity.

If the bands overlap, we have a metallic conductor because there is now in effect one partly-filled band. Compounds like CuZn represent this extreme case. Evidently there is a gradation among three-dimensional binary compounds between the extremes of ionic and metallic solids. As we shall see in Section 7.6, there is a similar gradation among elemental substances between extended covalent structures and typical metals.

(iv) Many 'subvalent' binary halides, oxides, sulphides etc. have metallic properties. These are compounds M_aX_b where M is of lower electronegativity than X and M is in an unusually low (often fractional) oxidation state. Examples include Cs_3O, Ag_2F, Gd_2Cl_3 and NbO. These usually have complex structures in which there appears to be considerable M–M interaction as well as M–X bonding, and the highest occupied MOs constitute a partly-filled band, leading to metallic behaviour. In some cases, discrete M_n clusters can be identified; for example, NbO contains Nb_6 octahedra which are also found in subhalides of Nb (see also

Section 8.5). There is no simple bonding theory to account for the existence of such compounds and their stoichiometries. You will often see reference made to materials such as ZrCl, and you must suspect a complex metallic structure rather than a simple ionic array in such cases.

Some metallic compounds in this category have structures which might imply an ionic description, and do not reveal significant M–M interaction. An example is CeS, which has the NaCl structure adopted by ionic monosulphides like BaS. If we imagine Ce^{2+} in CeS to have the configuration $[Xe]4f^15d^1$, the 4f electrons remain localised on the Ce atoms but the 5d electrons are delocalised in a conduction band which has substantial contributions from the relatively high-lying sulphur 3p orbitals. Disproportionation to Ce_2S_3 and Ce is discouraged by the low lattice energies of sulphides with relatively large cations (see Section 5.2). Most of the other lanthanides form similar monosulphides which have metallic conductivity. These are often formulated as containing Ln^{3+} and S^{2-} ions with one free electron per formula unit in a conduction band. However, EuS and YbS are ionic solids, similar to BaS; Eu(II) and Yb(II) are 'normal' oxidation states. Similar behaviour is exhibited by the lanthanide diiodides LnI_2. Thus EuI_2 behaves as an ionic solid, but GdI_2 has metallic behaviour and is formulated as $Gd^{3+}(I^-)_2(e^-)$.

(v) Mixed-valence compounds frequently exhibit high electrical conductivities, and sometimes other metallic properties as well. The presence in a crystalline substance of an element in two different oxidation states provides an easy mechanism whereby an electron can pass through the material, hopping from one atom to another with consequent changes in oxidation state. A good example is found in the tungsten bronzes, M_xWO_3, where M is a Group 1 atom (usually Na) and $0 < x < 1$. Tungsten(VI) oxide WO_3 has a three-dimensional structure with 6:2 coordination, i.e. each W is octahedrally surrounded by 6O, and each O is linearly coordinated by 2W. The unit cell has a large cavity at its centre, capable of accommodating a large M^+ cation:

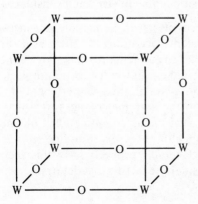

An MO treatment of a WO_3 crystal leads to a filled band of bonding and nonbonding MOs (mostly comprised of oxygen AOs). Not much higher in energy is an empty band of approximately nonbonding MOs, mainly 5d(W) in character but with some admixture of 2p(O). These are related to the t_{2g} level of CF theory which can engage in π overlap with oxygen 2p orbitals (see Section 8.2). The introduction of Na atoms into some of the holes at the centres of the unit cells leads to the formation of Na^+ ions; the extra electrons go into the conduction band just described and are responsible for the metallic characteristics of the tungsten bronzes. Thus Na_xWO_3 might be formulated as $Na_xW(V)_xW(VI)_{1-x}O_3$, i.e. each Na atom reduces one W(VI) to W(V). Alternatively, we might consider $Na_xW(VI)O_3(e^-)_x$, analogous to the formulation of CeS as $Ce(III)S(e^-)$.

Thousands of mixed-valence oxides have interesting electrical properties. Of particular topical interest are the high-temperature superconductors containing Cu(II) and Cu(III). In a typical metal, resistance to an electric current is caused by vibration of the atoms in the lattice. This will obviously increase with increasing temperature, but should not completely vanish at absolute zero because the zero-point energy ensures that the nuclei are still in motion. Some materials, however, have nil resistance, and infinite conductivity, at temperatures above absolute zero. Such materials – superconductors – have obvious technological possibilities, especially if the critical temperature T_c at which the electrical resistance vanishes is high enough to be attainable without difficulty. In 1986 it was discovered that $YBa_2Cu_3O_{7-x}$ ($x = 0.2$–0.5) were superconductors with critical temperatures as high as 90–100 K, much higher than those of the various alloys previously known to be superconductors and accessible by means of liquid nitrogen (b.p. 77 K). Since Y and Ba are always found in the III and II oxidation states respectively, Cu(II) and Cu(III) are present. The mechanism of the superconductivity is under intensive study, but it must involve a conduction band consisting mainly of 3d(Cu) orbitals.

7.6   Metallic versus covalent bonding in elemental substances

Apart from the atomic noble gases, elemental substances may be classified as metallic or covalent, according to their structures and properties at room temperature and ambient pressure. Covalent elemental substances may be subdivided as molecular or non-molecular, the latter category including one-, two- or three-dimensional structures. There is a 'grey area' between the extremes of the three-dimensional covalent structure and the typical metal; 'semi-metallic' or 'metalloid' behaviour is found in a number of cases. Even iodine, *prima facie* a molecular solid, has incipient metallic properties. In this section, we explore this grey area and consider the factors that determine which type of structure is adopted. The subtleties involved are underlined by the fact that metal–nonmetal allotropy is found in some cases.

Semi-metallic behaviour and semiconductors

In order to illustrate the gradation from covalent to metallic behaviour we look at the structures and properties of the Group 14 elemental substances.

Elemental carbon is found in two allotropic forms under ordinary conditions: graphite and diamond, the former being marginally the more stable at room temperature and atmospheric pressure. The diamond structure represents the most obvious manner in which a collection of gaseous carbon atoms might coalesce into a solid, corresponding to scheme (1) of Table 6.5 with each atom tetrahedrally bonded (sp^3) to four other C atoms. There are two C–C single bonds per atom, consistent with the atomisation enthalpy of $715 \, kJ \, mol^{-1}$, which is greater than that of any elemental substance save Nb, Ta, W, Re and Os, plus diamond's allotrope, graphite. Diamond certainly rivals these metals in its hardness and refractory behaviour. Graphite has a planar, two-dimensional structure based upon hexagonal benzene-like rings, with sp^2 hybridisation at the C atom. The mean C–C bond energy in graphite is $478 \, kJ \, mol^{-1}$, i.e. between a single and double bond. The relative stabilities of diamond and graphite are delicately balanced. The simplest VB description of graphite gives a C–C bond order of just 1.33 but the bonding is greatly enhanced by resonance. Although thermodynamically more stable than diamond, graphite is chemically more reactive for kinetic reasons; π bonds are more labile than σ bonds. The lubricity of graphite can be readily explained in terms of the ease with which the two-dimensional layers can slip over each other, overcoming the weak Van der Waals attraction between layers.

Silicon and germanium as elemental substances are found only in the diamond-type form. The reluctance of Si and Ge to enter into p_π–p_π bonding prohibits a graphite-type structure as a plausible allotrope. These are rather more reactive than diamond; the weaker Si–Si and Ge–Ge bonds make disruption of the lattice kinetically easier. Tin occurs in both a metallic form (white tin) and a covalent (diamond-type) form; the latter is slightly more stable at low temperatures. Lead forms only a metallic elemental substance.

The diamond structure exhibited by C, Si, Ge and Sn shows increasing metallic characteristics as we move down the Group. Elemental silicon and germanium have a metallic lustre, and although not metallic conductors have the property of *semiconductivity*. An MO treatment of a diamond crystal C_n, considering the 2s and 2p orbitals of each C atom and allowing for interactions between next-nearest neighbours, leads to a band of $2n$ bonding MOs, separated by a gap of about $500 \, kJ \, mol^{-1}$ from a band of $2n$ antibonding MOs. The difference between this situation and the band structure described in the previous section for Li and Be arises from the much stronger orbital overlap among carbon 2s and 2p orbitals, together with the different crystal structure. Given four valence electrons per atom, the bonding MOs – the valence band – are filled, consistent with the simple VB picture in which we have two C–C bonds per atom. At high temperatures, or on application of an electric field, electrons may be excited into the antibonding MOs; partial occupancy of this band leads to electrical conductivity, hence the term conduction band for the antibonding MOs. In the case of silicon, the gap between the top of the valence band and the bottom of the conduction band is only $106\,kJ\,mol^{-1}$. This can be explained in terms of the weaker Si–Si bonding compared with C–C; the situation is shown schematically below:

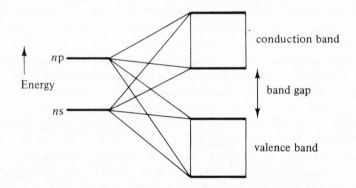

Thus the conduction band in silicon is much more accessible than that in diamond, and under an applied potential the population of the conduction band in silicon is sufficient to endow the elemental substance with an electrical conductivity some orders of magnitude better than that of diamond. This increases rapidly with increasing temperature, a characteristic of semiconductivity, as opposed to metallic conductivity, which decreases with increasing temperature. In the case of germanium, the band gap is reduced to $58\,kJ\,mol^{-1}$, and in grey tin it is a mere $8\,kJ\,mol^{-1}$; grey tin is on the brink of being a metallic conductor. The trend can be related to the decreasing homopolar bond energies down the Group; these reflect a smaller difference in energy between bonding and antibonding MOs as the atoms become larger. Another factor – which may be connected with the dramatic drop in the band gap in going from diamond to silicon – is the intervention of nd orbitals. These have the effect of depressing the energies of the lowest MOs in the conduction band, via d–p and d–s mixing. Their effect on the valence band is much smaller, since the MOs in this band are further away in energy from the nd orbitals. The effect is therefore to reduce the band gap.

Semiconductivity in the diamond-type lattice can be much enhanced by replacing some of the Group 14 atoms by Group 13 or Group 15 atoms. Thus if some Ga atoms are 'doped' into a crystal of germanium, the valence band will not be completely filled, and the conductivity is increased. This is known as p-type semiconductivity; the p stands for positive, a number of negative electrons being missing from the valence band. If P atoms are substituted for Si or Ge in a diamond-type lattice, additional electrons must go into the conduction band with consequent enhancement of conductivity. Such a crystal is an n-type semiconductor (n for negative, since we have additional electrons to fit in). A transistor is an electronic device which exploits the characteristics of junctions between n- and p-type semiconductors.

The metallic lustre of the elemental substances formed by the heavier Group 14 elements in the diamond structure can be interpreted in terms of the valence band/conduction band picture. The spectrum of excited states which can arise from promotion of an electron from the valence band to the conduction band covers the whole of the visible region, leading to opaqueness and specular reflectance. In the case of diamond itself, the lowest electronic excited state lies well into the ultraviolet.

The reader is invited to consider other possible allotropic modifications of the Group 14 elemental substances, and to become satisfied that none is likely to be more stable than the structures observed.

An allotrope of carbon based upon linear chains:

$$=\!C\!-\!C\!\equiv\!C\!-\!C\!\equiv\!C\!-\!C\!\equiv\!C\!-\!C\!=$$

has been described, but it appears to be thermodynamically uncompetitive with diamond and graphite. This should be apparent from the bond energies listed in Table 6.2.

Metallic versus molecular structures

What factors determine whether an elemental substance adopts a metallic or a covalent structure? From the simple model for metallic bonding, which views a metal as a lattice of cations embedded in a 'sea' of delocalised electrons, it may be supposed that atoms having low ionisation potentials are most likely to become assembled as metallic substances. This correlation is far from perfect, however. Thus the first and second ionisation energies of mercury are comparable with those of sulphur, but the alchemists viewed elemental mercury and sulphur as the quintessential metal and nonmetal respectively. A closely-related correlation can be found with electronegativity.

For any elemental substance, we can envisage both a metallic form and a covalent form; perhaps several covalent forms. The most stable structure under ordinary conditions will be the winner in a thermodynamic contest among all possible structures. For some elements – e.g. Sn and As – there is little to choose between the two. In order to shed some light on this problem, let us pose a specific question and seek a convincing answer. Why does elemental lithium (in common with the other members of Group 1) adopt a metallic structure in preference to molecular Li_2, whereas hydrogen and the halogens in their elemental states consist of diatomic molecules?

MO calculations should be helpful. The methods used by chemists to deal with finite molecules of modest size are not wholly appropriate for macroscopic metallic crystals. However, we can perform calculations on fragments of metallic lattices. A convenient fragment is the M_{35} molecule depicted in Fig. 7.14. This is a bcc structure, as adopted by the Group 1 elemental substances (excepting hydrogen), described in Section 7.5. An MO calculation on H_{35} leads to 35 MOs, constructed from the 1s atomic orbitals. It might be thought that, when plotted on an energy scale, these should be symmetric about the energy of the 1s orbital of a free H atom, with equal numbers of bonding and antibonding MOs with perhaps a few essentially nonbonding MOs in the middle. Thus we might expect the highest occupied MO (HOMO) of H_{35} to be approximately nonbonding. But this simplistic expectation is not fulfilled by even rather crude numerical calculations. It was noted in Section 7.2 that an antibonding MO is more antibonding than its bonding counterpart is bonding: the 'antibonding effect'. Thus in the MO energy diagram for H_2 (Fig. 7.1) the two MOs are not symmetric about the energy of the free 1s(H) AO.

In a more complex molecule, an MO which, *prima facie*, contains about equal amounts of in-phase (positive) and out-of-phase (negative) overlaps between AOs on neighbouring atoms will turn out to be distinctly antibonding rather than bonding. Thus a calculation on H_{35} (at a fairly low level of sophistication) leads to only eight bonding MOs and no fewer than 25 antibonding MOs, with two essentially nonbonding. The HOMO is extremely antibonding, and it is apparent that H_{35} has little or no cohesive energy. The same conclusion can be drawn from the results for any H_n $(n > 2)$ molecule.

If a similar calculation is performed on Li_{35}, considering first only the 2s orbitals, the 'antibonding effect' is less marked, on account of the weaker interactions among the relatively diffuse 2s orbitals. But the HOMO is still appreciably antibonding and it is not immediately obvious that Li_{35} should be stable with respect to Li_2. If the calculation is repeated with the lithium 2p orbitals included, s–p mixing has the effect of stabilising most of the MOs in the erstwhile s-band and the HOMO is now roughly nonbonding. This is illustrated in Fig. 7.15, where we show the effect of s–p interaction on the occupied MOs of Li_{35}. Evidently, such mixing makes an important contribution to the stability of the metallic lattice for lithium in its elemental state. We may say that the 2p orbitals – unoc-

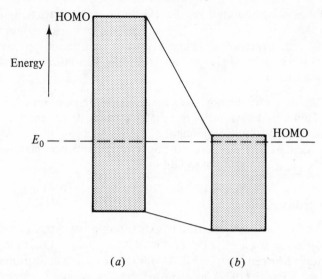

Fig. 7.15. Effect of including 2p orbitals in an MO calculation on lithium metal: (*a*) shows the energy range of occupied MOs without 2p orbitals in calculation, (*b*) shows the range with 2p included in the calculation. E_0 is the energy of 2s(Li) in the free Li atom.

cupied in the atomic ground state, and often ignored in simple MO treatments of bonding in molecules containing Li atoms – have a 'buffering effect' which tends to counteract the antibonding effect. Hydrogen has no 1p orbitals to exert a similar effect in H_{35}; the inclusion of 2s(H) and 2p(H) is of little help because these orbitals are too high in energy and too diffuse. Thus we can account for the preferences of elemental hydrogen and lithium for molecular and metallic structures respectively.

The antibonding effect is also apparent if we perform MO calculations on hypothetical molecules such as F_{35} and Cl_{35}. In the latter case, inclusion of the 3d orbitals exerts a small buffering influence but with a relatively large number of valence electrons compared with the number of valence orbitals, the HOMO is still strongly antibonding.

We may now summarise the features of an atom E which will favour the formation of a metallic lattice for the elemental substance in preference to a molecular structure. A metallic structure is favoured if:

(i) The atom E is large, in which case its orbitals will be rather diffuse and will overlap weakly with one another; this tends to reduce the antibonding effect in the metallic lattice, and makes a covalent structure less competitive since bond energies will be weak.

(ii)   E has a large number of valence orbitals, relative to its number of valence electrons. This encourages the buffering effect in the metallic structure, and helps to ensure a continuous, partly-filled band structure which endows the solid substance with metallic properties.

These features are obviously connected with the position of E in the Periodic Table, and such properties as electronegativity and ionisation potentials. Thus it is not surprising that the tendency to adopt a metallic structure increases as we descend any Group, and becomes minimal towards the right-hand side of the p block.

7.7   Further reading

All the inorganic chemistry texts listed in the Appendix discuss MO theory at varying levels. The best book to supplement the material in this chapter is Murrell, Kettle and Tedder (1979), listed in Section A.3 of the Appendix. Parish (1977), listed in Section A.4, is valuable for metallic bonding.

8

Coordination and organometallic compounds of the transition elements

8.1   Introduction: ligands

This chapter encompasses the largest and currently the most active area in inorganic chemistry; in a random sample of several recent issues of journals devoted to inorganic chemistry, 70% of the articles dealt with studies which could fit into the title of this chapter.

Coordination and organometallic chemistry are often regarded as distinct tribal divisions in the discipline of inorganic chemistry. However, the historical reasons behind the distinction are no longer applicable; the term 'coordination compound' may have outlived its usefulness, and the term 'organometallic compound' is often interpreted rather liberally. Accordingly, the two will be taken together as far as transition element chemistry is concerned.

Central to the orderly study of coordination/organometallic chemistry is the classification of ligands and an understanding of their functions. Ligands are generally molecules which exist independently in molecular substances (e.g. NH_3, CO) or ions which are found as entities in ionic solids (e.g. Cl^-, NO_2^-), although cases are known where the ligand is unstable unless bound to the central atom in a complex (e.g. cyclobutadiene C_4H_4). Ligands are sometimes classified as 'classical' or 'nonclassical'. 'Classical' ligands are those investigated by Werner and his disciples in the late nineteenth/early twentieth centuries, for example halide ions and amines. These form complexes which can usually be prepared and handled without difficulty on the open bench, and their bonding to the central atom can be described quite simply in terms of coordinate bonds involving σ donation of their lone pairs. 'Nonclassical' ligands (e.g. CO, PX_3, H^-) received relatively little attention until the middle of the present century, when techniques for handling air- and water-sensitive substances had been perfected. The bonding in their complexes often requires a consideration of rather more than simple σ donation. Another class of ligand – represented by olefins, aromatic

273

hydrocarbons and other species containing π bonds – is bonded via overlap between its π MOs (both bonding and antibonding) and suitable orbitals on the central atom. In such cases, it may be inappropriate to speak of 'donor atoms' on the ligands, since donation comes from π bonds (or, more strictly, from π bonding MOs) rather than from nonbonding orbitals localised on specific atoms.

Notwithstanding the previous sentence, all potential ligands possess one or more pairs of electrons which are available for donation to a central atom by one mechanism or another. A systematic survey of all known and potential ligands would be a formidable undertaking. In this section we will look at a number of representative ligands which illustrate the various modes of bonding to the central atom.

Ammonia NH₃

The ammonia molecule owes its capacity as a ligand to the lone-pair orbital on the N atom. In the language of VB theory (which is of limited value in d block and f block chemistry), the bond simply involves the σ overlap of the lone-pair orbital with an empty hybrid orbital of the central atom. In MO language, a complex ion $M(NH_3)_m^{n+}$ has m filled σ bonding MOs. There is no reason to suspect that any other orbitals on the NH_3 molecule are involved in its bonding to a central atom in a complex.

1,2-diaminoethane (ethylenediamine = en = NH₂CH₂CH₂NH₂)

Each of the two N atoms in $NH_2CH_2CH_2NH_2$ has the same function as the N atom of an ammonia molecule with respect to complex formation. However, en can take up two positions in the coordination sphere of the same central atom, e.g.:

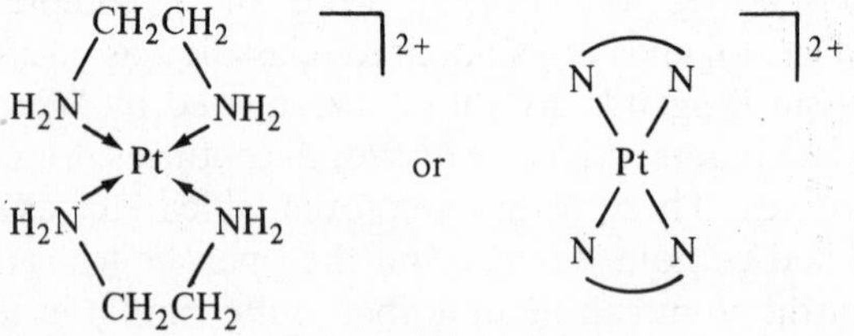

One en molecule can replace two NH_3 ligands, and is described as a chelating ligand (from the Greek word for the claw of a crab). Chelating ligands having two, three, four, five and six donor atoms are labelled respectively as bidentate, tridentate, quadridentate, quinquedentate and sexadentate, according to the number of 'teeth' with which they can 'bite'

the central atom. Chelating ligands are of special importance since they tend to form particularly stable complexes. Thus for the replacement:

$$Cd(CH_3NH_2)_4^{2+}(aq) + 2en(aq) \rightarrow Cd(en)_2^{2+}(aq) + 4CH_3NH_2(aq)$$

ΔG^0 is $-23\,kJ\,mol^{-1}$. This 'chelate effect' is further discussed in Section 8.4.

Chloride Cl⁻

In a complex ion such as $Cr(NH_3)_5Cl^{2+}$, the Cr–Cl bond can be described in terms of σ overlap between a filled 3p orbital (or 3s–3p hybrid) of Cl^- with an empty orbital of Cr^{3+}. In addition to this σ lone pair, Cl^- has at right angles to the M–Cl bond two filled 3p orbitals which might engage in π overlap. In an octahedral complex, the central atom d_{xy}, d_{xz} and d_{yz} orbitals (the t_{2g} level of ligand field theory) have the correct symmetry for π overlap (cf. d_π–p_π bonding in Main Group species discussed in Chapter 6):

Thus chloride is primarily a σ donor, but it has a secondary function as a π donor as well; the π overlap is rather weaker than the σ.

The chloride ion has four pairs of valence electrons, all of which can in principle form coordinate bonds, but their directional properties prohibit Cl^- from taking up two or more positions in the coordination sphere of the same central atom. The anion can, however, form bonds to two or more different central atoms; in other words, Cl^- can act as a *bridging ligand*, between two central atoms. We have already encountered bridging chlorine atoms in binary halides such as crystalline $BeCl_2$ and Ga_2Cl_6 (see Sections 6.3 and 9.2). Such compounds could be viewed as complexes, and their structures would certainly have been rejected by pre-Wernerian chemists. Examples of bridging Cl^- ligands include:

Triply-bridging chloride – Cl^- bonded to three central atoms – is known but is much less common. Dichlorides such as $MnCl_2$ and $NiCl_2$ having two-dimensional polymeric structures with $6:3$ coordination could be regarded as complexes in which Cl^- ions bridge three M^{2+} ions in trigonal fashion (see Sections 3.3 and 9.2). Quadruply-bridging Cl^- could be perceived in the tetrahedral $4:4$ structure of CuCl.

The other halide ions $- F^-$, Br^-, $I^- -$ can be viewed in their functions as ligands in much the same way as Cl^-. The extent of electron donation – which can be measured by various spectroscopic methods, such as the nephelauxetic effect – follows the expected sequence, $F^- < Cl^- < Br^- < I^-$; the lower the electronegativity of X, the more covalent the M–X bond becomes, i.e. the more electron density is donated by X^-. The amount of π donation as a fraction of the total $(\sigma + \pi)$ follows the reverse order, in keeping with the decreasing tendency of np orbitals to enter into π bonding as n increases. As noted elsewhere (Sections 6.3 and 9.2), F^- as a bridging ligand has some tendency to form linear (as opposed to bent) bridges.

Thiocyanate NCS⁻

The electronic structure of NCS^- can be described in terms of resonance:

$$N\equiv C-S^- \leftrightarrow {}^-N=C=S$$

That both forms make significant contributions demands sp hybridisation at the N atom and sp^2 at the S atom. There are thus two lone pairs on S and one on N. Five modes of coordination of NCS^- are found:

$$M \leftarrow \bar{N}CS \qquad \bar{N}CS \overset{M}{\nearrow} \qquad M \leftarrow \bar{N}CS \overset{M}{\nearrow}$$

$$\text{(i)} \qquad\qquad \text{(ii)} \qquad\qquad \text{(iii)}$$

$$\begin{matrix} M \\ \searrow \bar{N}CS \\ M \nearrow \end{matrix} \qquad \begin{matrix} M \\ \searrow \bar{S}CN \rightarrow M \\ M \nearrow \end{matrix}$$

$$\text{(iv)} \qquad\qquad \text{(v)}$$

Of these, types (i), (ii) and (iii) are most common. Type (iv) is rare and implies sp^2, rather than sp, hybridisation at the N atom. Thiocyanate is described as an *ambidentate* ligand, being able to coordinate via either or both of two atoms of different atomic number. It is not, however, a chelating ligand; a bidentate function would require that the lone pairs on both N and S point to the same central atom, which is not feasible. Other examples of ambidentate ligands include NO_2^-, CN^- and sulphoxides $R_2S=O$.

Carbon monoxide CO

CO acting as a ligand is known as carbonyl, and its complexes are called carbonyls. The lone pair on the C atom can form a dative bond, but CO in Main Group chemistry is a weak Lewis base and it is apparent that a simple representation:

$$M \leftarrow \overset{-}{C} \equiv \overset{+}{O} \quad \text{or} \quad M - \overset{-}{C} \equiv \overset{+}{O}$$

does not convey the full story. A wealth of experimental evidence supports the view that the bonding also involves π overlap between filled d orbitals on M and the empty π antibonding MOs of the CO molecule. This implies 'back-donation' from M to CO, and in VB language we could describe the bonding in terms of resonance:

$$M - \overset{-}{C} \equiv \overset{+}{O} \leftrightarrow M = C = O \leftrightarrow M \equiv \overset{-}{C} - \overset{+}{O}$$

The bonding is sometimes described as synergistic, implying that the donation M → CO enhances the Lewis basicity of CO towards M. Carbon monoxide tends to stabilise low oxidation states of the d block elements, since its bonding requires the presence of filled nd orbitals on M.

Other modes of bonding in CO complexes have been identified. The most important is the symmetrical bridging CO, in a three-membered ring:

$$\begin{array}{c} M \\ | \\ M \end{array} \!\! >C = O$$

The 'ketonic' description of the C–O bond is supported by its length and stretching frequency; but the M–C bonds should not be seen as simple σ bonds. An MO description involving three-centre bonding is required, and it is noteworthy that a carbonyl bridge between two atoms always involves a direct M–M bond. One, two or three CO molecules may bridge two M atoms, as shown in the examples below:

A single CO bridging two M atoms – as in the case of the rather unstable $Os_2(CO)_9$ shown above – is uncommon unless the M atoms form additional bonds, apart from the terminal M–C bonds. An example is $Co_4(CO)_{12}$ (Fig. 8.8 in Section 8.5). In a number of cases, 'semi-bridging' carbonyl is encountered; two M atoms (directly bonded to each other) are bridged unsymmetrically by one or more CO groups so that one M–C distance is appreciably shorter than the other. Triply-bridging CO is also found, i.e. the C atom is bonded to three M atoms.

Phosphine and its derivatives

PR_3 molecules where R is an organic group of relatively low electronegativity (e.g. alkyl) are quite good Lewis bases and might be expected to form complexes, like the analogous amines. Molecules PX_3 where X is an atom or group of high electronegativity (e.g. Cl, F, OR) have little or no Lewis basicity towards Main Group Lewis acids such as BX_3. That they do form a great number of complexes with atoms of the d block elements suggests a mode of bonding similar to that of CO, where P$\rightarrow$M donation is supplemented by P$\leftarrow$M back-donation. The latter is made possible by the availability of empty 3d orbitals on the P atom, which can overlap with filled d orbitals on M (d_π–d_π overlap). Remembering from Chapter 6 that highly-electronegative groups are necessary in order that a P atom may bring its 3d orbitals into play, and that electron-withdrawing groups should enhance the π-acceptor function of PX_3, it is understandable that π back-bonding should be particularly important in complexes of the phosphorus trihalides, and of phosphite esters $P(OR)_3$. Such bonding is less important, but is far from negligible, in many complexes of organic phosphine derivatives PR_3 and in their bidentate analogues such as $(C_6H_5)_2PCH_2CH_2P(C_6H_5)_2$. Ligands based upon three-coordinate P atoms can stabilise quite a variety of oxidation states, for example:

Ni(0): $Ni(PF_3)_4$
Ni(I): $Ni(PPh_3)_3Cl$
Ni(II): $Ni(PPh_3)_2Cl_2$
Ni(III): $Ni(PEt_3)_2Br_3$
(Ph = C_6H_5; Et = C_2H_5)

Anions such as PR_2^- appear as bridging ligands, while PF_4^- has recently been identified as a non-bridging ligand in a few complexes.

Ethylene C_2H_4

One of the earliest organometallic compounds to be isolated was Zeise's salt, $K[PtCl_3(C_2H_4)]$, which contains the anion:

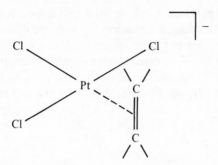

The C–C bond is perpendicular to the $PtCl_3$ plane, and its mid-point lies at a corner of a square where one might expect a ligand atom to be in a four-coordinate Pt(II) complex. The bonding of the C_2H_4 molecule to the Pt is believed to involve:

(i) Overlap of the filled π bonding MO of C_2H_4 with a Pt hybrid orbital of σ symmetry, leading to donation from ligand to Pt.

(ii) Overlap of a filled Pt 5d orbital with the empty π antibonding MO of C_2H_4, which amounts to π donation from Pt to C_2H_4:

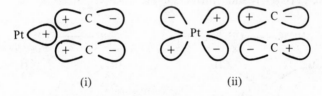

(i)   (ii)

In drawing structural formulae, the bonding of an olefin to a d block atom may be represented in two ways:

$$M \leftarrow \begin{matrix} C \\ \| \\ C \end{matrix} \qquad \text{or} \qquad M \begin{matrix} C \\ | \\ C \end{matrix}$$

Either is acceptable, although neither adequately conveys the nature of the bond; the observed C–C distance may distinguish one from the other.

Most other unsaturated hydrocarbon molecules – diolefins, acetylenes etc. – can function as ligands via bonds similar to that portrayed for ethylene. Unsaturated carbanions – such as allyl $CH_2{=}CH{—}CH_2^-$, or acetylides $RC{\equiv}C^-$ – can bind to central atoms via their π bonding and antibonding MOs; they can also function as simple σ donor ligands, using the lone pair on the C atom which bears the negative charge. Unsaturated organic ligands can often bind in a great many ways – over 20 have been identified for acetylenes and acetylides – and it is convenient to specify the number of carbon atoms which can be considered as bonded to the central atom in terms of the ligand's *hapticity*. Ligands are described as monohapto-, dihapto- etc., written as η^1, η^2 etc. For example, an allyl complex in which there are three approximately equal M–C distances, as in (i) below, is described as containing η^3-allyl, while one in which the M atom is uniquely bonded to only one C atom, as in (ii) below, is a complex of η^1-allyl.

(i)   (ii)

Aromatic molecules and ions (e.g. $C_5H_5^-$, C_6H_6 and $C_7H_7^+$) are found as ligands. Cyclopentadienyl $C_5H_5^-$ is usually η^5, as in complexes such as ferrocene $Fe(C_5H_5)_2$ and $Mn(CO)_3C_5H_5$:

It can, however, be η^3 or η^1 in some cases.

Unsaturated molecules/ions other than hydrocarbons and their simplest derivatives can function as ligands, using their π bonding and antibonding MOs in the manner (*mutatis mutandis*) described above for ethylene. CO_2, R_2CO, CS_2 and SO_2 provide examples:

$$M \underset{O}{\overset{\displaystyle C=O}{<}} \qquad M \underset{S}{\overset{\displaystyle C=S}{<}} \qquad M \underset{O}{\overset{\displaystyle S=O}{<}}$$

Note that CO_2 and CS_2 as free molecules are linear; but when complexed as shown above, they become bent.

Hard and soft acids and bases

Complex formation can usually be viewed as a Lewis acid–Lewis base reaction. The Lewis acid (or acceptor) in the present context is an atom or ion M^{n+} (where M is an atom of a d or f block element and n is its oxidation number) and the Lewis base (donor) is a ligand. Lewis acids and bases can be classified as 'hard' or 'soft'. A 'hard' base is one which donates via an atom (or atoms) of high electronegativity and low polarisability. The following are sequences of increasing hardness:

$$CH_3^- < NH_3 < OH_2 < F^-$$
$$I^- < Br^- < Cl^- < F^-$$

Generally, H^-, the heavier halides, and ligands which can be described as C-, P- or S-donors tend to be soft; N- and O-donors, and F^-, are characteristically hard bases. Small cations of high charge are the hardest acids. The softest acids are cations such as Ag^+, Hg^{2+} and Au^{3+}; lacking a noble gas configuration, they are quite polarisable. The principle which makes this classification useful is that hard acids and hard bases, or soft acids and soft bases, form the strongest complexes; 'like' tends to go with 'like'.

This classification has its origins in the observation that cations fall into two categories with respect to the thermodynamic stabilities of their complexes formed in aqueous solution with halide ions. Class (a) cations $M^{n+}(aq)$ form complexes $MX^{(n-1)+}(aq)$ such that the equilibrium constant $K = [MX^{(n-1)+}(aq)]/[M^{n+}(aq)][X^-(aq)]$ follows the sequence:

$$F^- > Cl^- > Br^- > I^-$$

Class (b) cations follow the reverse sequence of stability for halide complexes. The 'Class (a)/Class (b)' terminology is still to be found, but the broader concept of soft and hard acids and bases has found widespread acceptance.

The utility of the concept may be illustrated by comparing the chemistries in the III oxidation state of cobalt, rhodium and iridium. Co^{3+} is classified as a relatively hard acid, and the very large number of Co(III) complexes mostly involve N- and O-donor ligands. Attempts to prepare Co(III) complexes with P- or S-donor ligands often lead to reduction to

Co(II) or Co(I) species; Co^{2+} and Co^+ are softer acids than Co^{3+}. Cobalt(III) does form complexes with the soft base CN^-, and coordination of cyanide has a softening effect upon the cobalt(III). Thus the ambidentate ligand NCS^- usually binds to Co(III) via its hard end, i.e. the N atom, as in $Co(NH_3)_5NCS^{2+}$. But in $Co(CN)_5SCN^{3-}$, the thiocyanate is S-bonded. Rh^{3+} and Ir^{3+} are appreciably softer than Co^{3+}; as well as forming complexes with 'classical' hard bases, they will bind soft ligands such as phosphine derivatives and hydride. Complexes such as $[IrH_3(PPh_3)_3]$ have no analogue in cobalt(III) chemistry, and the thiocyanate ligands in $Rh(SCN)_6^{3-}$ are all S-bonded.

8.2   Bonding theories in coordination/organometallic chemistry

The reader may have been wondering which theory of bonding was being invoked in the previous section to deal with the binding of ligands to the central atom(s) in coordination/organometallic compounds of the d block elements. Both VB and MO language were employed, and later in this chapter the terminology of CF/LF theory will be encountered. Students and practitioners of this field have to develop some facility in changing from one language to another, often in the same breath. You will often read that VB theory is of limited value in this field; yet even the most enthusiastic supporters of MO theory lapse into VB terminology from time to time. We observed in Section 7.4 that the concept of the three-centre bond in Main Group chemistry was essentially an attempt to graft onto the electron-pair bond idea of Lewis (and its direct descendant, VB theory) some situations that are best tackled by means of MO theory. Coordination/organometallic chemists, proficient though most of them are in at least qualitative MO approaches, are still in their hearts unwilling to abandon the (2c, 2e) bond implicit in Lewis theory. To speak of, e.g., a π bond being formed by overlap between a filled cobalt 3d orbital and an empty phosphorus 3d orbital in a phosphine complex is to invoke VB theory. But, as in Main Group chemistry, there are situations where only MO theory will do.

CF theory is not really a theory of bonding at all; the assumption that the ligands are held by electrostatic forces is certainly erroneous. But the spirit and language of CF theory provide valuable simplifications where we have to focus our attention on the orbitals which constitute the partly-filled shell; very often, the frontier orbitals of MO theory are the d orbitals of CF theory.

Most of this section is devoted to a discussion of the MOs of an octahedral complex ML_6. This exercise is performed in most inorganic texts; the emphasis here is on its relationships with VB and CF treatments.

MO treatment of an octahedral complex ML_6

Here, M is taken to be a member of the 3d series. The 3d, 4s and 4p orbitals are deemed to constitute its valence orbitals. L is a ligand which has available for bonding to M only a lone pair, oriented such that σ overlap is possible with M orbitals having non-zero electron density along the M–L axes. Fig. 8.1 shows the labelling of the Cartesian axes, with respect to which the M atomic orbitals are labelled. The L lone pairs are

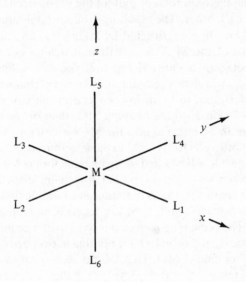

M atomic orbitals	Symmetry	L σ group orbitals[a]
$3d_{z^2}$ $\Big\}$ $3d_{x^2-y^2}$	e_g	$2\sigma_5 + 2\sigma_6 - \sigma_1 - \sigma_2 - \sigma_3 - \sigma_4$ $\sigma_1 - \sigma_2 + \sigma_3 - \sigma_4$
$3d_{xy}, 3d_{xz}, 3d_{yz}$	t_{2g}	–
4s	a_{1g}	$\sigma_1 + \sigma_2 + \sigma_3 + \sigma_4 + \sigma_5 + \sigma_6$
$4p_x$ $\Big\}$ $4p_y$ $4p_z$	t_{1u}	$\sigma_1 - \sigma_3$ $\sigma_4 - \sigma_2$ $\sigma_5 - \sigma_6$

[a] Normalising constants omitted

Fig. 8.1. M AOs and L σ group orbitals for MO treatment of octahedral complex ML_6, where M is a member of the 3d series.

labelled $\sigma_1, \sigma_2, \ldots, \sigma_6$ according to the numbering of the donor atoms. The group orbitals for a system of such high symmetry are very easily determined, without the need for any elegant group-theoretical techniques. All you have to do is to draw a picture of any M orbital, and see whether you can devise a matching combination of σ orbitals such that the overlaps are all equal in sign and magnitude. The only difficulty arises with d_{z^2}, whose large lobes along the z axis are not equivalent to the central annulus. The matching group orbital for d_{z^2} is best obtained by imposing the requirement that all the group orbitals have to be orthogonal to each other. The resulting group orbitals are listed in Fig. 8.1.

MOs are then constructed by taking bonding and antibonding combinations of the M AOs and their matching group orbitals. The MO energy diagram is shown in Fig. 8.2. The six bonding MOs – labelled a_{1g}, e_g and t_{1u} – will all be predominantly ligand in character, since we presume the donor atom to be of lower electronegativity than M so that the σ orbitals lie closer to the bonding MOs than the M AOs. The number of electrons to be inserted into the MOs is equal to $12 + n$, where n is the number of 3d electrons on M in the appropriate oxidation state. Thus the bonding MOs will be filled; if we imagine for book-keeping purposes that the 12 electrons originally in the L lone pairs are now in MOs which have some M character, this is tantamount to stating that some net transfer of electron density from L to M has occurred and L may be described as a σ donor. The remaining n electrons will have to go into the nonbonding t_{2g} level – the $d_{xy,xz,yz}$ orbitals have nothing to overlap with – and perhaps also the antibonding e_g^* MO. This, of course, is what we do in CF/LF theory; the difference is that the e_g level of CF theory comprises the pure d_{z^2} and $d_{x^2-y^2}$ orbitals, while the e_g^* level of MO theory is a twofold-degenerate antibonding MO which will be predominantly $d_{z^2}/d_{x^2-y^2}$ in character (especially if the overlap is relatively weak and if the ligand σ orbitals lie much lower in energy than the M 3d orbitals) but which will to some extent be delocalised over the ligands.

We now consider the possibility of π bonding. If the ligands have a full set of orbitals for this purpose – i.e. if there are two suitable orbitals on each L atom, p or d orbitals each with a nodal surface containing the M–L axis – it is easy to show that the $d_{xy,xz,yz}$ orbitals can now engage in π overlap, thus:

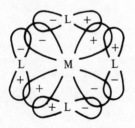

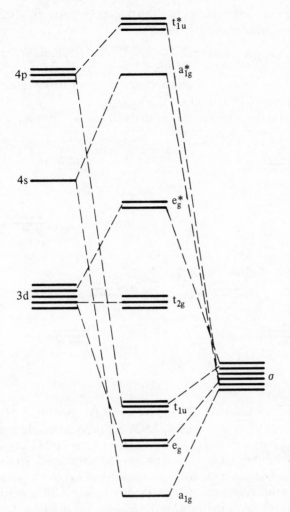

Fig. 8.2. MO energy diagram for octahedral ML_6, considering only σ bonding.

The 4p orbitals can also engage in π overlap, but because these are preoccupied with the more important σ bonding such overlap is of little consequence. The effect on the t_{2g} levels will be to produce a bonding set – labelled t_{2g} – and an antibonding set, t_{2g}^*. Which of these is to be identified with the t_{2g} level of CF/LF theory depends on the relative energies of the ligand π orbitals and the M 3d orbitals. We can envisage two situations:

(i) The ligand π orbitals lie lower in energy than the M 3d orbitals, in which case they will be filled in the free ligand L.

(ii) The ligand π orbitals lie higher in energy than the M 3d orbitals, and are empty in the free ligand.

Situation (i) will apply for ligands like F^-, OH^- etc. Situation (ii) holds for ligands where the capacity for π bonding stems from the postulated availability of, e.g., the 3d orbitals of the P atom in PX_3, or the π antibonding MOs of CO and other such ligands.

The effect upon the relevant part of the MO energy diagram is shown below:

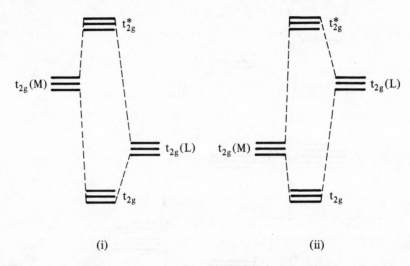

(i)   (ii)

In case (i), the t_{2g} level of CF theory is to be identified with the antibonding t_{2g}^* MOs; the t_{2g} bonding MOs are predominantly ligand in composition, and will be filled regardless of the number of d electrons on M. The e_g^* antibonding MOs will, of course, be unaffected, so that the energy separation between the MOs corresponding to the CF parameter Δ_0 is reduced by the π interaction. The bonding t_{2g} level will have some M 3d character, and its filling by erstwhile L electrons constitutes a degree of electron transfer from L to M; the ligands can be described as π donors.

In case (ii), it is now the bonding level t_{2g} which is the MO counterpart of the CF t_{2g} orbitals. The number of electrons to be placed in it will depend on the number of d electrons possessed by M in the appropriate oxidation state. But since electrons in t_{2g} will have some L character, and will be to some extent delocalised over the ligands, such occupancy leads to electron transfer from M to L. Thus ligands in this category are known as π acceptors, and we say that there is some degree of π back-bonding or back-donation. Note that the stabilisation of the CF t_{2g} level as a consequence of such interaction will increase the splitting Δ_0.

The above observations enable us to rationalise the spectrochemical series (Section 2.6), which is quite mysterious within an electrostatic description of the ligand field. π donor ligands (e.g. halide) occur at the beginning of the series, and produce the smallest splittings. Purely σ donor ligands like NH_3 occur around the middle, and the largest splittings are produced by π acceptors like CO and CN^-. (Note, however, that H^- and CH_3^- – pure σ donors surely – produce large splittings, comparable with those for π acceptors).

The octahedral splitting parameter Δ_0 can be written as the difference:

$$\Delta_0 = \Delta(\sigma) - \Delta(\pi)$$

where $\Delta(\sigma)$ and $\Delta(\pi)$ are respectively the destabilisations suffered by the (predominantly) $d_{z^2}/d_{x^2-y^2}$ and $d_{xy}/d_{xz}/d_{yz}$ orbitals, in CF language. Analysis of the d–d spectra of regularly octahedral complexes can lead to only one parameter; but studies of complexes where the ligands are not all the same can lead to the separation of splitting parameters into σ and π contributions. We can then draw up a two-dimensional spectrochemical series. Thus in increasing order of $\Delta(\sigma)$ we have:

$$I^- < Br^- < Cl^- < CH_3NH_2 < NH_3 < \text{ en} < H_2O < F^- < OH^-$$

For $\Delta(\pi)$, a partial ordering (data are sparse for many ligands) is:

$$PPh_3 < NO_2^- < C_5H_5N < NH_3 < I^- < Br^- < Cl^- < H_2O < F^- < OH^-$$

It should be noted that although the sign of $\Delta(\pi)$ (ligands to the left of NH_3 have negative values) tells us whether a ligand is a π donor or acceptor, the magnitude of $\Delta(\sigma)$ or $\Delta(\pi)$ is not to be taken as a quantitative measure of the extent of electron transfer. Thus F^- has large values for both parameters, but it would be absurd to suggest that F^- was a better donor than I^-. It is probably valid to say that the π donor component of fluoride's capacity as a ligand relative to its σ donor power is greater than for the other halide ions. But its position in the nephelauxetic series (see Section 2.6) shows that F^- is overall a relatively feeble electron donor.

MO treatments for other coordination geometries

MO treatments can, of course, be given for geometries other than octahedral. The results can be reconciled with CF theory in much the same way. In one important respect, however, the octahedral case is unique. The MOs arising from M AOs of different subshells all have distinct symmetry properties; there is no mixing of 3d with 4s or 4p. Each MO can be identified with a unique M AO as a contributor. But in lower symmetries, this is no longer the case. In a tetrahedral complex, the M orbitals $3d_{xz,yz,xy}$ belong to the same symmetry species as the $4p_{x,y,z}$ set

(t_2). This means that in all MOs of t_2 symmetry, both must make some contribution (not equal, of course). In a tetrahedral ML_4 complex, starting with a basis set of the M 3d, 4s and 4p orbitals and the L σ orbitals only, we obtain three t_2 sets of MOs, one bonding, one antibonding and one weakly antibonding; this last is the counterpart of the CF t_2 level. It is less antibonding than might have been expected on account of the 'buffering' effect of the antibonding t_2 level, and has considerable p character. The fact that d–d transitions are much more intense in tetrahedral complexes compared with octahedral can be attributed (in part) to the fact that in atomic spectroscopy d–d transitions are forbidden but d–p transitions are allowed. Thus a transition that involves an electron jump from the e to the t_2 level in a tetrahedral complex will have some d → p character. The situation can be illustrated by looking at a partial MO diagram which shows the effect of including d–p mixing in a calculation on tetrahedral ML_4:

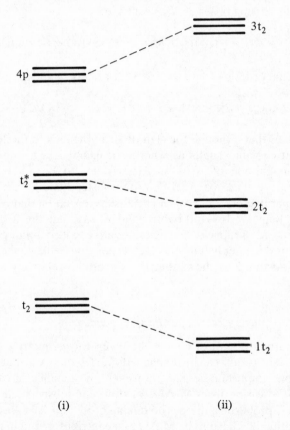

(i)   (ii)

In (i), the 4p orbitals of M (a 3d series atom) are excluded from the calculation, and the antibonding MO labelled t_2^* corresponds to the CF t_2

level (d_{xz}, d_{yz}, d_{xy}). In (ii), the calculation now includes 4p(M) in the basis set. The effect (cf. the inclusion of 2s orbitals in the calculation for He_2 in Section 7.2) is to stabilise the other t_2 MOs, now labelled $1t_2$ and $2t_2$, the latter being the MO counterpart of the CF t_2 level. Thus if we were to perform an MO calculation on tetrahedral ML_4 without the 4p subshell in the basis set, the effect of later 'switching on' the 4p orbitals would be to stabilise the MO counterpart of the CF t_2 level. In other words, this MO is antibonding with respect to 3d(M)–σ(L) overlap, but bonding with respect to 4p(M)–σ(L) interactions. Similar d–p mixing occurs with most other coordination geometries, except for linear two-coordination and square planar four-coordination. In these, we have mixing between nd_{z^2} and $(n + 1)$s, however. Thus the MO corresponding to the d_{z^2} orbital of CF theory may be appreciably bonding if the $(n + 1)$s orbital interacts more strongly with the ligand σ lone-pair orbitals than the M nd orbitals do. This is connected with the fact that d_{z^2} is usually the lowest d orbital in a CF analysis of the d–d spectrum of a square planar complex. It may also provide much of the driving force for the (so-called) Jahn–Teller distortions observed in six-coordinate copper(II) complexes. The most common coordination geometry for Cu(II) is the axially elongated octahedron or $(4 + 2)$ coordination shown in Fig. 8.3 for $Cu(NH_3)_4(NO_2)_2$. As two *trans* ligands are withdrawn along the z axis, the CF d_{z^2} orbital is rapidly stabilised, partly because (reverting now to MO terminology) it is less antibonding with respect to 3d(M)–σ(L) overlap and partly because it becomes bonding with respect to 4s(M)–σ(L) overlap. This orbital is filled, of course, in a d^9 complex. An axial compression, or an equatorial elongation, would not have nearly the same stabilising effect. Why do the two axial ligands not disappear altogether, leaving a square planar, four-coordinate complex? Sometimes this happens in copper(II) systems but more often a compromise is struck between the effects of stabilising the CF d_{z^2} orbital and destabilising the bonding MOs to which the withdrawn ligands contributed.

Where the nd subshell is full or nearly full (d^8, d^9, d^{10}), there is a tendency to avoid coordination geometries in which antibonding MOs of predominantly M nd character would necessarily be occupied, unless these MOs are appreciably stabilised via d–s and/or d–p mixing. Thus octahedral six-coordination is relatively uncommon for d^{10} systems, in which there are four electrons in the antibonding MOs corresponding to the CF e_g level. Octahedral Ni(II) complexes (d^8) are common enough, although other coordination geometries – square planar, tetrahedral, trigonal bipyramidal and square pyramidal – are competitive. For Pd(II) and Pt(II), the octahedral e_g^* MOs are more strongly antibonding – or, in CF language, the d-orbital splitting is much larger, as is always the case in comparing 3d ions with their 4d and 5d congeners – and six-coordination is very rare. Comparing the structural chemistry of zinc with that of magnesium, we might expect close similarities because Mg^{2+} and Zn^{2+} are very similar in size. But whereas MgO, MgS and $MgCl_2$ all have octahedral six-coordination about the Mg atom, the zinc analogues prefer tetrahedral four-coordination. In the case of linear two-coordination – uncommon in the zinc Group 12d (except for Hg), but common for Cu(I), Ag(I) and Au(I) – simple CF considerations place the d_{z^2} orbital high in energy, and its MO counterpart is appreciably antibonding with respect to the involvement of d_{z^2}. But this orbital is appreciably stabilised by mixing with $(n + 1)$s. In the next section we take a more systematic look at the occurrence of coordination numbers and their associated geometries.

VB theory in d block chemistry

In Chapters 6 and 7, it was observed that VB theory works well in Main Group chemistry provided that we are dealing with closed-shell systems. It runs into difficulties with open-shell systems, especially those where – in an MO treatment – we have partial occupancy of an antibonding level. The same constraints are valid when we come to apply VB theory in d block chemistry; they are necessarily more severe.

VB theory is quite satisfactory as a basis for describing the bonding in d^0 or d^{10} systems, which are treated in the same way – *mutatis mutandis* – as analogous Main Group systems. Otherwise, its performance is decidedly patchy. One difficulty is that for certain coordination geometries, there are ambiguities as to the type of hybridisation to be invoked and indeed the application of hybridisation is much more difficult where d orbitals are involved. Let us consider first the case of an octahedral complex ML_6. The formation of six coordinate bonds requires the construction of six d^2sp^3 hybrid orbitals using (in the case of a 3d series central atom) $3d_{z^2}$, $3d_{x^2-y^2}$, 4s, $4p_x$, $4p_y$ and $4p_z$, these being the central atom M orbitals having the correct directional properties for overlap with L σ lone pairs. These must be left empty, so that for a d^n central atom/ion, n electrons have to be accommodated elsewhere. For $n \leqslant 6$, this is no problem – the n electrons are placed in $3d_{xy,xz,yz}$ – unless we have a high-spin d^4, d^5 or d^6 system, in which case it is difficult to account for the number of unpaired electrons revealed by the experimental magnetic moment. The solution to this problem is to invoke the use of 4d rather than 3d orbitals in the hybridisation scheme; thus, for example, in the case of an octahedral iron(III) complex evidently having five unpaired electrons per Fe, use of the $4d_{z^2}$ and $4d_{x^2-y^2}$ orbitals for hybridisation leaves all five 3d orbitals available for the accommodation of nonbonding electrons, and interelectron repulsions will be minimised if they are singly occupied by electrons of the same spin. If we have a low-spin complex of

Fe(III), with just one unpaired electron, we can use the $3d_{z^2}$ and $3d_{x^2-y^2}$ orbitals for bonding; and the observed magnetic properties can be accounted for by placing five nonbonding electrons in the three nonbonding orbitals $3d_{xy,xz,yz}$. But there is no principle to guide us as to which of 3d or 4d is favoured in a particular case, other than the experimental evidence provided by the magnetic properties. For d^7, d^8, d^9 and d^{10} complexes, it is, of course, necessary to use the 4d orbitals rather than 3d for bonding, in order to find room for the electrons which constitute the partly-filled shell.

Other hybridisation schemes are fairly clear-cut, e.g. sp^2d for square planar four-coordination (using $d_{x^2-y^2}$), sd^3 for tetrahedral four-coordination (using $d_{xy,xz,yz}$) or dsp^3 for trigonal bipyramidal five-coordination (with d_{z^2}). But because of d–s and d–p mixing in these coordination geometries, there is an element of ambiguity. Thus we could substitute nd_{z^2} for $(n + 1)$s in a square planar system, or the three $(n + 1)$p orbitals for the t_{2g} nd orbitals in the tetrahedral case. The 'real' hybridisation will be somewhere between the extremes, and it is in these situations that MO theory is much more suitable for quantitative calculations. It is quite useful and legitimate to describe the σ bonding in the square complex $Ni(CN)_4^{2-}$ in terms of dsp^2 (or sp^2d) hybrids, with the eight valence electrons of Ni(II) in the four remaining 3d orbitals after $3d_{x^2-y^2}$ has been committed to the fray. One might then proceed to form π bonds by looking at the overlaps between filled 3d orbitals and suitable orbitals on the CN⁻ ligands. But in a square copper(II) complex, we run into trouble. In a square planar d^9 system, where is the ninth electron to go? All five 3d orbitals are in use. The next available orbital in energy is $4p_z$, which is not involved in the dsp^2 hybridisation. But ESR evidence shows unequivocally that the odd electron is in an orbital resembling $d_{x^2-y^2}$. This is not to be taken as a defeat for VB theory; rather, the theory is over-reaching itself in trying to pin down odd electrons in open-shell situations.

It might be thought that VB theory should come into its own in d block organometallic chemistry, where we are usually dealing with closed-shell species. But, as already noted, there are problems in specifying the hybridisation schemes. Pauling has published a number of interesting papers since the mid-1970s on this topic, but few organometallic chemists are likely to be seduced from MO theory. In Main Group and organic chemistry, it is often helpful to dispose of the σ framework in terms of hybridisation, and then to focus our attention – often using MO theory – on the delocalised π bonding, or on how two or more molecular fragments fit together to form a large molecule. This approach has its uses in d block chemistry too. For example, molecular substances such as $(NR_2)_3Mo{\equiv}Mo(NR_2)_3$ are rendered as having a triple bond between Mo atoms. Apart from the simple fact that Mo has a valency of six and can therefore form six bonds, can this formulation be justified? A VB treatment is much simpler than a full-blown MO calculation. The arrangement about each Mo is approximately tetrahedral, so we invoke four sd^3 hybrids. In the valence state, these will all be singly occupied and poised to form four 'ordinary' (as opposed to coordinate) σ bonds, one of which will be an Mo–Mo bond. The Mo atom (ground state configuration [Kr]$4d^55s$) thus has two electrons to spare after dealing with the σ bonding. If the Mo–Mo axis is labelled z, the 4d orbitals involved in hybridisation are $4d_{z^2}$, $4d_{xy}$ and $4d_{x^2-y^2}$. Thus the two remaining Mo electrons are placed in $4d_{xz}$ and $4d_{yz}$, which have the correct symmetries to overlap with their counterparts on the other Mo atom to form two Mo–Mo π bonds.

8.3   Coordination numbers and geometries

In this section we summarise the most common coordination numbers and the associated geometries in mononuclear complexes, and also in bridged polynuclear complexes where there is no direct M–M bond; the situation in cluster compounds (dealt with in Section 8.5) is rather more involved.

Coordination number two

This is quite rare, and is almost entirely restricted to complexes whose oxidation states correspond to the d^{10} configuration. The geometry is usually linear (or nearly so). If a filled nd subshell interacts at all with the ligands, there will result at least one filled antibonding MO. In linear ML_2, there is just one σ antibonding MO of largely nd character, the counterpart of d_{z^2} in CF theory. As mentioned in the previous section, this is appreciably stabilised by $nd_{z^2}-(n + 1)$s mixing. Higher coordination numbers are certainly found for d^{10} systems, but they are discouraged unless the filled antibonding MO(s) are stabilised by d–s or d–p mixing, as discussed in Section 8.2. Linear, two-coordinate complexes are mainly formed by Pd(0) and Pt(0) (e.g. M(PPh$_3$)$_2$), by Cu, Ag and Au in the I oxidation state (e.g. CuCl$_2^-$, Ag(NH$_3$)$_2^+$, Au(PPh$_3$)Cl) and by Hg(II) (e.g. Hg(NH$_3$)$_2^{2+}$).

Coordination number three

This is also rare, and is found mainly for complexes where the central atom has the d^{10} configuration. As for linear two-coordination, triangular three-coordination leads to just one weakly-antibonding occupied MO and is thus favoured by a filled nd subshell. Three-coordinate, planar complexes include HgI_3^-, $Au(PPh_3)_2Cl$ and $Cu(CN)_3^{2-}$. Mention may also be made of the molecular compounds $M[N(SiMe_3)_2]_3$ where $M = Ti$, V, Cr, Mn, Fe and the square brackets do not enclose a discrete complex species! These are not strictly coordination compounds, and the bulky trimethylsilylamide groups presumably prevent higher coordination numbers.

Coordination number four

At an early stage in the development of coordination chemistry, tetrahedral and square coplanar geometries were identified. More recently, geometries intermediate between these extremes have been observed.

Tetrahedral complexes are common among the 3d series. CFSE considerations favour the d^2 and d^7 configurations as being particularly auspicious for tetrahedral coordination geometries. Cobalt(II) certainly forms a great number of tetrahedral complexes, but many other geometries are represented in Co(II) chemistry. A number of otherwise rare oxidation states are represented by tetrahedral MO_4^{n-} ions where M has formally the d^2 configuration, e.g. CrO_4^{4-}, MnO_4^{3-}, FeO_4^{2-}. All the 3d elements, and all d^n configurations among the 3d series, are represented by tetrahedral complexes. In the 4d and 5d series, however, tetrahedral coordination geometry is much less common. The larger sizes of the heavier d block atoms favour higher coordination numbers than four, except where square coplanar four-coordination is favoured electronically.

CFSE arguments suggest that square coplanar coordination should be particularly favoured by the low-spin d^8 configuration, since we have four relatively stable filled orbitals. In MO language, the lowest empty MO in a square coplanar d^8 complex (predominantly $d_{x^2-y^2}$) is quite strongly antibonding, especially for 4d/5d atoms. Among the 3d series, square planar complexes are particularly common for Co(II), Ni(II) and Cu(II), i.e. d^7, d^8 and d^9 respectively. These systems all have some tendency to add additional ligands, in order to attain octahedral six-coordination (or $4 + 2$) in the case of Cu(II)). Among the heavier transition elements, square coplanar four-coordination is dominant for d^8 systems, e.g. Rh(I), Pt(II), Au(III).

Configurations intermediate between a square and a tetrahedron are known. An important example is Wilkinson's catalyst, $Rh(PPh_3)_3Cl$. Four-coordinate rhodium(I) would normally adopt a square planar configuration, but the bulk of the phosphine ligands causes a distortion from coplanarity of the four donor atoms, increasing the distances between the PPh_3 groups. The conflict of interest between the central atom and the ligands is no doubt important in determining the reactivity of this compound (see Section 9.8). The $CuCl_4^{2-}$ ion in several solid complexes is described as a flattened tetrahedron. This distortion is commonly blamed on the Jahn–Teller effect which is operative for d^9 systems; the square planar configuration can be regarded as an extreme case of flattening a tetrahedron, as shown below:

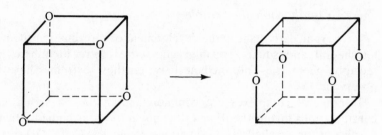

Square coplanar $CuCl_4^{2-}$ does occur, for example in $[C_6H_5CH_2CH_2NH_2CH_3]_2[CuCl_4]$. It would seem that, for electronic reasons, the square planar configuration is preferred to tetrahedral. However, Cl–Cl repulsion favours the tetrahedron. Where N—H....Cl hydrogen bonding is strong, the fractional negative charge on the Cl atoms is diminished, and the consequent lowering of interligand repulsion encourages the square planar configuration. Thus tetrachloro-cuprates(II) with cations such as Cs^+ or NR_4^+, offering no possibility of hydrogen bonding, usually contain the flattened-tetrahedral anion. Secondary ammonium cations $NH_2R_2^+$ appear to favour the square planar configuration for $CuCl_4^{2-}$. With NH_4^+ or RNH_3^+, the square planar anions are stacked in such a way that the Cu now has distorted octahedral 4 + 2 coordination. Chlorocuprates(II) which contain the genuinely four-coordinate, square planar $CuCl_4^{2-}$ ion are characteristically thermochromic; on mild heating, their colour changes from emerald green to yellow-brown. The high-temperature form contains flattened-tetrahedral anions; presumably heating disrupts the hydrogen bonding that is necessary to stabilise the square planar configuration.

Coordination number five

Until the 1950s, this was thought to be quite rare. Now, however, five-coordinate complexes are known for practically all the d block elements, and for all d^n configurations. Two idealised geometries are found in five-coordination: the trigonal bipyramid (TBP) and the square pyramid (SP), shown below:

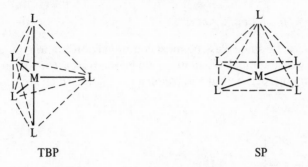

TBP   SP

A number of ligands favour TBP coordination geometry. A ligand of the type $N(CH_2CH_2NR_2)_3$ is quadridentate, and its stereochemistry dictates a 'tripod' arrangement about the central atom; a fifth ligand completes TBP coordination geometry in, e.g., $[CoLBr]^+$, where L is $N(CH_2CH_2NMe_2)_3$.

Where the ligands have no obvious influence, it appears that there is little to choose energetically between TBP and SP configurations for five-coordinate complexes. In the solid state, crystal packing effects may be dominant. For example, the SP $MnCl_5^{2-}$ ion is found in at least one crystalline solid. But the fact that the corresponding In(III) and Tl(III) compounds adopt the same structure (in conflict with the VSEPR rules) suggests that the preference for SP geometry has nothing to do with the d^4 configuration of Mn(III) and is probably dictated by the packing of ions in the particular crystal structure. In general, attempts to correlate the question of SP versus TBP coordination geometry in terms of CF arguments are rather unconvincing. Both SP and TBP configurations are found for $Ni(CN)_5^{3-}$ in the same compound in:

$$[Co(NH_2CH_2CH_2NH_2)_3][Ni(CN)_5]1.5H_2O,$$

Emphasising the subtleties of the factors that govern the geometries of five-coordinate species.

CF arguments do, however, predict that the d^8 (low-spin) configuration should be stable for TBP geometry, and this is indeed common for complexes of Fe(0), Co(I) and Ni(II) in the 3d series. In the 4d and 5d series, analogous compounds are formed, except that five-coordination is very rare for Pd(II), Pt(II), Ag(III) and Au(III); these prefer square coplanar four-coordination.

SP coordination is sometimes adopted in cases where the central atom would normally prefer octahedral six-coordination, but the sixth site is blocked. An example is $Ru(PPh_3)_3Cl_2$, where it is apparent that an H atom of a phenyl group lies in the way of any incoming ligand and the octahedral coordination usually adopted by Ru(II) (d^6) is sterically prevented. This compound is very reactive and has important catalytic properties; see also Section 9.8.

Coordination number six

This is by some way the most common coordination number for complexes of the d block elements. Two geometries are associated with six-coordination: octahedral and trigonal prismatic, which are compared below:

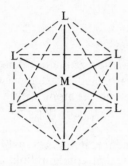

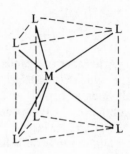

octahedron   trigonal prism

The trigonal prism is quite rare, and is mainly restricted to complexes ML_3 where L is a bidentate ligand with S atoms as donors. The trigonal prism allows closer ligand–ligand contact than the octahedron. For most ligands, this favours the octahedron since it means less ligand–ligand repulsion. But in S-donor complexes, the relatively short S–S distances imply some weak bonding (as is often seen in Main Group sulphur compounds) and the trigonal prismatic arrangement favours such bonding.

Octahedral six-coordination is especially favoured by the low-spin d^6 configuration. This can be understood in terms of simple CFSE considerations. For M(III) (M = Co, Rh, Ir) and Pt(IV), hardly any complexes other than octahedral ones are known. These complexes are kinetically fairly inert, in the sense that they undergo ligand exchange reactions slowly. For this reason, much of our knowledge of kinetics and mechanism in transition element chemistry has come from studies of low-spin d^6 octahedral complexes (see Sections 9.4 and 9.5).

The d^3 and d^8 configurations also give rise to particularly high CFSEs with octahedral coordination. Chromium(III) certainly forms a vast number of kinetically inert octahedral complexes, and is rarely found in any other coordination geometry. However, analogous complexes of Mo(III) and W(III) are rare; these tend to disproportionate unless stabilised by M–M bonding. Nickel(II) is the most familiar d^8 species in the 3d series and it forms a considerable range of octahedral complexes. However, this configuration is in competition with the four-coordinate square planar and five-coordinate geometries for d^8 species. Palladium

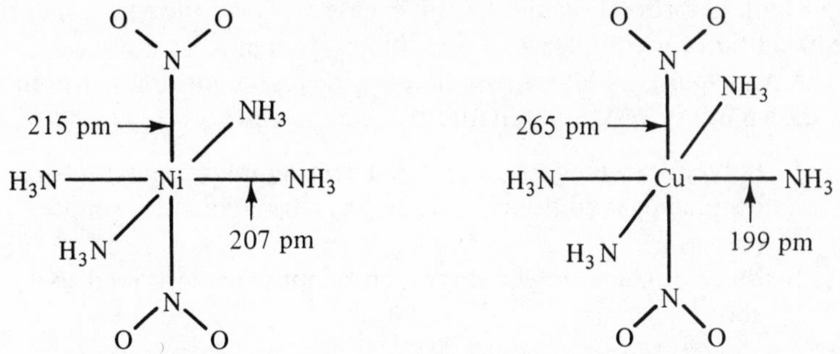

Fig. 8.3. Illustration of Jahn–Teller effect: comparison of axial and equatorial M–N distances between $M(NH_3)_4(NO_2)_2$ (M = Ni, Cu).

and platinum form very few octahedral complexes in the II oxidation state; the CF e_g level is now strongly antibonding, and planar four-coordination is preferred.

The d^4 and d^9 configurations are subject to the so-called Jahn–Teller effect, which prohibits regularly octahedral geometry. The Jahn–Teller distortion usually takes the form of an axial elongation, i.e. the partial removal of two *trans* ligands to give a configuration intermediate between an octahedron and a square plane, i.e. (4 + 2) coordination. In Fig. 8.3 we compare the structures of $[Ni(NH_3)_4(NO_2)_2]$ and $[Cu(NH_3)_4(NO_2)_2]$; note that the Cu–NH$_3$ bond length is slightly shorter than Ni–NH$_3$, while the Cu–NO$_2$ distance is much greater than in the Ni(II) complex. The removal of two *trans* (axial) ligands relieves interligand repulsion somewhat and allows the equatorial NH$_3$ ligands to approach the Cu atom more closely.

Few (if any) authentic cases of an axial compression or equatorial elongation – leading to 2 + 4 coordination – as opposed to axial elongation are known. The reader may find reference to K_2CuF_4 as an example, but a redetermination of this structure reveals the usual elongation, with 4F at 192 pm and 2F at 222 pm. The reasons for the apparent preference for elongation rather than compression are now quite well understood in terms of the MO arguments set out in Section 8.2.

Coordination numbers higher than six

These are unusual but by no means unknown for central atoms belonging to the 3d series. They are more common for the larger 4d and 5d central atoms, especially towards the left-hand side; large central atoms/ions allow the binding of more ligands because the interligand repulsions which tend to discourage high coordination numbers are less important. Coordination numbers in excess of six are the rule rather than the exception for complexes of the f block elements.

The principal geometries associated with high coordination numbers are shown in Fig. 8.4, and fall into two categories:

(a) Polyhedra which can be derived from simpler, regular polyhedra by placing additional ligands over triangular or square faces (capping).

(b) More regular polyhedra which cannot be described as in (a) above.

In category (b) we have the pentagonal bipyramid for seven-coordination; the cube, square antiprism and dodecahedron for eight-coordination; and the icosahedron for 12-coordination.

Seven-coordination can also be achieved by capping a triangular face of an octahedron, or a rectangular face of a trigonal prism. Seven-coordinate complexes exhibit no marked preference for any of the three known geometries. For example, in $Na_3[ZrF_7]$ the anion has pentagonal bipyramidal geometry, while in $(NH_4)_3[ZrF_7]$ it prefers the capped trigonal prism, for no obvious reason. The capped octahedron is much rarer than its two rivals.

In the case of eight-coordination, the cube is fairly rare; the compounds $Na_3[MF_8]$ (M = Pa, U, Np) constitute the best-established examples. As for the trigonal prism versus the octahedron, interligand repulsion can be reduced by distortion of the cube to yield the square antiprism or dodecahedron (this is best demonstrated by playing with models). These two are about equally common. For example, in the case of $Mo(CN)_8^{4-}$ both square antiprismatic and dodecahedral coordination are found in the solid state; in solution, the ion is stereochemically nonrigid (or fluxional) and appears to convert rapidly from one geometry to the other.

Nine-coordination is most commonly achieved by the tricapped trigonal prism, and is found in ReH_9^{2-} and $M(H_2O)_9^{3+}$ (M = lanthanide). Rarer is the capped square antiprism.

Higher coordination numbers (up to 16) are encountered but are neither common nor important, except in the cases of organometallic complexes involving ligands of high hapticity. The icosahedron is a very regular figure, and a distorted variant best describes the coordination geometry of $Ce(NO_3)_6^{2-}$, the nitrate ligands being bidentate.

8.4   Some thermodynamic aspects

This section is largely devoted to the results of thermodynamic studies of equilibria in aqueous solution involving hydrated cations $M^{n+}(aq)$, ligands and the complexes formed by these. Some of the thermodynamic properties of $M^{n+}(aq)$ ions have already been discussed

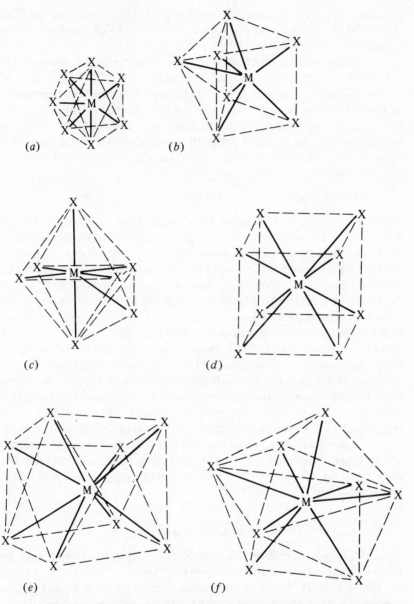

Fig. 8.4. Coordination geometries for seven- and eight-coordination: (a) pentagonal bipyramid, (b) capped trigonal prism, (c) capped octahedron, (d) cube, (e) square antiprism and (f) triangulated dodecahedron. Central atom–ligand bonds are shown by full lines; dashed lines indicate the edges of the polyhedral figure.

(Chapter 5); they, of course, are complex ions, and salts such as $NiSO_4.7H_2O$ (which contains the octahedral $Ni(H_2O)_6^{2+}$ ion) qualify as coordination compounds.

The reader may wonder at this point whether bond energies for coordination compounds can be tabulated and utilised in chemical arguments, in the same way as in Main Group chemistry. The complexity of most coordination/organometallic compounds makes this very difficult, and the ways in which the formation of a coordinate bond affects the bonding within the ligand renders the bond energy concept distinctly dubious. Consider, for example, the relatively simple case of hexacarbonylchromium(0). For the reaction:

$$Cr(g) + 6CO(g) \rightarrow Cr(CO)_6(g)$$

ΔH^0 is $-740\,kJ\,mol^{-1}$, leading to a mean Cr–CO bond energy of $123\,kJ\,mol^{-1}$. However, the coordination chemist is more interested in the Cr–C bond energy. This is expected to be greater than $123\,kJ\,mol^{-1}$ because π donation from Cr to CO effectively transfers electron density into the ligand π antibonding orbitals, weakening the C–O bond. Thus the figure of $123\,kJ\,mol^{-1}$ would represent the M–C bond energy if the carbon–oxygen triple bond remained unaffected, but a correction needs to be made since there is irrefutable evidence that the carbon–oxygen bond order is less than three in the complex. The Cr–C bond energy is then estimated to be about $360\,kJ\,mol^{-1}$. This kind of calculation is of dubious validity even for relatively simple systems, and is quite impracticable for most coordination/organometallic compounds. Most inorganic chemists prefer to assess the relative strengths of bonds in related compounds by means of spectroscopic measurements and crystallographically-determined bond lengths, rather than attempt calculations of absolute bond energies. However, progress is being made towards the determination of reliable M–M bond energies in cluster and other compounds (see Section 8.5).

Complex formation in aqueous solution

Consider a hydrated cation M and a ligand L (which may be neutral or anionic) in aqueous solution (for clarity and simplicity, charges are not specified). If the coordination number of M is 6, and if L is unidentate, we can form complexes ML_n ($n = 1$–6). The equilibria can be written in two ways:

(i) $ML_{n-1} + L \rightleftharpoons ML_n$ $k_n = [ML_n]/[ML_{n-1}][L]$
(ii) $M + nL \rightleftharpoons ML_n$ $K_n = [ML_n]/[M][L]^n$

The constants k_n are *stepwise formation constants*, while the K_n are *overall formation constants*. They are simply related by:

$$K_n = k_1 k_2 \ldots k_n$$

or:

$$\log_{10} K_n = \log_{10} k_1 + \log_{10} k_2 + \ldots \log_{10} k_n$$

Since formation constants are often very large, it is more convenient to adopt the logarithmic formation. (Stepwise and overall formation constants are often given the symbols K and β respectively.)

Let us now look at some representative results. Consider the situation where $M = Ni^{2+}(aq)$ and $L = NH_3(aq)$. The experimental values of k_n are given below:

n	1	2	3	4	5	6
$\log_{10} k_n$	2.79	2.26	1.69	1.25	0.74	0.03

Remember that we are not dealing with addition of NH_3 molecules to a naked cation; each step involves the displacement of a water molecule by NH_3. Evidently, the substitution of the nth H_2O by NH_3 becomes progressively more difficult with increasing n. Values of ΔH_n^0 and ΔS_n^0 for each step can be obtained, and it turns out that ΔH_n^0 is roughly constant (in the range -16 to $-18 \, \text{kJ mol}^{-1}$). A progressively less favourable entropy change ΔS_n^0 with increasing n is responsible for the variation of $\log k_n$. This can be understood (qualitatively at least) in terms of the statistical interpretation of entropy. The more ways there are of describing any change, the more positive will be ΔS^0 for the change. The first ammonia molecule to displace a water molecule from its site in $Ni^{2+}(aq)$, i.e. $Ni(H_2O)_6^{2+}$, has a choice of six positions. The last NH_3 whose coordination produces $Ni(NH_3)_6^{2+}$ has no choice at all. Thus ΔS_n^0 becomes less favourable as n increases.

A vast number of formation constants have been measured, and many correlations have been made in comparing their relative magnitudes. For a given M, formation constant data for a range of ligands will reflect its status as a hard or soft acid. Thus in the case of a soft acid such as Hg^{2+}, we find that the formation constants with halide ions increase in the order $F^- < Cl^- < Br^- < I^-$, while the reverse sequence is found for harder acids. With a hard base such as NH_3, there is a good correlation between k_n values and the radius of M^{n+}, suggesting that an ionic model has some value (if not absolute validity) in the interpretation of such data. Soft ligands, however, give rise to very high formation constants with the large, polarisable cations such as Hg^{2+} and Pb^{2+}. The high toxicities of these ions – a serious environmental problem in many places – arise from their strong binding to sulphur atoms in enzymes, often displacing other cations (e.g. Fe^{2+}, Cu^{2+}, Zn^{2+}) from their rightful places or otherwise disturbing the delicate three-dimensional structures of proteins and inhibiting enzyme action. The formation constant $\log k_1$ for some complexes between $Hg^{2+}(aq)$ and sulphur-donor (soft) ligands can be in excess of 40.

The chelate and macrocyclic effects

Chelating ligands tend to form thermodynamically more stable complexes than comparable non-chelating ligands. For example, in the $Ni^{2+}(aq)/en$ system (en = ethylenediamine $NH_2CH_2CH_2NH_2$) we find $\log_{10} K_3 = 18.4$, compared with $\log_{10} K_6 = 8.7$ for $Ni^{2+}(aq)/NH_3$. Thus for the equilibrium:

$$Ni(NH_3)_6^{2+}(aq) + 3en(aq) \rightleftharpoons Ni(en)_3^{2+}(aq) + 6NH_3(aq)$$

$\log_{10} K$ is equal to 9.7, which can be re-expressed as $\Delta G^0 = -55\,kJ\,mol^{-1}$. Further analysis leads to $\Delta H^0 = -12\,kJ\,mol^{-1}$ and $T\Delta S^0 = 44\,kJ\,mol^{-1}$ (at 25 °C). Thus the 'chelate effect' in this case is largely due to entropy, with some contribution from a favourable enthalpy term. ΔH^0 for such reactions is sometimes positive and sometimes negative, but it is usually close to zero and much smaller than the entropy term $T\Delta S^0$. Various attempts have been made to rationalise the sign and magnitude of ΔH^0 but it is so small that such efforts are probably inappropriate. The important point is that the entropy term leads to the greater stability of the chelate complex.

The conventional explanation runs as follows. A process in which the number of independent species increases is generally accompanied by an increase in entropy. For the reaction:

$$Ni(NH_3)_6^{2+}(aq) + 3en(aq) \rightarrow Ni(en)_3^{2+}(aq) + 6NH_3(aq)$$

we count seven species on the right-hand side and only four on the left-hand side, so a positive ΔS^0 is expected. This interpretation has been challenged for thermodynamic reasons relating to units and standard states. The detailed argument need not trouble us, but some authors prefer to interpret the chelate effect in a more pictorial way to obviate any objections from pedantic thermodynamicists. Imagine a solution containing (say) ammonia and ethylenediamine in roughly equal concentrations; the ligands are competing for coordination sites on M^{n+} ions. Once one end of an en ligand has attached itself to an ion, its other donor atom is necessarily close to the ion and is more likely to take up a second site than an ammonia molecule; in effect, the local concentration of —NH_2 groups is greatly increased in the immediate vicinity of the ion. This gives the chelating ligand an edge in the thermodynamic contest for coordination sites.

The coordination of a chelating ligand necessarily leads to the formation of rings; examples are shown in Fig. 8.5. Five- or six-membered rings are most common. Larger rings are rather rare, and it is often suggested

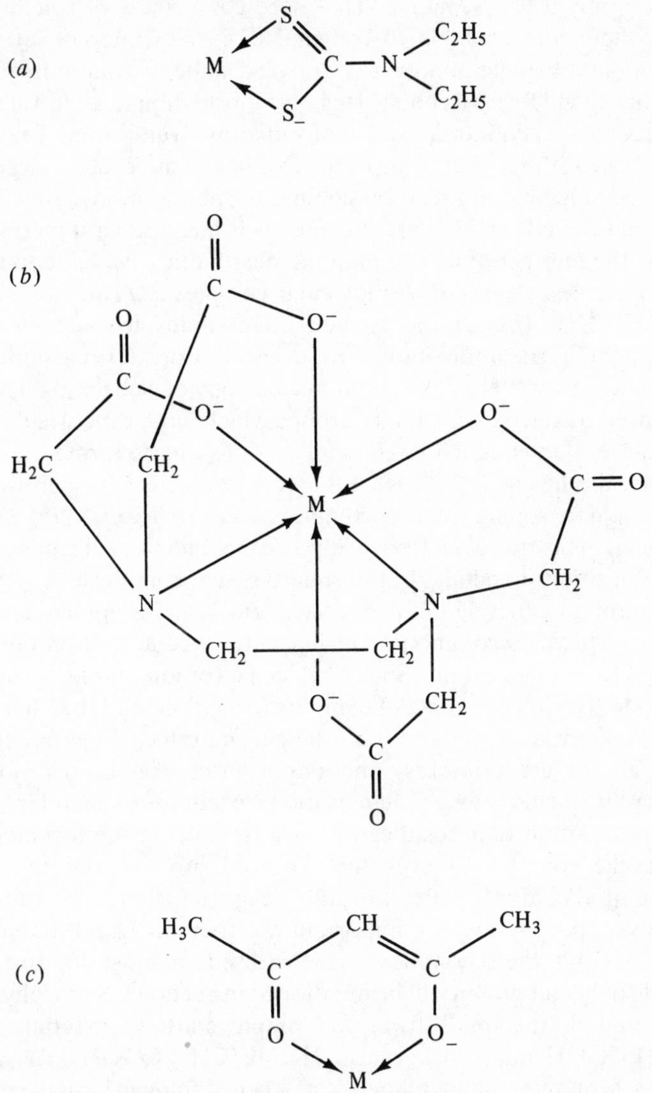

(a)

(b)

(c)

Fig. 8.5. Examples of complex formation by chelating ligands, showing (a) a four-membered ring with diethyldithiocarbamate (dtc), (b) formation of five-membered rings with ethylenediaminetetraacetate EDTA and (c) a six-membered ring with acetylacetonate (acac).

that steric effects, or 'ring strain', can be blamed. A better rationale comes from a consideration of rotational entropy. A molecule such as $NH_2(CH_2)_nNH_2$ can adopt a number of conformations; given the essentially free rotation which is permissible about C–C and C–N bonds, these differ by only a few $kJ\,mol^{-1}$. The more conformers of roughly equal energy there are available to a ligand molecule (interconvertible via rotations about single bonds), the greater will be its rotational entropy. When the ligand forms a chelate ring on coordination to a central atom, it may become locked into a fixed conformation; even if some flexibility is allowed, the rotational entropy of the free ligand molecule is largely lost. This effect appears on the debit column of the thermodynamic balance sheet, and for $NH_2(CH_2)_nNH_2$ ligands, will increase with increasing n, because the number of conformational possibilities available to the free ligand increases rapidly with molecular complexity. Thus for diamines $NH_2(CH_2)_nNH_2$ ($n > 2$) the chelate effect tends to fade away with increasing n, as the unfavourable rotational entropy term assumes more significance. However, if the rotational entropy of a chelating ligand can be reduced by introducing bulky groups which hinder the 'free' rotation about single bonds, much larger chelate rings can be formed. Thus, for example, the ligand $R_2P(CH_2)_{12}PR_2$ ($R = C(CH_3)_3$) can span two *trans* positions in a square planar complex, e.g. $IrCl(CO)L$ (cf. Vaska's compound in Section 3.2). Practically all other bidentate ligands can span only *cis* positions in octahedral or square planar complexes.

The term 'macrocycle' implies a 'large ring', and is applied to a cyclic ligand which can accommodate at its centre a central atom bonded to (usually) four donor atoms. Such a molecule (or ion) has little rotational entropy to lose in the course of complex formation, and the 'macrocycle effect' is often more marked than the chelate effect. Macrocyclic complexes are of great interest and importance. The preparation of a macrocyclic ligand – via condensation reactions of simpler organic molecules – often requires a cation as a 'template'; the formation of a macrocyclic complex induces the required product and the correct configuration/conformation. Porphin (Fig. 8.6) and its substituted derivatives (porphyrins) are important macrocyclic ligands in biological processes, and their complexes are utilised in most life-forms; the example of haemoglobin will be mentioned in Section 9.8. Porphyrins are synthesised in mammals from the simple starting materials glycine NH_2CH_2COOH and succinic acid $HOOC(CH_2)_2COOH$. The process involves a number of metalloenzymes (i.e. biological catalysts which contain atoms of relatively low electronegativity) and may well be an example of a template synthesis.

8.5   'Metal–metal' bonds and cluster compounds

Until the 1950s, it was supposed that atoms of the d and f block elements – in common with other atoms of relatively low electronegativity belonging to the Main Groups – would not form strong covalent bonds with their own kind. The reasons behind this prejudice are difficult

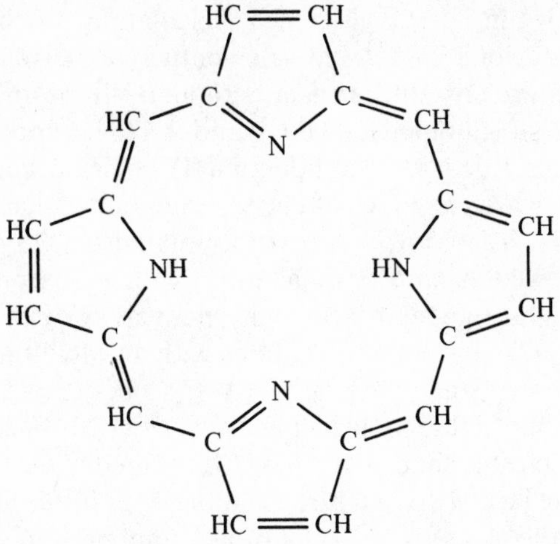

HC ═══ CH

Fig. 8.6. The porphin molecule. The NH hydrogen atoms are readily removed as protons and the resulting dianion acts as a quadridentate ligand.

to explain at this distance of time and the contemporary student may wonder why the confirmation of 'metal–metal' bonded systems c. 1960 aroused so much surprise. After all, the mercurous ion Hg_2^{2+} was thoroughly established by the 1930s.

The C–C bond in organic chemistry makes possible the formation of chains and rings. However, a feature of d block chemistry is the formation of M_n clusters ($n \geqslant 3$) in which each M atom is bonded to at least two others, as well as to other ligands. Clusters may be divided into two categories:

(a)   Halides of the early transition elements in relatively low oxidation states;

(b)   Carbonyl clusters, mostly formed by the later transition elements, again in low oxidation states.

Cluster compounds, and M–M bonds in general, tend to be particularly common among compounds of the 4d and 5d series elements, although they are by no means rare among the 3d elements.

Cluster halides

In Section 5.2, we noted that ZrF_3 and HfF_3 were unstable with respect to disproportionation to the tetrafluoride and the elemental substance, and that the chlorides, bromides and iodides of an element in a low oxidation state were less unstable than the fluorides to such disproportionation. In Group 6d (Cr, Mo, W), calculations of this kind suggest that MoF_3 and $MoCl_3$ are stable as essentially ionic solids, but $MoCl_2$, WF_3 and WCl_3 are unstable. In fact, although WF_3 is unknown, compounds having the stoichiometries $MoCl_2$ and WCl_3 are known. But they should not be described as molybdenum(II) chloride and tungsten(III) chloride. They are examples of cluster compounds, having the structures shown in Fig. 8.7. '$MoCl_2$' is best formulated as $[Mo_6Cl_8]Cl_4$. The cation may be viewed as an Mo_6 octahedron with a Cl atom triply bridging each of the eight triangular faces. Alternatively, we can see the cation as a cubic arrangement of eight Cl atoms with an Mo atom at the centre of each of the six square faces. In 'WCl_3', better written as $[W_6Cl_{12}]Cl_6$, we have one Cl doubly bridging each of the 12 edges of the W_6 octahedron. Other cluster halides based on M_6 octahedra are Nb_6Cl_{14}, Ta_6Cl_{15} and Nb_6F_{15}, the last of these a rare example of a cluster fluoride. M_6 clusters appear in the 4f series, in the curious compounds having the empirical formulae Gd_2Cl_3 and Yb_2Cl_3.

In all of these, the bonding evidently requires strong M–M bonding within the M_6 clusters, and the structure is not held together solely by M–Cl–M bridge bonds. The stoichiometries are difficult to rationalise, and full-blown MO treatments are necessary in order to rationalise the 'right to exist' of such a species. $PdCl_2$ and $PtCl_2$ both crystallise in two forms, one of which has the molecular structure M_6Cl_{12}. This differs from the $W_6Cl_{12}^{6+}$ ion depicted in Fig. 8.7 in that strictly planar MCl_4 units can be discerned; the WCl_4 moieties in 'WCl_3' are square pyramidal. It would seem that M–M bonding is not important in the Pd and Pt compounds, and the structure adopted is dictated by the stability of square planar four-coordination for these d^8 systems. The other form of MCl_2 (M = Pd, Pt) is a one-dimensional chain, again with square four-coordination about each M.

Bromides and iodides analogous to the chlorides discussed so far can be prepared; but Nb_6F_{15} is the only well-authenticated example of a fluoride containing an M_6 cluster. The propensity to disproportionation of lower fluorides is probably the main reason for the scarcity of cluster fluorides; M–M bonding rescues chlorides, etc. from this fate but is not enough to stabilise, e.g., WF_3 (although MoF_3 is known, having the ReO_3 structure with 6:2 coordination). Another factor is that M–M bonding appears to involve principally the M nd orbitals. In low oxidation states, these are sufficiently extended radially that nd–nd overlap is considerable at typical M–M distances. In higher oxidation states, the nd orbitals are contracted and M–M overlap is weak. The same effect can be achieved by having highly-electronegative atoms attached to the M atoms, so that nd–nd overlap is likely to be poor in cluster fluorides.

The redox behaviour of octahedral clusters and their coordination chemistry have been extensively studied.

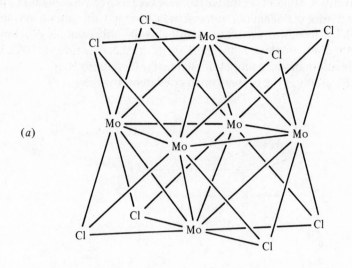

(a)

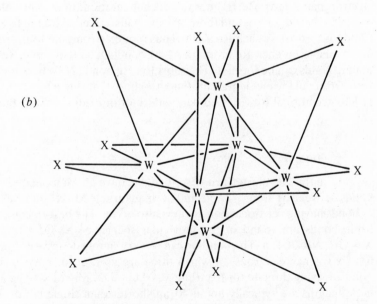

(b)

Fig. 8.7. M_6 cluster structures; (a) the $Mo_6Cl_8^{4+}$ ion in 'MoCl$_2$', and (b) the $W_6X_{12}^{6+}$ cation (X = Cl, Br) in 'WX$_3$'.

The other type of cluster halide is represented by Re_3Cl_9 and its derivatives. Students in the early 1960s were told of a compound $CsReCl_4$ which was diamagnetic; since it was assumed to contain tetrahedral $[ReCl_4]^-$ ions, this was cited as probably the only example of a low-spin tetrahedral complex. The report of its crystal structure in 1963 was a landmark in modern inorganic chemistry. It contains $Re_3Cl_{12}^{3-}$ ions, with the Re atoms at the corners of an equilateral triangle:

The bonding can be quite simply described in terms of two-centre, electron-pair bonds. Re(III) has the d^4 configuration; of its nine valence orbitals – five 5d, one 6s and three 6p – five will be needed for the five Re–Cl bonds formed by each Re atom. Thus four electrons and four orbitals are available on each Re atom for M–M bonding, so that three Re=Re double bonds can be formed. The short Re–Re bond (248 pm, compared with 276 pm in the elemental substance) is consistent with a bond order of 2. The structure of Re_3Cl_9 is similar, with chlorine bridges joining the Re_3 clusters.

Multiple M–M bonds

In the Re_3 cluster just described, we have an example of double bonds between M atoms. Triple and even quadruple M–M bonds (as well as bonds having fractional bond order) also occur. The best-documented triple bonds are found in the molecular species M_2X_6 (M = Mo, W; X = OR, NR_2, CR_3). The bonding can be very simply described. Each M has six valence electrons, of which three are used to form M–X single bonds, leaving three to form a triple M–M bond, $X_3M\equiv MX_3$. The M–M bond lengths are typically about 40 pm shorter than single bonds. (See also Section 8.2.)

Quadruple bonds are found in dimeric molecules Mo_2L_4 (L = RCOO) and the anions $M_2(CH_3)_8^{4-}$ (M = Cr, Mo, W). Among the best-known cases are $Mo_2Cl_8^{4-}$ and $Re_2Cl_8^{2-}$. In all of these, each M atom is bonded to four ligands L in the form of a square; the two L_4 squares are eclipsed, and are both perpendicular to the M–M axis:

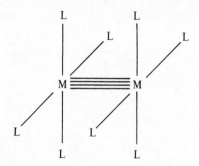

The M–M bonding can be described by considering only the nd orbitals. After dealing with the M–L bonding, four of these are available for M–M bonding. If we take as an example $Mo_2Cl_8^{4-}$, Mo(II) is a d^4 species, so that a quadruple bond can be formed. This comprises one σ, two degenerate π and one δ bond. The last of these is formed by the overlap of the d_{xy} orbitals (the M–L bonds define the x and y axes). For a more elaborate MO treatment, see more advanced texts. Although the δ overlap is relatively weak, there is now general agreement that the evidence – both experimental and theoretical – supports the idea of a quadruple bond. The Mo–Mo distance in $Mo_2Cl_8^{4-}$ is 214 pm, significantly shorter than Mo≡Mo.

A bond order of 3.5 is postulated for the paramagnetic $Tc_2Cl_8^{3-}$ ion.

Carbonyl clusters

Binuclear and polynuclear carbonyl complexes were among the first compounds containing M–M bonds to be prepared early in the present century, but it was not until the crystal structure determination of $Mn_2(CO)_{10}$, i.e. $(CO)_5Mn-Mn(CO)_5$, in 1957 that such bonds were authenticated in carbonyls. The study of polynuclear carbonyl complexes has been a major growth area of chemistry since then. Clusters containing more than 50 M atoms have been characterised. Some examples are illustrated in Fig. 8.8.

As well as neutral molecular carbonyls $M_n(CO)_m$, carbonyl anions are found. Hetero-atoms such as H, C, N and S appear in many cases. Hydrogen atoms – which may be present in both neutral and anionic species – are often difficult to locate with precision because they tend to be invisible to X-ray crystallographers in the presence of heavy atoms. They are sometimes terminal, for example in $Mn(CO)_5H$, an octahedral

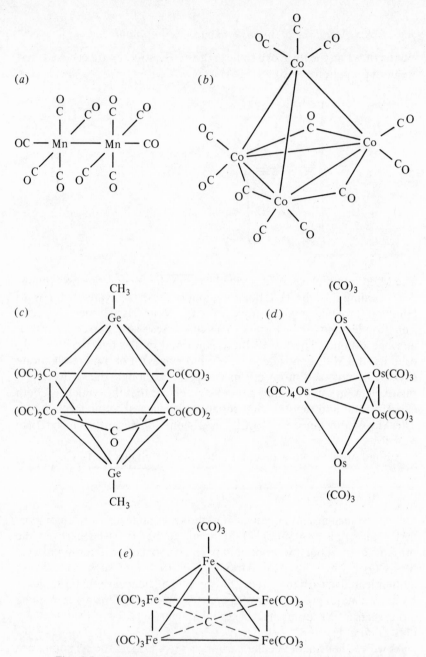

Fig. 8.8. Structures of some carbonyl complexes with M–M bonds, showing (a) a simple unbridged single bond, (b) a tetrahedral M_4 cluster with both terminal and bridging CO ligands, (c) a square planar M_4 cluster with two $GeCH_3$ groups completing an octahedron, (d) a trigonal–bipyramidal M_5 cluster with no bridging CO ligands and (e) a square pyramidal M_5 cluster with a carbon atom at the centre of the basal plane.

complex. They may also be doubly or triply bridging, i.e. an H atom may lie equidistant from two M atoms constituting an edge of the polyhedral figure which describes the cluster, or equidistant from the three M atoms which define a triangular face of, e.g., an octahedron. An H atom may also be encapsulated within the M_n cluster. A heavier hetero-atom may be similarly encapsulated – often at the centre of an octahedral M_6 cluster – or it may lie on the surface, bridging three or four M atoms. Ligands other than CO may be present.

At first glance, the structures and stoichiometries of polynuclear carbonyls and their derivatives are of baffling complexity. As we shall see in the next section, some very simple electron book-keeping devices help to bring some order from the apparent chaos.

8.6 Electron-counting and isolobal relationships

The development of organic chemistry in the latter part of the nineteenth century was greatly aided by elementary valency number theory. In classical structure determinations (by degradation and synthesis), it was important to know what structures, and what structural units, were possible. Organic chemists – long before crystallographic and spectroscopic methods became available – were able to write down structural formulae for some quite large and complex molecules, and they got most of them right.

Contemporary inorganic chemistry – at any rate in the area of organometallic and cluster compounds of the transition elements – is in a very different situation. A formidable arsenal of techniques for characterisation and structure determination is available, but the structural principles which govern which species are stable and which are not are only slowly emerging. Electron-counting procedures – which sometimes, to the uninitiated, may verge upon numerology – have proved extremely useful in elucidating these principles. Thus we have an 18-electron rule, which serves the same purpose as the octet rule for small Main Group atoms. The isolobal concept develops analogies between organic and inorganic chemistry.

The 18-electron rule (or EAN rule)

The octet rule (discussed in Section 6.1) is nearly always followed by C, N, O and F atoms; it is very often followed by other Main Group atoms. This eight-electron rule – of inestimable value to organic chemists – arises from the fact that these atoms have four valence orbitals (one ns, three np). If all of these are being used in bonding, or for the accommodation of lone pairs, the atom will necessarily have an octet configuration. Does an analogous rule apply to compounds of the d block elements?

It was recognised at an early stage in the development of the electronic theory of bonding that many complexes of the transition elements do obey an 'effective atomic number (EAN) rule'. Consider, for example, $Co(NH_3)_6^{3+}$. Cobalt(III) has the d^6 configuration. If we suppose that each NH_3 ligand contributes two electrons to the bonding by donation of its lone pair, we arrive at a total of 18 electrons in which the Co atom has at least a share, giving it an effective atomic number equivalent to the noble gas Kr. This was thought to be analogous to the Ne configuration attained by C, N, O and F in most of their compounds. The EAN rule was by no means universally obeyed. Yet, it was noticed in the 1920s that a high proportion of complexes do obey the rule and this remains true today.

We can now explain the restricted applicability of the EAN rule by means of VB, CF and MO arguments. The Co atom in $Co(NH_3)_6^{3+}$ has nine valence orbitals at its disposal – five 3d, one 4s and three 4p. Six of these are used for bonding (d^2sp^3 hybridisation, in VB language); since Co(III) has a low-spin d^6 configuration, the three nonbonding 3d orbitals are just filled. In MO language, a d block atom will have an 18-electron configuration, and will obey the EAN rule, if all bonding and nonbonding MOs are filled, no σ antibonding MO is occupied, and no valence orbital is left wholly or partially unemployed.

It is easy to see why the rule is regularly violated, and to identify the circumstances in which it is likely to be followed. $Co(NH_3)_6^{3+}$ can be reduced to $Co(NH_3)_6^{2+}$; this has the high-spin d^7 configuration and is reasonably stable to oxidation, disproportionation etc. Although we now have a $(t_{2g})^5(e_g)^2$ configuration, the e_g level is only weakly antibonding and its occupation is tolerated. Octahedral Ir(III) complexes, however, have no tendency to undergo reduction to Ir(II), which is a very rare oxidation state. In 4d and 5d complexes, the splitting parameter Δ_0 is much larger than for analogous 3d complexes; this is attributable to the inherently better overlap characteristics of 4d and 5d orbitals, so that the e_g level is now quite strongly antibonding. Octahedral complexes of the 4d and 5d elements where the central atom has more than six d electrons are very rare, since occupation of e_g has a powerful destabilising effect.

Octahedral complexes where the central atom has an incomplete t_{2g} level are common enough in all three transition series, unless at least some of the ligands are π acceptors, such as CO, PR_3 etc. The t_{2g} orbitals are stabilised by π back-bonding, becoming quite strongly-bonding MOs. In the absence of such interactions, the t_{2g} level is essentially nonbonding. But there will naturally be a tendency to fill all bonding MOs. Thus octahedral complexes with π acceptor ligands nearly always have a d^6 central atom which is equivalent to saying that the 18-electron or EAN rule is obeyed.

Other geometries than octahedral will also obey the rule – given π acceptor ligands – if nine bonding MOs can be constructed, although VB theory is often adequate. For example, in a tetrahedral complex we would invoke sp³ hybrids on the central atom to deal with the σ bonds. Given a full set of π acceptor orbitals on the ligands, all five d orbitals can form π bonds, so that 18 electrons are required. In MO language, we have five bonding MOs of predominantly d character. This implies a d^{10} configuration, which is always associated with tetrahedral geometry when the 18-electron rule is obeyed. Likewise, trigonal bipyramidal complexes where the central atom has the d^8 configuration will obey the rule; here (in VB terminology) we need dsp³ hybrids to form the σ bonds between the central atom and the ligands, and the remaining four d orbitals can all form π bonds. In short, nine bonds can be formed requiring 18 electrons. In MO language, nine bonding MOs are constructed.

In the application of the 18-electron rule, it is convenient to ignore the formal oxidation number on the central atom, which in many cases is ambiguous or indeterminate. The sum of the number of valence electrons on the neutral central atom, plus the number of ligand electrons involved in bonding to the central atom, should (after adjustment for net charge in the case of an ion) add up to 18. Thus H, CH_3, Cl etc. are one-electron ligands. If we wished to assign an oxidation number to the central atom in, e.g., $Co(CO)_4H$, we would consider an H^- ion as an electron-pair donor. But since we propose to count nine electrons in respect of the Co atom, and not eight for Co^+, the H contributes just one electron. Ligands such as CO and PR_3 – neutral molecules with just one lone pair available for bonding to a central atom – are two-electron ligands. Three-electron ligands include NO and η^3-allyl CH_2CHCH_2. (For the purpose of oxidation state assessment, NO is regarded as being bound as NO^+, isoelectronic with CO. But in electron-counting, we are looking at the three-electron ligand NO.) The cyclopentadienyl ion $C_5H_5^-$ is deemed to contribute its six π electrons to the bonding in its complexes, but for present purposes we are dealing with the five-electron radical C_5H_5. We now consider some examples of the application of the rule.

$Cr(CO)_5H^-$ The tally is as follows:

Valence electrons on Cr atom: 6 ($3d^54s$)
Contribution from 5 CO   : 10
Contribution from H   : 1
Negative charge   : 1

Total   : 18

$Mn_2(CO)_{10}$ Each Mn atom is bonded to 5 CO groups (see Fig. 8.8). Thus it has a share in $7 + (5 \times 2) = 17$ electrons, one short of the required 18. This is remedied by forming a single Mn–Mn bond, just as the 7-electron CH_3 radical will tend to dimerise to give C_2H_6 in which the C atoms obey the octet rule.

$(C_5H_5)Fe(CO)_2^-$ This anion is of some importance as a strong nucleophile. We count:

8(FE atom) + 4(2CO) +5 (C_5H_5) + 1 (charge) = 18.

The 16- and 14-electron rules

Like most rules in bonding theory and structural chemistry, the 18-electron rule is often violated, even among the nonclassical complexes containing π acceptor ligands where, we have suggested, it should apply. Square planar complexes follow a 16-electron rule; examples include Wilkinson's catalyst $Rh(PPh_3)_3Cl$ (see also Section 9.8) and the vast number of square Pt(II) complexes.

The best way of looking at the 16-electron rule is to recall the analogy between the 18-electron rule and the eight-electron rule followed by many Main Group species. As explained in Section 6.1, the octet rule is followed if all four valence orbitals – one ns and three np – of a Main Group atom are fully employed, either for bonding or for the accommodation of nonbonding pairs. In a square complex ML_4, only eight of the nine M valence orbitals are accounted for if – using VB language for the moment – we use dsp^2 hybrids for σ bonding and fill the four remaining nd orbitals. An extra pair of electrons would destabilise the system, because they would go into the strongly antibonding MO, the counterpart of $nd_{x^2-y^2}$ in CF theory (here we have managed to invoke three theories in two sentences; cf. Section 8.2). The 'unemployed' orbital is $(n + 1)p_z$, i.e. the p orbital whose lobes are directed perpendicular to the ML_4 plane. This orbital can in fact take part in the bonding via overlap with a filled ligand π orbital (cf. the apparent adherence to a 'sextet' rule of boron in, e.g., BCl_3, R_2NBR_2 discussed in Section 6.3). It is noteworthy that in the vast majority of 16-electron square complexes, at least one of the ligands is a π donor such as Cl^-. But the electrons involved are not considered in the electron count. For example, a Cl atom (or Cl^- ion) will have a filled 3p orbital of suitable symmetry for π overlap with p_z; but non-bridging Cl counts as a one-electron ligand (or Cl^- as a two-electron ligand). In our tally, we count only:

(a)   ligand electrons donated to the central atom via σ overlap, and
(b)   central atom electrons which are donated to the ligands via π back-bonding, or which are in stable nonbonding d orbitals.

Linear complexes where the central atom (e.g. Cu(I), Ag(I) or Au(I)) has a d^{10} configuration follow a 14-electron rule. Here, the σ bonding can be described by using sp hybrids, so that two np orbitals on the central atom can be regarded as out of action, and we require 14 electrons to fill up the d subshell and provide for two bonds.

Electron-counting in polynuclear carbonyls

The 18-electron rule can be applied – with some adjustments – to the structures and stoichiometries of polynuclear carbonyl complexes. To illustrate the principles, let us go back to the simple case of $Mn_2(CO)_{10}$. We found that the Mn atom was one electron short of 18 and that formation of the Mn–Mn bond – giving each Mn atom a share in an extra electron – would make up the deficit. In the general case of an M_n cluster, we can say that N, the number of two-centre, electron-pair M–M bonds, is given by:

$$N = \tfrac{1}{2}(18n - P)$$

where P is the conventional electron count, the sum of M and ligand valence electrons. Here are some examples; the structures of the complexes cited are shown in Fig. 8.8.

$Co_4(CO)_{12}$ Here, P is found to be $(4 \times 9) + (12 \times 2) = 60$, so that $N = 6$. A tetrahedral arrangement of Co atoms has six edges, and hence six M–M bonds. This is indeed the structure.

$Co_4(CO)_{11}(GeMe)_2$ The GeMe group is a three-electron ligand, with the methyl group taking up one of the four Ge valence electrons. Thus P is 64 and $N = 4$. The most plausible figure whereby four Co atoms form just four Co–Co bonds is a square plane, as observed experimentally.

$Os_5(CO)_{16}$ N is found to be 9 for this complex, and the only five-vertex figure with nine edges is a trigonal bipyramid; the observed structure has a slightly-distorted trigonal bipyramidal arrangement of Os atoms.

$Fe_5(CO)_{15}C$ Remembering to count four electrons as a contribution from the carbido-C atom (which sits in the middle of the basal plane of a square pyramidal Fe_5 cluster), we obtain $N = 8$, consistent with the eight-sided figure.

An alternative structure for an M_5 cluster, consistent with $N = 8$, is the edge-bridged tetrahedron. Such a structure is found in the case of $H_2Os_5(CO)_{16}$, for which the electron count is the same as for $Fe_5(CO)_{15}C$. There is no simple principle to distinguish between the two.

Other M_5 possibilities are shown below:

(a)   (b)

The 'hinged butterfly' structure (a) is exhibited by $Ru_5C(CO)_{16}$, while the 'bowtie' structure (b) is found for $Os_5(CO)_{19}$. These have N values of 7 and 6 respectively, consistent with the number of M–M bonds in each case.

When we come to M_6 clusters, the 18-electron rule and the concept of simple two-centre, electron-pair bonds reveal inadequacies. Consider $Os_6(CO)_{18}$, for which N is found to be 12. This is consistent with an octahedral Os_6 cluster, having 12 edges. As it turns out, $Os_6(CO)_{18}$ has an M_6 skeleton which can be described as a capped trigonal bipyramid (or, alternatively, as a bicapped tetrahedron, since a tetrahedron can be changed into a trigonal bipyramid by placing a fifth atom above one triangular face). This figure also has 12 edges, so that the structure is still consistent with the 18-electron rule and $(2c, 2e)$ M–M bonds. However, the anion $Os_6(CO)_{18}^{2-}$ does have an octahedral Os_6 cluster, despite the fact that N is now only 11. What structure might we expect for an M_6, $N = 11$ cluster? An obvious possibility would be a square pyramid M_5, with a sixth M atom capping one of the triangular faces, to yield three additional M–M bonds. This structure is exhibited by $H_2Os_6(CO)_{18}$ which, of course, is isoelectronic with $Os_6(CO)_{18}^{2-}$; no simple electron-counting theory could explain the difference.

The apparently anomalous structure of $Os_6(CO)_{18}^{2-}$ can be explained by MO arguments. In order to construct an octahedral M_6 cluster, using $(2c, 2e)$ bonds, we would require four orbitals of suitable symmetry and four electrons on each M atom. An MO calculation might be expected to yield 12 bonding and 12 antibonding MOs; but, in fact, we obtain 13 stable (not necessarily all bonding) MOs from such a calculation. Thus an octahedral cluster demands 26 electrons, not 24, for M–M bonding. This means that P, the total number of M and ligand valence electrons, is 86 for octahedral clusters and not 84 as might have been expected. A large number of M_6 systems conform to this '86-electron' count. Examples include $Rh_6(CO)_{16}$, $Fe_6C(CO)_{16}^{2-}$ and $Ru_6N(CO)_{16}^{-}$.

More elaborate electron-counting procedures have been established, and these are usually more appropriate for the rationalisation of high-nuclearity clusters. An approach which was originally applied to complex boranes has been extended to carbonyl clusters; and a number of topological theories have been described. Although there is some way to go, the structures, stoichiometries and relative stabilities of carbonyl clusters are now fairly well understood. As in other areas, however, experiment is still far ahead of theory.

Isolobal relationships

The isolobal concept was largely developed in the 1970s by R. Hoffmann, who won the 1981 Nobel Prize for Chemistry. His Nobel Lecture was entitled: 'Building Bridges between Inorganic and Organic Chemistry', a fitting description of isolobal relationships. A fragment or radical in which a d block atom is bonded to a number of π acceptor ligands (such as CO, PR_3, C_5H_5) is said to be isolobal with a simple organic radical such as CH, CH_2 or CH_3 if the two fragments have similar bonding capabilities. These are assessed by means of MO calculations and inspection of the resulting 'frontier orbitals', i.e. the highest occupied and lowest unoccupied MOs (HOMO and LUMO) whose characteristics will determine the capacity of the fragment to form further bonds. An isolobal connection is established if the HOMOs and LUMOs are similar in energy, shape and symmetry. It is often possible to predict such relationships by simple electron-counting procedures. For example, consider the fragment $Mn(CO)_5$. Counting seven valence electrons on Mn and two for each CO, the Mn is one electron short of 18. This is analogous to CH_3, where the C atom is one electron short of its octet. Both behave as univalent radicals, attaining the required electron counts by formation of a single bond. Similarly, $Mn(CO)_5^+$, and the isoelectronic $Cr(CO)_5$, are isolobal with CH_3^+. Other 17-electron species isolobal with CH_3 include $Co(CO)_4$, $Fe(C_5H_5)(CO)_2$ and $Cr(C_5H_5)(CO)_3$. Since Au tends to follow a 14-electron rule, the 13-electron group $AuPPh_3$ is also isolobal with CH_3 (and with the 17-electron groups mentioned above). Sixteen-electron groups such as $Fe(CO)_4$, $Mn(C_5H_5)(CO)_2$ and $Cr(C_5H_5)(CO)(NO)$ are isolobal with CH_2. The 15-electron groups $Co(CO)_3$, $Ni(C_5H_5)$ and $Cr(C_5H_5)(CO)_2$ are isolobal with CH.

The analogies are far from perfect. For example, CH dimerises to give acetylene HC≡CH but the isolobal $Co(CO)_3$ does not dimerise. However, $Co(CO)_3$ does behave as a trivalent radical. The cluster carbonyl $Co_4(CO)_{12}$ (Fig. 8.8) can be seen as analogous to tetrahedrane C_4H_4 (note, however, that three of the CO groups are bridging). The compounds $Co_n(CO)_{3n}(CR)_{4-n}$ ($n = 1,2,3$) have been prepared, all containing tetrahedral Co_nC_{4-n} clusters.

Isolobal relationships have done much to place a branch of inorganic chemistry – complexes of d block atoms with π acceptor ligands – on the same footing as organic chemistry, in the sense that one can determine at a glance whether a given structure has a 'right to exist'. As well as helping to rationalise much existing chemistry, isolobal considerations are proving valuable in the strategic thinking of synthetic chemists. For example, consider the species (i) and (ii) below:

$$[(CO)_5Cr \overset{H}{-\!\!-\!\!-} Cr(CO)_5]^- \qquad (CO)_5Cr \overset{H}{-\!\!-\!\!-} AuPPh_3$$

         (i)   (ii)

The anion (i) is isoelectronic with $Mn_2(CO)_{10}$ and thus can be said to obey the 18-electron rule. It was the first of many species found to contain an M–H–M bridge, analogous to the boranes. The Cr–Cr distance of 340 pm is rather large for a single bond (cf. 293 pm for the Mn–Mn bond in $Mn_2(CO)_{10}$). One way in which this anion has been viewed is as an adduct between $Cr(CO)_5H^-$ and the 16-electron species $Cr(CO)_5$ which, being isolobal with CH_3^+, acts as an electrophile and accepts an electron pair. Now $AuPPh_3$ is isolobal with CH_3, and $AuPPh_3^+$ with CH_3^+. This suggested to a team of synthetic chemists that the neutral molecule (ii) – a Lewis acid/Lewis base adduct of $Cr(CO)_5H^-$ with $AuPPh_3^+$ – should be stable. The compound was duly prepared, by the reaction between $Cr(CO)_5H^-$ and $Au(PPh_3)Cl$.

8.7   Further reading

All the standard texts listed in Section A.3 of the Appendix have lengthy sections on coordination and organometallic compounds of the transition elements. See also the books listed in Sections A.10 and A.11. Bell (1977) gives the fullest account of the chelate effect. Cotton and Wilkinson (1988) (Section A.3) is best for the catalytic applications of complexes. Jolly (1984) (Section A.3) discusses electron-counting in polynuclear carbonyls in some depth.

9

Inorganic reactions and their mechanisms

9.1 Introductory remarks

In this chapter we try to classify the more important types of reactions encountered in inorganic chemistry, and describe some of their mechanisms. The emphasis is placed upon the principles which determine the stability or instability, existence and nonexistence of inorganic substances from the viewpoint of the ease or otherwise of preparing a compound, and the tendency a compound – once prepared – may have to react spontaneously to give other products. Both thermodynamic and kinetic considerations are obviously involved here. The division of material between this chapter and the next has not been easy, and there is inevitably a good deal of overlap. Coupling reactions, which might have deserved a section in this chapter, are discussed in Sections 10.5 and 10.6.

Organic chemists are usually concerned with homogeneous – i.e. single-phase – reactions of molecular substances in inert solvents. The classification of such reactions and their mechanisms is relatively simple. Inorganic chemists, however, have to deal with reactions involving gases, solids and liquids/solutions and often have to consider heterogeneous equilibria. In solution, the solvent may play an active part, both in the thermodynamic and kinetic aspects of the reaction. Nor is the inorganic chemist dealing almost exclusively with molecular substances. For these reasons, inorganic reactions and their mechanisms are not susceptible to any really satisfactory scheme of classification. Inorganic chemistry has a long way to go before its practitioners can emulate their organic colleagues in armchair strategies for synthesis. Serendipity and 'green fingers' still play an important part; but our knowledge of inorganic reaction mechanisms is expanding rapidly, and only a very brief sketch can be presented here.

9.2 Donor–acceptor reactions

One of the most common types of reaction is that between A and B where A can be viewed as an electron-pair acceptor, and B an electron-

319

pair donor. In VB language, the bond formed between A and B arises from the overlap of a filled orbital on B with an empty orbital on A: a dative or coordinate bond. A simple example is:

$$BCl_3 + PH_3 \rightarrow Cl_3B \leftarrow PH_3$$

PH_3 is classed as a Lewis base, having a lone pair localised on the P atom, while BCl_3 has an empty valence orbital – a boron 2p orbital, directed perpendicular to the molecular plane – which, if not wholly unemployed (see Section 6.3), is certainly underemployed. The resulting compound can be classed as a coordination compound or complex, and is often called a Lewis acid/Lewis base adduct. Reactions of this type can be studied in the gas phase or in solution. In the former case, and in the latter as well if the donor and acceptor are molecular species in an inert solvent, the question of the reaction mechanism is trivial. The kinetics, as might be expected, are second order with a straightforward bimolecular rate-determining step in which donor and acceptor simply add together. Interest in such reactions is focussed more upon the thermodynamics. A very useful analysis depends upon an empirical formula proposed by Drago, whereby ΔH^0 for a reaction between an acceptor A and a donor B is given by:

$$-\Delta H^0 = E_A E_B + C_A C_B$$

The parameters E_A and E_B are deemed to represent the tendencies of A and B respectively to undergo reaction via electrostatic interaction, where C_A and C_B represent the propensities of A and B towards covalent bonding by formation of a coordinate bond. The equation is valid only for reactions between neutral molecules, in the gas phase or in an inert, non-polar solvent. It is useful in predicting the enthalpy change for such a reaction, and the parameters (arguably) provide some insights into the nature of the bonding in adducts. As might be expected, donors/acceptors having highly polar bonds, and hence large fractional charges on the atoms involved in the formation of adducts, have large E values. A large value of C_A tends to be associated with a Lewis acid having a low-energy empty orbital, while a Lewis base having a filled donor orbital lying relatively high in energy will tend to have a large C_B. In other words, a high-energy HOMO and a low-lying LUMO (see Section 7.2) lead to large values of C_B and C_A respectively. A strong covalent interaction occurs where the energies of the donor and acceptor orbitals are about equal. A useful example of how the parameters vary among a series of related molecules is afforded by the bases NH_3, CH_3NH_2, $(CH_3)_2NH$ and $(CH_3)_3N$. As H is replaced by CH_3, the fractional negative charge on the N atom decreases, C being more electronegative than H. This leads to a decrease in E_B as we go from NH_3 to $(CH_3)_3N$. C_B increases with increasing substitution, probably because the lone-pair orbital is raised in energy by successive alkyl substitution. Experimental evidence for this comes from the ionisation potential (measured by photoelectron spectroscopy) which falls from $10.2\,eV$ for NH_3 to $7.8\,eV$ for $(CH_3)_3N$. In other words, the energy of the nitrogen lone-pair orbital (the HOMO) increases from $-10.2\,eV$ to $-7.8\,eV$ as the H atoms are replaced by methyl.

It is tempting to correlate these E and C parameters with the concept of hard and soft acids and bases (see Section 8.1). This is difficult, however. The idea of 'hardness' implies a single-parameter scale (although no satisfactory numerical scale of hardness/softness has yet been established) while the Drago approach requires two parameters for each acid or base (acceptor or donor).

Many reactions which could be viewed in terms of donor/acceptor functions involve the intimate participation of the solvent; these are dealt with in Section 9.3. Others are strictly substitution reactions, e.g.:

$$Zn^{2+}(aq) + Cl^{-}(aq) \rightarrow ZnCl^{+}(aq)$$

Here, we can imagine the formation of a coordinate bond between Cl^- and Zn^{2+}, but a water molecule is displaced from the coordination sphere of the cation. Reactions of this kind are covered in Section 9.4.

An important type of reaction is one between two or more molecules to form an adduct in which coordinate bonds are formed between ligand atoms on one Lewis acid (acceptor) molecule and the central atom on another, e.g.:

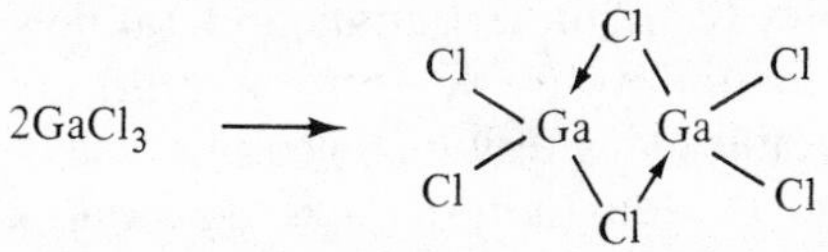

The formation of polymeric structures such as those of $BeCl_2$, $CdCl_2$, $CrCl_3$ etc. (see Section 3.3) can be regarded as reactions of this type. Adducts may be formed between (or among) two or more Lewis acids, e.g.:

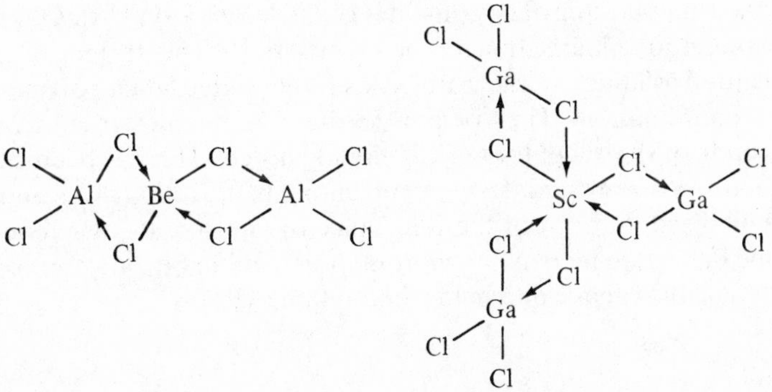

In both of these, the Group 13 atom has tetrahedral four-coordination. The other central atom has the same coordination as in the binary chloride. As discussed in Section 3.3, beryllium(II) chloride has a one-dimensional chain polymeric structure under ordinary conditions. This can be regarded as a donor/acceptor complex, like Ga_2Cl_6. Likewise, $ScCl_3$ (which has the $CrCl_3$ structure; see Section 3.3) can be regarded as a polymeric complex formed by $ScCl_3$ molecules via donor/acceptor interactions. When we speak of, say, beryllium(II) chloride as a Lewis acid, we are perhaps thinking of the gas-phase molecule $BeCl_2$ and the formation of molecular complexes such as $[Be(NH_3)_2Cl_2]$. However, solid $BeCl_2$ does react with Lewis bases to form complexes. Such reactions are evidently substitutions rather than additions, and to speak of $[Be(NH_3)_2Cl_2]$ as an adduct could be misleading.

What factors favour the formation of polynuclear bridged species like these from simple molecules? As with other types of adduct (e.g. $Cl_3B \leftarrow PH_3$), we have to consider the tendencies of the atoms involved in the formation of coordinate bonds to increase their coordination numbers. The fact that additional bonds are formed on adduct formation has to be set against the increased promotion energy which may accompany the change in hybridisation, and the increased repulsion among nonbonded atoms. These considerations are illustrated by looking at binary halide molecules of Groups 2 and 13 and their relative propensities towards dimerisation or polymerisation. In Section 3.3, we looked at the case of $BeCl_2$. Although the mean Be–Cl bond energy in $Be_2Cl_4(g)$ or $BeCl_2(s)$ is less than in $BeCl_2(g)$, dimerisation and (at low enough temperatures to outweigh the unfavourable entropy term) polymerisation are energetically favourable. Magnesium(II) chloride has the $CdCl_2$ structure with octahedral six-coordination of each Mg atom. This two-dimensional polymer formation reflects the higher coordination numbers attainable by Mg, given its greater size and the availability of 3d orbitals. In Group 13, the boron trihalides do not dimerise (but see Section 9.5). This may be attributed to the stabilisation of the planar monomer by p_π–p_π bonding (lost upon dimerisation and the change to sp^3 hybridisation) and to the importance of repulsive terms between halogen atoms bonded to the small B atom. The trihalides of the larger atoms Al, Ga, In and Tl all undergo dimerisation or polymerisation. The trifluorides may be regarded as ionic, or alternatively as three-dimensional polymers having $6:2$ coordination. The trichlorides have the same structure as $CrCl_3$ (Section 3.3), except for Ga_2Cl_6 already noted. This has been cited as an example of the 'middle element anomaly' (Section 4.1) and can be rationalised in terms of the reluctance of Ga (like its neighbours in the same Period) to increase its coordination number beyond four, compared with its allies above and below in the same Group.

Ion-transfer reactions

The donor/acceptor reactions considered so far all involve the formation of new bonds, using erstwhile nonbonding electron pairs. A closely-related type of reaction is one in which an ion is transferred from one molecule to another, to form an ionic solid, e.g. the proton-transfer reaction:

$$NH_3(g) + HCl(g) \rightarrow NH_4Cl(s)$$

Ion transfer may determine the structure of a compound as ionic or molecular, e.g.:

$$2PCl_5(g) \rightarrow PCl_4^+PCl_6^-(s)$$

The ionic structure adopted by phosphorus(V) chloride at room temperature can be rationalised as the product of Cl^- transfer between PCl_5 molecules. A high proportion of known noble gas compounds contain cations such as XeF^+.

The formation of an oxoacid salt from the binary oxides can be seen as an O^{2-} transfer reaction. For example, the equation:

$$CaO(s) + CO_2(g) \rightarrow CaCO_3(s)$$

involves the formation of the ionic solid $Ca^{2+}CO_3^{2-}$ from the ionic solid $Ca^{2+}O^{2-}$ and the molecular gas CO_2. In effect, the formation of one mole of $CaCO_3$ requires the transfer of one mole of oxide ion from CaO to CO_2. Reactions of this type always involve the transfer of O^{2-} from a basic oxide (which can be viewed as containing discrete oxide ion) to an acidic oxide (usually molecular but often polymeric).

We now consider the factors which favour ion transfer between *molecules*, with particular reference to halide ion transfer. Thus we are concerned with the reaction:

$$AX_m + BX_n \rightarrow AX_{m-1}^+ BX_{n+1}^-$$

The principal exothermic term in a stepwise thermochemical analysis (such as we found useful in Chapter 5) will be the lattice energy of the product. This is not readily obtainable experimentally (unlike the lattice energies of simple ionic solids) and its magnitude is not amenable to any simple analysis. As we shall see a little later, a purely ionic description of such products is often inappropriate anyway. Let us focus attention on the ease of formation of AX_{m-1}^+ and BX_{n+1}^-. The removal of X^- from AX_m will be favoured by:

(a)   A weak A–X bond (which must be broken to form AX_{m-1}^+).

(b)   A small atom A and a large halogen atom X, so that the lowering of the coordination number of A is accompanied by a significant relief in X–X repulsion.

(c)   A change in hybridisation of A in going from AX_m to AN_{m-1}^+, which leads to a large reduction in promotion energy.

For the formation of BX_{n+1}^-, the converse will hold: the acceptance of X^- will be favoured by a strong B–X bond, a large atom B and small X, with little or no increase in promotion energy required to achieve the valence state for the higher coordination number. It will be apparent that for the reaction:

$$2AX_m \rightarrow AX_{m-1}^+ AX_{m+1}^-$$

there is a conflict of interest between cation and anion formation.

These considerations are well illustrated by looking at the Group 15 pentahalides. The bonding in the trigonal bipyramidal molecules has been discussed in Chapter 6, and the postulate of nd orbital participation helps to rationalise their occurrence. No nitrogen(V) halides are known, although NF_5 may have been detected as a short-lived species at low temperatures. The cation NF_4^+ occurs in solids, prepared by reactions such as:

$$NF_3(g) + BF_3(g) + F_2(g) \rightarrow NF_4^+ BF_4^-$$

and we can speculate that NF_5 is formed as a transient intermediate. This should be quite a strong F^- donor, given its very low bond energy. The anion NF_6^- is not known. At room temperature, phosphorus(V) fluoride, chloride and bromide are stable, but they have very different structures. PF_5 is a molecular gas. PCl_5 is a solid consisting of PCl_4^+ and PCl_6^- ions, while PBr_5 is $PBr_4^+ Br^-$. Molecular PCl_5 and PBr_5 can be obtained in the gas phase, although the pentabromide is thermally rather unstable. PI_5 has been obtained at low temperatures and is formulated as $PI_4^+ I^-$; the molecule is unknown. The structural observations can be rationalised as follows. The tendency of a PX_5 molecule to form PX_4^+ should increase in the order $PF_5 < PCl_5 < PBr_5 < PI_5$; with increasing size of the halogen atom X, the P–X bond energies decrease and the increasing bulk of the X atom favours donation of X^-. Moreover, as discussed in Section 6.1, the accessibility of phosphorus 3d orbitals for bonding decreases from PF_5 to PI_5; since these are not required in PX_4^+, the formation of the latter is favoured for the heavier halogens. Conversely, the tendency to accept a halide ion will follow the order $PF_5 > PCl_5 > PBr_5 > PI_5$. Thus although the PF_5 molecule is a powerful fluoride acceptor – the PF_6^- ion is well known in a large number of solids – PF_4^+ is very rare, and phosphorus pentafluoride is molecular in all phases. The PCl_5 molecule is a better halide donor but a poorer halide acceptor than PF_5; the fact that its donor and acceptor capabilities are well-matched accounts for the formation of

$PCl_4^+PCl_6^-$. PBr_5 and PI_5 are very poor X^- acceptors, but excellent donors; hence the formation of $PX_4^+X^-$.

Turning to arsenic in Group 15, we find molecular AsF_5, similar to PF_5, and $AsCl_5$, unstable above $-50\,°C$ but apparently molecular in all phases. It would appear that, given the generally lower As–X bond energies, the formation of $AsCl_4^+$ is favoured compared with PCl_4^+, and compounds such as $AsCl_4^+AsF_6^-$ were known long before the elusive $AsCl_5$ was finally prepared. The $AsCl_6^-$ ion is apparently unknown.

Antimony(V) fluoride is a viscous liquid at room temperature. The SbF_5 molecule is a powerful fluoride acceptor, but has no tendency to donate fluoride at all. As well as SbF_6^-, we find $Sb_2F_{11}^-$, i.e. $[F_5Sb\text{—}F\text{—}SbF_5]^-$. This can be seen as a complex between the powerful Lewis acid SbF_5 and SbF_6^-, in which a coordinate bond is formed between an F atom of the anion and the Sb atom of the molecule (cf. molecules like Ga_2Cl_6 discussed above). The very strong tendency of the large Sb atom to attain octahedral six-coordination (at least when bonded to relatively small atoms like F) is also apparent in the structure of antimony(V) fluoride itself. ^{19}F NMR spectra indicate that the viscous liquid contains polymeric chains, with *cis*-bridging fluorines, so that the environment of the Sb atom can be written as:

which is essentially the same as in $Sb_2F_{11}^-$ except that the Sb—F—Sb bridge is almost linear in the anion. In crystalline SbF_5 (m.p. $8\,°C$) tetrameric molecules Sb_4F_{20} are formed, again with approximately octahedral coordination about each Sb:

A similar structure is adopted by the pentafluorides of the d block elements.

The marked propensity of the xenon fluorides towards formation of cations is related to the weakness of the Xe–F bond. We find cations such as XeF^+, XeF_3^+, XeF_5^+, $Xe_2F_3^+$ (i.e. $[F\text{—}Xe\text{—}F\text{—}Xe\text{—}F]^+$, with a bent Xe–F–Xe bridge) and $Xe_2F_{11}^+$ (an adduct of XeF_6 with XeF_5^+; cf. $Sb_2F_{11}^-$).

These are mostly associated with anions such as SbF_6^-, $Sb_2F_{11}^-$ and fluoroanions of the d block elements in high oxidation states, e.g. PtF_6^-, AuF_6^-.

There are cases where it is not easy to draw a clear distinction between molecular Lewis acid/Lewis base complexes and ionic substances of the kind discussed immediately above. For example, in the compound formulated as $[ICl_2][AsF_6]$, each I atom has as near neighbours two F atoms from the 'anion' at a distance of only 265 pm, compared with 350 pm for the sum of the Van der Waals radii. The compound might be alternatively formulated as a covalent polymer with (admittedly rather long) I–F covalent bonds.

It was suggested earlier in this section that oxoacid salts such as $CaCO_3$ could be viewed as products of reactions between basic oxides (containing O^{2-} discrete ions) and covalent (molecular/polymeric) oxides in which oxide ions are transferred to form oxo-anions. Analysis of the thermochemistry of such reactions has led to the formulation of a numerical scale of acidity for oxides. In Table 9.1 the acidity parameter a is listed for the most important binary oxides. Highly-negative values indicate a basic oxide, while acidic oxides have positive values.

Table 9.1 *Acidity parameter a for binary oxides*

Oxide	a	Oxide	a	Oxide	a	Oxide	a
H_2O	0	CO_2	5.5	TiO_2	0.7	Y_2O_3	−6.5
		SiO_2	0.9	ZrO_2	0.1	La_2O_3	−6.1
Li_2O	−9.2	SnO_2	2.2	V_2O_5	3.0	Ce_2O_3	−5.8
Na_2O	−12.5	PbO	−4.5	CrO_3	6.6	CeO_2	−2.7
K_2O	−14.6			MoO_3	5.2	Pr_2O_3	−5.8
Rb_2O	−15.0	N_2O_3	6.6	WO_3	4.7	Nd_2O_3	−5.7
Cs_2O	−15.2	N_2O_5	9.3	MnO	−4.8	Sm_2O_3	−5.1
		P_4O_{10}	7.5	Mn_2O_7	9.6	Eu_2O_3	−5.1
BeO	−2.2	As_2O_5	5.4	Tc_2O_7	9.6	Gd_2O_3	−5.0
MgO	−4.5	Sb_2O_3	0	Re_2O_7	9.0	Tb_2O_3	−4.3
CaO	−7.5	Bi_2O_3	−3.7	FeO	−3.4	Dy_2O_3	−4.7
SrO	−9.4			Fe_2O_3	−1.7	Ho_2O_3	−4.5
BaO	−10.8	SO_2	7.1	CoO	−3.8	Er_2O_3	−4.3
RaO	−11.5	SO_3	10.5	NiO	−2.4	Tm_2O_3	−4.2
		SeO_2	5.2	Cu_2O	−1.0	Yb_2O_3	−4.5
B_2O_3	1.5	SeO_3	9.8	CuO	−2.5	Lu_2O_3	−3.3
Al_2O_3	−2.0	TeO_2	3.8	Ag_2O	−5.0		
Ga_2O_3	−1.6			ZnO	−3.2	ThO_2	−3.8
In_2O_3	−2.4	Cl_2O_7	11.5	CdO	−4.4		
Tl_2O	−6.8	I_2O_5	7.1	HgO	−3.5		

Amphoteric oxides (oxides which dissolve in acids to give cationic species, and also in alkalis to give anionic hydroxo- or oxo-anions) have slightly negative values (between -5 and 0). The a value for H_2O is arbitrarily set equal to zero. The enthalpy change ΔH^0 for the reaction between the basic oxide B and the acidic oxide A is given approximately by $-(a_A - a_B)^2$ in kJ mol^{-1}, where the stoichiometric unit corresponds to the transfer of one mole of oxide ion from B to A. For example, from Table 9.1 we would predict ΔH^0 for the reaction:

$$CaO(s) + CO_2(g) \rightarrow CaCO_3(s)$$

to be -169 kJ mol^{-1}; the experimental value is -178 kJ mol^{-1}. The formation of a salt from a strongly-basic oxide and a strongly-acidic oxide is highly exothermic, i.e. if $a_A - a_B$ is large, while salt formation is only feebly exothermic if $a_A - a_B$ is small. Thus iron(III) carbonate is unknown; it is probably unstable with respect to decomposition to the binary oxides $Fe_2O_3(s)$ and $CO_2(g)$ because the favourable entropy term (at room temperature) outweighs the feebly endothermic ΔH^0 for the decomposition. The ease with which carbonates (and other oxoacid salts) undergo thermal decomposition correlates well with the a values involved. The a values themselves can be correlated empirically with electronegativities. The higher the electronegativity of A, the more acidic is the binary oxide A_xO_y. For a given element A, the acidities of its oxides increase with increasing oxidation state of A.

9.3 Acids, bases and solvents

All the reactions discussed in the previous section could be described as acid/base phenomena, defining acids and bases quite liberally. The importance of ionic equilibria in aqueous solution was recognised in the 1880s by Arrhenius, who proposed that acids were sources of $H^+(aq)$ while bases were sources of $OH^-(aq)$, and it was soon realised that this definition was closely related to the self-dissociation of water:

$$H_2O \leftrightharpoons H^+(aq) + OH^-(aq)$$

for which the equilibrium constant K is 10^{-14} at $25\,°C$. A more general definition (Lowry and Brønsted, 1923) views an acid as a donor of protons and a base as an acceptor of protons. This allows the acid/base concept to be extended to other protonic solvents such as liquid ammonia, sulphuric acid, hydrogen fluoride etc., all of which undergo self-dissociation via proton-exchange between solvent molecules:

$$2NH_3 \leftrightharpoons NH_4^+ + NH_2^-$$
$$2H_2SO_4 \leftrightharpoons H_3SO_4^+ + HSO_4^-$$
$$2HF \leftrightharpoons H_2F^+ + F^-$$

All these equilibria involve processes in which a proton accepts a lone pair from an atom X, or the reverse, i.e.:

$$H^+ \curvearrowleft :X \longrightarrow H-X^+$$

or: $$H \overset{\frown}{-} X \longrightarrow H^+ \; X^-$$

It is easy to see how the acid/base concept can be further generalised to include reactions/equilibria in which ions other than H^+ are transferred, e.g.:

$$XeF_4 + AsF_5 \rightarrow [XeF_3][AsF_6]$$
$$CaO + SO_3 \rightarrow CaSO_4$$

These two examples each have a crystalline solid as the product. Analogous reactions occur in solution, and some liquids undergo self-dissociation with exchange of an anion between solvent molecules, e.g.:

$$2BrF_3 \rightleftharpoons BrF_2^+ + BrF_4^-$$
$$2ICl_3 \rightleftharpoons ICl_2^+ + ICl_4^-$$

Even more general is the Lewis concept of acids and bases; a Lewis base has a lone pair available for formation of a coordinate bond, and a Lewis acid has a vacant acceptor orbital handy. This concept is applicable to reactions in the gas phase or in inert solvents (as discussed in the previous section) as well as to complex formation in solution and the acid/base phenomena studied by Arrhenius, Brønsted and Lowry.

In this section we are concerned with reactions and equilibria in polar solvents (not necessarily all protonic solvents) with emphasis on acid/base processes. In a protonic solvent HX which undergoes self-ionisation to H_2X^+ and X^-, we say that H_2X^+ is the *conjugate acid* of HX, which is the conjugate acid of X^-. Likewise HX is the *conjugate base* of H_2X^+ and X^- is the conjugate base of HX. In other words, an acid is converted into its conjugate base by loss of a proton, while a base becomes its conjugate acid on acceptance of a proton. A solute which protonates solvent molecules – thereby shifting the self-ionisation equilibrium in favour of the cationic species – behaves as an acid, while a solute which accepts protons from solvent molecules is a base. In the non-protonic (or aprotic) liquid ICl_3, a chloride acceptor such as $SbCl_5$ acts as an acid:

$$ICl_3 + SbCl_5 \rightarrow ICl_2^+ + SbCl_6^-$$

A chloride ion donor like KCl acts as a base:

$$ICl_3 + KCl \rightarrow K^+ + ICl_4^-$$

Chemists have always been interested in measuring and interpreting the relative strengths of acids and bases, especially the former. The acid strength of a solution in a protonic medium is obviously to be measured by its power to protonate a base. For a solvent HX, this will depend on both the concentration and the intrinsic proton-donating power of H_2X^+. Likewise, the base strength of such a solution depends on $[X^-]$ and on the proton-accepting power of X^-. If a solute is a strong proton donor – stronger than HX – it will be completely deprotonated by the solvent and will exist in solution in the form of its conjugate base. This means that H_2X^+ is the strongest acid that can exist in HX; likewise, X^- is the strongest base. In a given solvent HX, the strength of an acid HY is to be measured by the magnitude of the equilibrium constant K for the process:

$$HX + HY \leftrightharpoons H_2X^+ + Y^-$$

This can be construed as the competition between the bases HX and Y^- for protons. It should be apparent that the more powerful HY is as a proton donor, the weaker will be Y^- as a proton acceptor. In other words, a strong acid will have a weak conjugate base and vice versa. It is often more convenient to rationalise the relative strengths of acids in terms of the proton affinities of their conjugate bases. We look first at acids in aqueous solution, and then at acid/base and other equilibria in non-aqueous polar solvents.

The relative strengths of acids (mostly oxoacids) in aqueous solution

Apart from the binary hydrides of Groups 16 and 17, Lowry/Brønsted acids in aqueous solution are nearly all oxoacids, i.e. substances containing O–H bonds which ionise in aqueous solution to give oxo-anions and $H^+(aq)$ (or H_3O^+). Most oxoacids are molecular hydroxides $E(OH)_n$, such as $B(OH)_3$, $Ge(OH)_4$ and $Te(OH)_6$, or oxohydroxides $EO_m(OH)_n$. In addition, we have more complex species containing E–E bonds or E–O–E bridges. In $EO_m(OH)_n$ – for example, $NO_2(OH)$, $PO(OH)_3$, $SO_2(OH)_2$, $IO_3(OH)$ – the m O atoms are held to E by bonds having at least some double bond character, via $p_\pi-p_\pi$ or $d_\pi-p_\pi$ overlap. Oxohydroxides may be seen as being derived from hydroxides by elimination of H_2O, and are favoured by elements E whose atoms form double bonds to O atoms.

It is of interest to note that many of the 'familiar' oxoacids – familiar in the sense that their salts are well known – are difficult or impossible to isolate as pure substances. An example is nitrous acid $NO(OH)$ or HNO_2; the pure substance has not been isolated, but dilute solutions are readily obtained, and the molecule has been studied by electron diffraction in the gas phase. The nitrite ion NO_2^- is much more stable; many ionic nitrites are known, and NO_2^- (aq) is quite stable at alkaline pH. Here, the anion is stabilised by resonance:

Likewise, carbonic acid $CO(OH)_2$ is rather unstable with respect to CO_2 and water since the very strong bonds in CO_2 are resistant to addition. But the CO_3^{2-} and HCO_3^- ions are stabilised by resonance. The relative strengths of oxoacids in aqueous solution are evidently dependent on the stabilities of their conjugate bases, in which resonance (electron delocalisation) appears to play some part.

The strength of an oxoacid in aqueous solution is measured by its pK_a value. For oxoacids which may be formulated as $EO_m(OH)_n$, there is a close correspondence between observed pK_a values and m, viz.:

m	pK_a
0	7.5–9.2
1	1.8–3.5
2	(-3)–(-1)
3	(-10)–(-7)

From the discussion of HF in Section 1.5, it should be apparent that attempts to explain such trends must be treated with caution. A useful and generally-accepted rationalisation involves a consideration of the attraction of an O atom in an oxo-anion for a proton. If, for example, we compare ClO^-, ClO_2^-, ClO_3^- and ClO_4^-, the negative charge is increasingly delocalised as the number of O atoms is increased, there being respectively one, two, three and four resonance structures for the anion. The increasing electronegativity of the central Cl atom with increasing oxidation number also facilitates the reduction of the fractional negative charge on each O atom, and discourages recombination of the anion with a proton. The latter factor is helpful in interpreting trends in acid strength such as:

$$SO_2(OH)_2 > SeO_2(OH)_2 > CrO_2(OH)_2$$
$$ClOH > BrOH$$
$$ClO_2(OH) > BrO_2(OH) > IO_2(OH)$$

For oxoacids having more than one OH group, we can determine values of pK_1, pK_2 etc. The difference $(pK_n - pK_{n-1})$ is nearly always between 4 and 5; the second and successive ionisations are always more difficult to accomplish than the first. It should not be difficult to see that SO_4^{2-} is much more susceptible to protonation than $SO_3(OH)^-$; the former has four protonatable O atoms compared with only three in the latter, and each of these will be carrying a higher fractional negative charge than the O atoms in $SO_3(OH)^-$.

Acid/base and other reactions in non-aqueous solvents

A protonic solvent HX must be both a donor and an acceptor of protons, i.e. both an acid and a base. But some solvents are distinctly more acidic, and less basic, than others. Thus in aqueous solution, NH_3 is a base while H_2SO_4 is an acid. Clearly, in its capacity as a solvent, NH_3 is a better proton acceptor than a proton donor, while the reverse is true for H_2SO_4. Since NH_3 is more basic than H_2O, its conjugate acid NH_4^+ must be a weaker acid than H_3O^+. Since NH_4^+ is the strongest acid which can exist in liquid ammonia, this is a poor medium for performing reactions requiring a strong proton donor. On the other hand, NH_2^- is a stronger base than OH^- and cannot survive in aqueous solution. Liquid ammonia is an excellent solvent where a strongly-basic medium is desired. A more acidic solvent than water – anhydrous HF, pure sulphuric acid or glacial acetic acid – would be chosen for a reaction medium requiring a very strong proton donor.

Strong acids such as $HClO_4$, HNO_3 and H_2SO_4 are completely ionised in water and their relative strengths cannot be differentiated. This is the 'levelling effect', arising from the fact that no acid stronger than H_2X^+ can survive in HX. Acids that are strong in water may be much weaker in a more acidic solvent; they may even act as bases. Their relative strengths are then readily differentiated. Thus in H_2SO_4, $HClO_4$ is a weak acid, HNO_3 a weak base and CH_3COOH quite a strong base. If a very strongly-acidic medium is required – in order to protonate an extremely reluctant base – we need an acidic solvent like H_2SO_4 or HF and a proton donor which behaves as a strong acid in an acidic solvent, producing a high concentration of, e.g., $H_3SO_4^+$ or H_2F^+. In the case of H_2SO_4, few solutes behave as strong acids. One is pyrosulphuric acid $H_2S_2O_7$ – i.e. $(OH)O_2S$—O—$SO_2(OH)$ – which is produced by adding SO_3 to sulphuric acid. Another is '$HB(OSO_2OH)_4$' which has no independent existence; the solution formed by the reaction:

$$H_3BO_3 + 2H_2SO_4 + 3SO_3 \rightarrow H_3SO_4^+ + B(OSO_2OH)_4^-$$

is a very potent proton donor. The concentration of H_2F^+ in liquid HF can be greatly enhanced by the addition of a fluoride acceptor like SbF_5:

$$2HF + SbF_5 \rightarrow H_2F^+ + SbF_6^-$$

Mixtures of SbF_5 and fluorosulphonic acid $SO_2F(OH)$ are among the most powerful proton-donating media known. Here the SbF_5 acts as a Lewis acid, as usual, but not as a fluoride acceptor:

$$SbF_5 + 2SO_2F(OH) \rightarrow F_5Sb-O-SO_2F^- + SOF(OH)_2^+$$

Solutions such as those described above, having great potency as proton donors, are often called 'superacids'. They can protonate the most unlikely substrates, even hydrocarbons. Superacid media stabilise other improbable cationic species which would otherwise undergo solvolysis in a protonic solvent. Examples include the many cyclic polyatomic cations formed by the Group 16 elements, such as S_8^{2+}, Se_4^{2+}, Te_6^{2+}. Crystalline salts of these with anions such as SbF_6^- can be obtained from superacid solutions.

Acid/base reactions can be defined in non-protonic solvents. A popular solvent in this category is $POCl_3$, which is believed – although there has been some controversy over this – to undergo self-ionisation:

$$POCl_3 \rightarrow POCl_2^+ + Cl^-$$

Here we can say that $POCl_2^+$ is the characteristic acid, and Cl^- the characteristic base of the $POCl_3$ solvent system. Chloride donors are therefore bases, and chloride acceptors acids in $POCl_3$. Many covalent chlorides have the capacity both to donate and to accept Cl^-, and a scale of donor/acceptor strength can be set up from studies in $POCl_3$.

There are, of course, many other types of reaction that can profitably be studied in solvents other than water. Redox reactions will be discussed in Section 9.4, where we will look at analogies between reduction/oxidation and acid/base processes, and the use of non-aqueous solvents as media suitable for the use of strongly oxidising or reducing agents.

It should be noted that self-ionisation is not an essential prerequisite for a satisfactory polar solvent. Liquids such as acetonitrile CH_3CN or dimethylsulphoxide $SO(CH_3)_2$ appear not to ionise but they make very useful solvents for electrolytes as well as for polar molecular substances. As with H_2O, NH_3, H_2SO_4 etc., they owe their solvent powers to their polarity, leading to dipole–dipole interaction in the case of polar molecules as solutes and ion–dipole attraction in the case of electrolytes. There may in addition be considerable covalent bonding, via coordinate bond formation, in the case of cations. In solvents which do undergo appreciable self-ionisation, coordination often needs to be considered explicitly in discussing acid/base and other reactions and equilibria.

9.4   Redox reactions in polar solvents

There are close analogies between acid/base and reduction/oxidation reactions. Indeed, a very general definition of acids and bases due to Usanovich (not mentioned in Section 9.3) embraces redox processes as well. Every Brønsted acid has a conjugate base, and every base its conjugate acid, interchangeable via proton transfer. Likewise, we think in terms of redox couples, each comprising an oxidising agent and a reducing agent; these are interchangeable via ion–electron half-reactions in which electrons appear on the same side as the oxidising agent (the solvent, and its associated conjugate acid or base, may also appear). For example, there is a clear analogy between the reactions:

$$SeO_4{}^{2-}(aq) + 2H^+(aq) \rightarrow H_2SeO_4(aq)$$

and

$$SeO_4{}^{2-} + 4H^+(aq) + 2e \rightarrow H_2SeO_3(aq) + H_2O$$

In the second reaction, there is a change in the oxidation state of Se from VI to IV.

The strength of an acid in a given solvent is measured by the equilibrium constant K_a for its acid dissociation. The analogous quantity for a redox couple is the potential, relative to a standard electrode. The stronger an acid, the weaker is its conjugate base. Likewise, the more strongly oxidising is the oxidising half of a redox couple, the more feeble is the other half as a reducing agent. For example, consider the cases below:

$$U^{4+}(aq) + e \rightarrow U^{3+}(aq) \quad E^0 = -0.61\,V$$
$$MnO_4^-(aq) + 8H^+(aq) + 5e \rightarrow Mn^{2+}(aq) + 4H_2O \quad E^0 = +1.51\,V$$

Both of these redox potentials are applicable at a pH of zero; that the MnO_4^-/Mn^{2+} value applies in acid solution is obvious from the appearance of $H^+(aq)$ in the equation. Highly-charged cations like U^{3+} and U^{4+} only exist in acid solutions (see Section 5.5). The E^0 values are, as usual, relative to the standard hydrogen electrode. We can say immediately that MnO_4^- is a powerful oxidising agent in aqueous solution at acid pH, and is capable of oxidising gaseous hydrogen to $H^+(aq)$. It is also capable of oxidising U(III) to U(IV) at zero pH. $Mn^{2+}(aq)$ is a very weak reducing agent in so far as it is not easily oxidised to MnO_4^-. It will not reduce U(IV) to U(III) at zero pH, nor will it reduce proton to gaseous H_2. $U^{3+}(aq)$ is quite a strong reducing agent, and will reduce protons to $H_2(g)$, at unit concentration of $H^+(aq)$. But $U^{4+}(aq)$ is a feeble oxidising agent, being reduced to $U^{3+}(aq)$ only with difficulty.

The neutralisation of acids with bases provides many valuable volumetric methods of chemical analysis; and redox titrations are useful as well. But here we encounter an important difference between acid/base and redox reactions in solution. Acid/base reactions which involve the transfer of protons are very fast; indeed they are usually instantaneous for all practical purposes. In protonic solvents, polar H–X bonds are very labile and undergo rapid proton exchange. For example, if $B(OH)_3$ – a very weak acid – is recrystallised from D_2O, we obtain a fully-deuterated product. Redox reactions, on the other hand, are often very slow under ordinary conditions. To return to the analogy between acid/base and redox titrations, many readers will be familiar with the reaction between permanganate and oxalic acid; the reaction is very slow at room temperature and, for titrimetric purposes, should be carried out at about $60\,°C$. The mechanism whereby a redox reaction takes place tends to be more complicated than that for an acid/base neutralisation. This, of course, means that the rate of the reaction, and the factors which determine the rate, can be conveniently studied and a wealth of information has been amassed on this topic (see later in this section).

Like acid/base chemistry, redox chemistry is strongly dependent on the solvent. The susceptibility of the solvent to oxidation or reduction will obviously place restrictions on the scope of redox chemistry which can be studied in it. In the case of water, we have to consider:

(1)   The oxidation of $O(-II)$ to $O(-I)$, with formation of H_2O_2.
(2)   The oxidation of $O(-II)$ to $O(0)$, with formation of O_2.
(3)   The reduction of $H(I)$ to $H(0)$, with formation of H_2.
       The relevant potentials are as follows:

$$
\begin{aligned}
& & E^0\,(\text{V}) \\
2H^+(aq) + H_2O_2(aq) + 2e &\rightarrow 2H_2O & +1.78 \\
4H^+(aq) + O_2(g) + 4e &\rightarrow 2H_2O & +1.23 \\
H^+(aq) + e &\rightarrow \tfrac{1}{2}H_2(g) & 0.00 \\
HO_2^- + H_2O + 2e &\rightarrow 3OH^-(aq) & +0.87 \\
O_2(g) + 2H_2O + 4e &\rightarrow 4OH^-(aq) & +0.40 \\
2H_2O + 2e &\rightarrow H_2(g) + 2OH^-(aq) & -0.83
\end{aligned}
$$

The first three potentials are applicable to acid solutions (pH = 0) and the last three to alkaline solutions (pH = 14). In acid solution, any oxidising agent which forms part of a couple having a reduction potential greater than 1.23 V should oxidise water, with evolution of oxygen gas. In practice, many agents with higher potentials than this can be handled in aqueous solution because the oxidation of water is often rather slow. Water is rather more susceptible to reduction. Reducing agents where the reduction potential is negative in sign can survive at higher pH; at a pH of 7, the reduction potential for the reduction of water to $H_2(g)$ is -0.41 V. Again, some powerful reducing agents can be used in aqueous solution because they react slowly with water.

If a strong oxidising agent is needed, a solvent having less susceptibility to oxidation than water may be required. HF or H_2SO_4 would fit the bill. For reactions involving strong reducing agents, NH_3 is a good solvent; ammonia is not easily reduced although it is vulnerable to oxidation, to a variety of possible products: N_2H_4, N_2, NO, HNO_2, HNO_3 etc.

Apart from its own susceptibility to oxidation or reduction, a solvent can affect redox equilibria by modifying the relative stabilities of oxidation states of solutes. Thus Cu^+ is unstable in aqueous solution to disproportionation (Section 5.4); but it is quite stable in acetonitrile. This arises from the relative magnitudes of the solvation energies and entropies of Cu^+ and Cu^{2+} in the different solvents. In ammonia, cobalt(III) is much more stable relative to cobalt(II) than in water. The higher ligand field splittings produced by NH_3 favour the low-spin d^6 configuration.

Mechanisms of oxidation/reduction reactions in solution

As in organic chemistry, most of our detailed knowledge (as opposed to sheer speculation) about reaction mechanisms comes from kinetic studies. Such studies may suggest (but rarely prove) a particular mechanism, and further confirmation is usually necessary. This may come from isotopic labelling experiments – which show us where particular atoms in the products have come from – or from the identification of reaction intermediates and suggestive by-products.

We saw in the previous section how acid/base reactions can be viewed as ion-transfer processes, the ions in question being usually proton, halide or oxide, without any changes in oxidation state. Redox reactions may often be seen as *atom*-transfer reactions, in which H, O, halogen etc. are transferred from one ion/molecule to another, with concomitant changes in oxidation state, e.g.:

$$R_2CHOH + O_2 \rightarrow R_2CO + H_2O_2$$
$$NiO + C \rightarrow Ni + CO$$
$$XeF_2 + SF_4 \rightarrow Xe + SF_6$$

Since the transfer of an atom (as opposed to an ion) can effect changes in oxidation number, it is tempting – although dangerous – to speculate that such transfers actually occur in the course of redox reactions, and form the basis for their mechanisms. As we shall see, this is sometimes true but the situation is usually more complex.

We consider first reactions in aqueous solution which can be described in terms of electron transfer between complex ions of the d block elements, e.g.:

$$Cr^{2+}(aq) + Fe^{3+}(aq) \rightarrow Cr^{3+}(aq) + Fe^{2+}(aq)$$
$$Fe(CN)_6^{4-} + IrCl_6^{2-} \rightarrow Fe(CN)_6^{3-} + IrCl_6^{3-}$$

A simpler situation from the point of view of a theoretical treatment – although more difficult to study experimentally – is electron exchange between ions which constitute two halves of the same redox couple, e.g. MnO_4^-/MnO_4^{2-}, $Co(NH_3)_6^{2+}/Co(NH_3)_6^{3+}$ etc. Two distinct types of mechanism have been postulated. In the *outer-sphere* mechanism, the coordination spheres of both oxidant and reductant remain intact as electrons are transferred, and the oxidation numbers of the central atoms change. The *inner-sphere* mechanism describes a situation where a bridged binuclear complex is formed as an intermediate, and the bridging ligand – which may be Cl^-, OH^- etc. or an ambidentate ligand like NCS^- – provides a pathway for electron transfer.

It might be thought that the transfer of an electron from, say, MnO_4^{2-} to MnO_4^- should require no activation energy following collision between the ions. But the equilibrium bond lengths in these two anions are not the same; the atoms tend to contract in size with increasing oxidation number, so that the Mn–O distance in MnO_4^{2-} is about 10 pm greater than in MnO_4^-. If both ions at the instant of electron transfer are in their equilibrium positions, both MnO_4^- and MnO_4^{2-} will be formed in vibrationally excited states and there will be a significant activation barrier to electron transfer. This is minimised if both oxidant and reductant make the effort to meet half-way, by distorting towards identical configurations, the one having slightly longer and the other slightly shorter Mn–O bonds than at equilibrium.

The activation barrier is further minimised if the loss/gain of an electron has little effect on the bonding or CFSE of the species involved. For electron transfer between $Mn(CN)_6^{4-}$ and $Mn(CN)_6^{3-}$ (low-spin d^5 and low-spin d^4 respectively), the outer-sphere electron transfer is very rapid (the second-order rate constant at 25 °C is greater than $10^6 \, l \, mol^{-1} \, s^{-1}$). This applies wherever a t_{2g} electron in an octahedral complex is being exchanged, and especially where π acceptor ligands are present. Usually, the internuclear distance will decrease as the oxidation state of the central atom is increased. But with ligands like CN^-, the contraction which accompanies oxidation may be very small, because some π back-donation is lost. The fact that there are only four electrons in the t_{2g} level (π bonding) in $Mn(CN)_6^{3-}$ compared with five in $Mn(CN)_6^{4-}$ tends to counteract the effect of the higher oxidation state of Mn. However, the rate constant for electron exchange between $Co(NH_3)_6^{2+}$ and $Co(NH_3)_6^{3+}$ is lower by a factor of about 10^{10}. Here, the Co(II) species is high-spin d^7 $(t_{2g})^5(e_g)^2$ while the Co(III) ion is low-spin d^6 $(t_{2g})^6$. The electron transfer now involves a more drastic change, and the activation barrier is correspondingly greater.

The outer-sphere mechanism was thoroughly established by studies of reactions such as:

$$Cr(H_2O)_6^{2+} + Co(NH_3)_5Cl^{2+} \rightarrow Cr(H_2O)_5Cl^{2+} + Co(NH_3)_5(H_2O)^{2+}$$

The reaction proceeds via the formation of the chloro-bridged intermediate $[(H_2O)_5Cr\!-\!Cl\!-\!Co(NH_3)_5]^{4+}$, which breaks down to give the products. Octahedral Cr(III) complexes are *substitution-inert*, i.e. they undergo ligand exchange very slowly. This property arises from the d^3 configuration; the higher CFSE leads to large activation barriers to substitution reactions, a property shared to a lesser extent by low-spin d^6 complexes such as octahedral Co(III). Most chloro-complexes in water undergo rapid aquation (replacement of Cl^- by H_2O) but the $Cr(H_2O)_5Cl^{2+}$ ion survives long enough for its positive identification as a product, and provides firm evidence for the bridged intermediate. If the reaction is performed in a solution to which isotopically-labelled Cl^- has been added, practically no labelled chlorine turns up bonded to Cr(III). The reaction can be perceived as a chlorine atom transfer from Co(III) to Cr(III). Hydrogen atom transfer is postulated for the reaction:

$$Cr^{2+}(aq) + Fe^{3+}(aq) \rightarrow Cr^{3+}(aq) + Fe^{2+}(aq)$$

Here it is thought that the oxidising agent is $Fe(H_2O)_5(OH)^{2+}$ (this is one of the dominant species in dilute solution at a pH of about 2) while the reducing agent is $Cr(H_2O)_6^{2+}$. The bridged intermediate may be formulated as:

$$[(H_2O)_5Fe\!\!-\!\!O\ldots H\ldots O\!\!-\!\!Cr(H_2O)_5]^{4+}$$

with $|$ H and $|$ H below the two O atoms respectively.

and breaks down to give as the ultimate products $Fe(H_2O)_6^{2+}$ and $Cr(H_2O)_5OH^{2+}$. This can be viewed as a transfer of a hydrogen atom from H_2O coordinated to Cr(II) to OH coordinated to Fe(III), leading to oxidation of Cr and reduction of Fe.

In the oxidation of $Cr^{2+}(aq)$ by $IrCl_6^{2-}$, there is evidence for the formation of a chloro-bridged intermediate, but no overall chlorine transfer takes place: the products are $Cr^{3+}(aq)$ and $IrCl_6^{3-}$. Bridging does appear to assist the process of electron transfer, but ultimate atom transfer is not absolutely essential.

All the reactions discussed so far involve changes in oxidation state by only one unit, and can be described in terms of one-electron transfer processes. But many elements have stable oxidation states differing by two units, e.g. Pt(II)/Pt(IV), Tl(I)/Tl(III), Sn(II)/Sn(IV). Can two electrons be transferred from one ion to another in a single step? The answer is yes, provided that some latitude is exercised in the definition of a single step; some reactions, apparently with two-electron transfer steps, may in fact require two one-electron steps but the second is so fast that the process has the appearance of a single two-electron transfer. Remember too that the transfer of an O atom amounts to a change in oxidation state by two units. A useful generalisation is that species derived from d block or f block elements react with one another by one-electron steps; species derived from p block elements usually react with each other in two-electron steps. A species derived from a d block element may react with a species derived from a p block element by either a one- or two-electron step, the former being more common. Examples of two-electron redox reactions include:

$$Sn(II) + Tl(III) \rightarrow Sn(IV) + Tl(I)$$
$$V(II) + Tl(III) \rightarrow V(IV) + Tl(I)$$

However, when Tl(III) oxidises Fe(II) to Fe(III), there is evidence for Tl(II) as a short-lived intermediate.

Many reactions between oxo-anions appear to be oxygen transfers from one species to another, e.g.:

$$SO_3^{2-}(aq) + ClO_3^-(aq) \rightarrow SO_4^{2-}(aq) + ClO_2^-(aq)$$

If the reaction is carried out in water enriched with ^{18}O and Ba^{2+} added to precipitate $BaSO_4$, no ^{18}O is found in the sulphate, so that the extra O atoms must have come from ClO_3^-. But if the ClO_3^- is enriched with ^{18}O, we obtain sulphate ions in which the oxygen content is ^{18}O-enriched to a quarter of the extent of that in the original chlorate. Most reactions involving oxo-anions are in this category. The rate tends to increase with the hydrogen ion concentration; protonation of the O atoms in the oxo-anion tends to weaken the E–O bonds (which, with the O atoms unprotonated, usually have some double bond character) as well as making them more susceptible to nucleophilic attack; e.g.:

The effectiveness of such acceleration will, of course, depend on the basicity of the oxo-anion; the stronger the oxoacid, the more slowly will its conjugate base(s) react in redox processes. This helps to account for the sluggishness of ClO_4^- as an oxidising agent. Although thermodynamically quite a strong oxidising agent in acid solution:

$$ClO_4^-(aq) + 2H^+(aq) + 2e \rightarrow ClO_3^-(aq) + H_2O \quad E^0 = +1.23\,V$$

only a few reducing agents react with perchlorate at an appreciable rate. On the other hand, very concentrated solutions of perchloric acid – containing a high proportion of undissociated $HClO_4$ molecules – are dangerously reactive towards oxidisable material. Despite its thermodynamic classification as a strong oxidising agent, sodium perchlorate is widely used to provide a chemically-inert electrolyte solution of high and constant ionic strength for the study of equilibria in aqueous solution; ClO_4^- has little tendency to form complexes.

9.5   Substitution reactions

This section is concerned with reactions of the type:

$$EX_nY + Z \rightarrow EX_nZ + Y$$

Thermodynamically, the feasibility of such a reaction will depend on the relative strengths of the E–Y and E–Z bonds, on the solvation energies (in solution) of the species, and on entropy terms. If the equilibrium lies to the left, the reaction may still proceed to near-completion if, for example, one of the products is removed by precipitation. From the mechanistic viewpoint, two extreme situations are recognised:

(1) An *associative* process (often designated *A*), in which the rate-determining step is the formation of an intermediate:

$$EX_nY + Z \xrightarrow{\text{slow}} Y\!-\!EX_n\!-\!Z \xrightarrow{\text{fast}} EX_nZ + Y$$

The bimolecular rate-determining step implies a second-order reaction. The intermediate can often be identified (e.g. spectroscopically) or even isolated and characterised; in the case of a free radical, it can sometimes be 'trapped' by reagents which quickly bind such species to give a readily-identifiable product.

(2) A *dissociative* (*D*) process, in which the rate-determining step is the breaking of the E–Y bond. This rupture may be homolytic (the electron pair constituting the E–Y bond is equally shared by the two fragments, which will be free radicals) or heterolytic, i.e. the bonding pair remains on one of the two atoms.

A heterolytic rupture will, in the case of an 'ordinary' covalent bond, lead to formation of a cation and an anion; in the case of a coordinate bond, the ligand simply departs along with the electron pair it contributed to form the bond. A dissociative mechanism exhibits first-order kinetics; its rate is independent of the concentration of the incoming group Z. The intermediate EX_n in the overall reaction:

$$EX_nY \xrightarrow{\text{slow}} EX_n + Y \xrightarrow[Z]{\text{fast}} EX_nZ + Y$$

may be detected, trapped or isolated as proof of a dissociative mechanism. The acceleration of a reaction by a polar solvent is sometimes cited as evidence for a heterolytic dissociation; solvation is expected to favour the formation of ions.

An intermediate situation between the *A* and *D* extremes is an *interchange* (*I*) mechanism. Here, the breaking of the E–Y bond and the formation of the E–Z bond occur simultaneously as a concerted process; at no time is there an intermediate with both a fully-formed E–Y and an E–Z bond. The transition state is a complex which can be described as $Y \ldots EX_n \ldots Z$. Like the *A* mechanism, the kinetics are second order, but the intermediate (or activated complex) has no real existence as a static entity in the reaction mixture; unlike the intermediate in the *A* mechanism, there is no energy barrier to the breakdown of $Y \ldots EX_n \ldots Z$, which appears as a maximum on the profile of energy versus reaction coordinate which the reader will no doubt have seen in elementary treatments of reaction kinetics and mechanisms.

The labels I_a and I_d are used to denote mechanisms which approach, but fail to reach, the extremes of A and D character respectively. In the case I_d, the E–Y bond is perceptibly weakened prior to any significant bonding of Z to E, but EX_n is never present as a free entity. The label I_a is appropriate where there is evidence of incipient E–Z bond formation while the E–Y bond remains more or less intact. It is often impossible to decide whether a mechanism should be labelled A, I or I_a, on the strength of the experimental evidence alone. If the intermediate Y—EX_n—Z is a plausible species, having a 'right to exist', we may tend to favour A as opposed to I. If, on the other hand, the intermediate offends our notions about bonding and stability, we are inclined to postulate an I mechanism. Thus the well-known nucleophilic substitution reaction at a saturated carbon atom:

$$RX + Y^- \rightarrow RY + X^-$$

is classified as S_N1 – a dissociative mechanism with R^+ as an intermediate – or as S_N2, with a bimolecular rate-determining step which would be described as I rather than A, because the intermediate in a genuinely associative step requires the objectionable pentavalent carbon.

The $A/I/D$ terminology is, of course, useful and applicable to reactions other than substitutions. The formation of a typical Lewis acid/Lewis base adduct like $Cl_3B \leftarrow PR_3$ is clearly an associative process where the complex is kinetically and thermodynamically stable towards any further reaction. This is not necessarily true of all such adducts, however; the complex $Cl_3B \leftarrow NHR_2$ rapidly loses HCl to give Cl_2B—NR_2. As discussed in Section 6.3, there is appreciable p_π–p_π bonding between B and N and the large B–N bond energy (about $450\,kJ\,mol^{-1}$ in Cl_2BNR_2, compared with about $330\,kJ\,mol^{-1}$ in Cl_3BNR_3) provides much of the thermodynamic driving force. However, the rate-determining step is the bimolecular association of a Lewis acid with a Lewis base. Reactions which involve ion or atom transfer between molecules/ions may proceed via a relatively stable bridged intermediate (A mechanism) or the exchange may require a concerted process (I mechanism).

Substitution reactions of Main Group species

We now look at reactions which can be written stoichiometrically as:

$$EX_n + Y \rightarrow EX_{n-1}Y + X$$
$$EX_{n-1}Y + Y \rightarrow EX_{n-2}Y_2 + X$$
etc.

where the reactions may involve molecules in the gas phase, or molecules/ ions in solution. Inorganic chemists may also be concerned with hetero- geneous systems. Here we are mainly concerned with substitution reac- tions of molecular species. Whether the mechanism is associative or dissociative depends on the tendency or otherwise of E to increase its coordination number from n to $n + 1$. Some examples will suffice to illustrate this.

The boron trihalides are all molecular substances, consisting of planar BX_3 molecules in all phases. Substitution reactions such as:

$$BX_3 + OR^- \rightarrow ROBX_2 + X^-$$

appear to involve the formation of $ROBX_3^-$ as an intermediate; such anions can be stabilised in crystalline solids and clearly have a 'right to exist'. An associative mechanism is obviously preferable to a dissociative one unless species such as BX_2^+ can be stabilised by complex formation, e.g. with the solvent. But in a polar, coordinating solvent, the BX_3 will already be present in the form of the tetrahedral complex BX_3(solvent); thus in an ether $R_2'O$ the reaction is likely to follow the route:

$$BX_3(OR_2') + OR^- \rightarrow BX_3OR^- + OR_2' \rightarrow BX_2OR(OR_2') + X^-$$

A D or I process for both steps would be more plausible than A.

Reactions of the type:

$$X_3BNR_3 + NR_3' \rightarrow X_3BNR_3' + NR_3$$

have been studied in the gas phase and in non-polar solvents, and it is apparent that they proceed via dissociative mechanisms. The enthalpy and entropy of activation are both large and positive, implying the breaking of the B–N bond and the dissociation of the adduct as the rate- determining step. An associative mechanism would be improbable, given the reluctance of boron to increase its coordination number beyond four and the fact that three-coordinate boron is familiar in stable substances. However, similar studies of analogous aluminium and gallium com- pounds indicate an associative rate-determining step, with intermediate formation of $X_3E(NR_3)(NR_3')$. Here the ability of the larger atoms Al and Ga to increase their coordination numbers is crucial. If we look at substitution of the Lewis acid instead of the base in such complexes, e.g.:

$$X_3ENR_3 + E'X_3 \rightarrow X_3E'NR_3 + EX_3$$

we find the mechanism to be dissociative if the amines are secondary or tertiary, with bulky R groups; otherwise it is associative. Evidently steric effects are at work; bulky substituents discourage the associative step.

We have already mentioned substitution reactions at saturated carbon atoms: the familiar S_N1 and S_N2 mechanisms of the organic chemist. In the latter case, the five-coordinate intermediate scarcely has any real existence; it cannot be isolated, its formation is based on circumstantial evidence and it violates one of the most venerable principles of bonding theory: the inability of carbon to form more than four bonds. In the process:

$$R_3CX + Y \rightarrow X \ldots CR_3 \ldots Y \rightarrow CR_3Y + X$$

the transformation of the reactants into the products involves the simultaneous and concerted formation of the C–Y bond and the breaking of the C–X bond. This is the classic interchange mechanism, falling somewhere between the extremes of association and dissociation.

Reactions of silicon compounds are dominated by the ability of the Si atom to increase its coordination number to five or even six. We have already mentioned (Section 4.1) the rapid hydrolysis of $SiCl_4$ and SiH_4 compared with the relative inertness of the analogous carbon compounds. Catenated silanes Si_nH_{2n+2} are thermally unstable, decomposing on mild heating to elemental silicon and hydrogen. It would be expected that such molecules would decompose more readily than the analogous hydrocarbons, with rupture of the weaker Si–H and Si–Si bonds. However, it is believed that an easier route to decomposition of $RSiH_2SiH_3$ is available via elimination of the transient silene species SiH_2 and SiHR; a five-coordinate intermediate containing a three-centre Si–H–Si bond is formed:

Closely related to substitution reactions are exchange or redistribution reactions between molecules, e.g.:

$$EX_n + E'Y_m \rightarrow EX_{n-1}Y + E'Y_{m-1}X$$

In the gas phase or in non-polar solvents where ions are unlikely to be formed, dissociative routes, with the breaking of E–X and E′–Y bonds and the intermediate formation of EX_{n-1} and $E'Y_{m-1}$, will require high temperatures unless the bonds are very weak. But if a plausible intermediate complex can be found, an associative mechanism may allow rapid exchange at modest temperatures. Thus the boron trihalides undergo rapid exchange; if BF_3 and BCl_3 are mixed, the ^{19}F NMR spectrum quickly shows evidence of BF_2Cl and $BFCl_2$. These mixed halides cannot be isolated as pure substances because of the tendency for the halogens to undergo rapid exchange. Presumably the intermediate:

is formed, and may break down as indicated by the broken lines to form either BF_3 and BCl_3, or BF_2Cl and $BFCl_2$. The intermediate has never been isolated, or even positively identified; but it is a plausible species and the analogous Al_2Cl_6 etc. are well known.

Similar behaviour is observed for, e.g., SiF_4 and $SiCl_4$. Here we postulate a five-coordinate halogen-bridged intermediate:

$$
\begin{array}{ccc}
F & & Cl \\
| & \diagup F \diagdown & | \\
F \!-\! Si & & Si \!-\! Cl \\
| & \diagdown Cl \diagup & | \\
F & & Cl
\end{array}
$$

However, CF_4 and CCl_4 do not undergo halogen exchange, except at high temperatures where dissociative pathways become possible. Mixed halides of silicon can be isolated; the rate of exchange is slow at low temperatures. The activation barrier to formation of the bridged intermediate is evidently higher than for the boron trihalides; this can be rationalised by noting that the change from an sp^3 to an sp^3d valence state is more costly in promotion energy than the change from sp^2 to sp^3. Halogen-exchange reactions are of great preparative value. The trihalides of P, As and Sb undergo exchange readily. SbF_3 is a volatile molecular solid at room temperature, and is thus easily handled. Heating a molecular chloride with SbF_3 is a useful route to fluorides.

Substitution reactions of complexes in solution

Here we are concerned with ligand-replacement reactions of complexes where the central atom is one of the d block elements. Most studies have been carried out on octahedral complexes where the central atom has the d^3 or especially the low-spin d^6 configuration, or on square coplanar complexes where the central atom is d^8. These configurations are associated with maximum CFSE, and their complexes are kinetically inert (as opposed to labile), undergoing relatively slow ligand exchange in solution. Perhaps the simplest ligand-exchange process is that between coordinated water and solvent water in $M(H_2O)_6^{2+}$ ions. This appears to involve a dissociative mechanism, and the experimentally-determined activation energies for $M^{2+}(aq)$ correlate well with calculated values for the loss of CFSE which accompanies a change from octahedral six-coordination to square pyramidal five-coordination. In the case of $M^{2+}(aq)$, the exchange is always very fast, with first-order rate constants in the range 10^2–$10^9\,s^{-1}$. But $Cr(H_2O)_6^{3+}$ undergoes exchange very much more slowly. A chromium(III) salt containing $Cr(H_2O)_6^{3+}$ ions enriched with ^{18}O can be recrystallised from ordinary water without much loss of ^{18}O. Relatively slow reactions are, of course, preferable for the kineticist who wants to make accurate rate measurements under a variety of conditions.

It would appear that in most reactions in aqueous solution of the type:

$$ML_5X + Y \rightarrow ML_5Y + X$$

the first step is *aquation*, the replacement of X by H_2O. The next step is called *anation*, the replacement of water by Y; the term is derived from the fact that Y is usually an anion. The mechanism for anation appears to be dissociative (D or I_d) in most cases, i.e.:

$$ML_n H_2O \longrightarrow ML_n + H_2O$$
$$\downarrow +Y$$
$$ML_n Y$$

Most of the work on the kinetics and mechanism of aquation – the first step in octahedral substitution – has been done on cobalt(III) complexes, which are neither too inert nor too labile for exhaustive investigations. The aquation of $Co(NH_3)_5X^{2+/3+}$ (the charge depends on whether X is neutral or anionic) has been studied in great depth. The rate law for such a process is found to take the form:

$$\text{rate} = k_A c + k_B c[OH^-]$$

where c is the concentration of $Co(NH_3)_5X^{2+/3+}$. The subscripts A and B distinguish the rate constants as being those appropriate for acid and base hydrolysis respectively (use of the term hydrolysis in this context is questionable, because the bonds in water are not being broken; hydration or aquation would be preferable, but this terminology is now well entrenched). The evidence is somewhat confusing and equivocal, but the mechanism for acid hydrolysis is probably dissociative, somewhere between D and I_d. The form of the rate law would suggest that the base hydrolysis is an associative process, with formation of $Co(NH_3)_5X(OH)^{+/2+}$ as a seven-coordinate intermediate. This has been ruled out on a number of grounds, however. It is generally agreed that the role of OH^- in base hydrolysis is to remove a proton from a coordinated NH_3, to give $Co(NH_3)_4(NH_2)X^{+/2+}$. This then undergoes dissociative loss of X to give a five-coordinate inter-mediate, which takes up water readily. For reasons not as yet wholly explained, NH_2^- as opposed to NH_3 in the coordination sphere appears to render the leaving group X labile.

In the case of square coplanar d^8 complexes – Pd(II) and Pt(II) systems have been most studied – it is clear that an associative mechanism operates. This is not surprising, since the steric constraints which might discourage formation of a seven-coordinate Cr or Co complex will be less important for a five-coordinate complex of the larger Pd or Pt.

The mechanism is best illustrated by looking at the replacement of X by

Y in the complex $PtXL_2L'$, where the two identical ligands L are *trans* to each other:

$$
\begin{array}{ccccc}
\text{L} & & \text{L} & & \text{L} \\
| & & | & & | \\
\text{L'} - \text{Pt} - \text{X} & \xrightarrow{\text{Y}} & \underset{\text{L'}}{\overset{\text{Y}}{\diagdown}}\text{Pt} - \text{X} & \xrightarrow{-\text{X}} & \text{L'} - \text{Pt} - \text{Y} \\
| & & | & & | \\
\text{L} & & \text{L} & & \text{L}
\end{array}
$$

The intermediate is trigonal bipyramidal (a few stable trigonal bipyramidal Pt(II) complexes have been isolated), with both entering and leaving groups plus the group *trans* to the leaving group in equatorial positions. This ensures that the *trans* arrangement of the two L groups is retained in the product.

This brings us to a phenomenon which has inspired a great deal of experimental work and theoretical speculation: the '*trans* effect'. Given a square planar complex, which of the four ligands (assuming these to be unidentate) is most likely to undergo substitution by an incoming ligand? Apparently ligands tend to render the ligands *trans* to themselves labile in an order called the *trans-directing series*:

$$CN^- > CO > H^- > CH_3^- > PR_3 > NO_2^- > I^- > Br^- > Cl^- > NH_3 > H_2O$$

A ligand which lies towards the right-hand side of this series and which is *trans* to a ligand towards the left-hand side of the series is likely to suffer substitution. Recognition of the *trans* effect is of great value in preparative strategies. A simple example is the preparation of the *cis* and *trans* isomers of $[Pt(NH_3)_2Cl_2]$. If $PtCl_4^{2-}$ is treated with aqueous ammonia, we obtain $Pt(NH_3)Cl_3^-$. Since Cl^- has a stronger *trans*-directing power than NH_3, further substitution by NH_3 will lead to the *cis*-product:

$$
\begin{array}{ccccc}
\left[\begin{array}{c} \text{Cl} \\ | \\ \text{Cl} - \text{Pt} - \text{Cl} \\ | \\ \text{Cl} \end{array}\right]^{2-} & \xrightarrow{\text{NH}_3} & \left[\begin{array}{c} \text{Cl} \\ | \\ \text{Cl} - \text{Pt} - \text{NH}_3 \\ | \\ \text{Cl} \end{array}\right]^{-} & \xrightarrow{\text{NH}_3} & \begin{array}{c} \text{NH}_3 \\ | \\ \text{Cl} - \text{Pt} - \text{NH}_3 \\ | \\ \text{Cl} \end{array}
\end{array}
$$

If we start with the $Pt(NH_3)_4^{2+}$ ion and substitute chloride for ammonia in the coordination sphere, we first obtain $Pt(NH_3)_3Cl^+$. The NH_3 *trans* to Cl is now rendered labile so that the next product is *trans*-$Pt(NH_3)_2Cl_2$:

$$
\begin{array}{ccccc}
\left[\begin{array}{c} \text{NH}_3 \\ | \\ \text{H}_3\text{N} - \text{Pt} - \text{NH}_3 \\ | \\ \text{NH}_3 \end{array}\right]^{2+} & \xrightarrow{\text{Cl}^-} & \left[\begin{array}{c} \text{NH}_3 \\ | \\ \text{H}_3\text{N} - \text{Pt} - \text{Cl} \\ | \\ \text{NH}_3 \end{array}\right]^{+} & \xrightarrow{\text{Cl}^-} & \begin{array}{c} \text{NH}_3 \\ | \\ \text{Cl} - \text{Pt} - \text{Cl} \\ | \\ \text{NH}_3 \end{array}
\end{array}
$$

The formation of one geometrical isomer rather than another is of much more than academic interest; cis-$Pt(NH_3)_2Cl_2$ has important applications as an anti-cancer agent, but the *trans* isomer is inactive in this respect.

The theoretical basis for the *trans* effect has been the subject of much debate. Various interpretations – ranging from simple electrostatic considerations to elaborate MO calculations – show that, for example, an M–N bond is weaker if it is *trans* to a CO group than if it is *trans* to Cl. Experimental evidence for such weakening is apparent from measurements of bond lengths, force constants, NQR coupling constants etc. But such measurements tell us about thermodynamics, rather than kinetics. Many writers use the term *trans influence* to describe this thermodynamic phenomenon, distinct from the *trans effect*, a kinetic phenomenon. Although there is a correlation between the two, the connection is indirect because an associative mechanism applies in most substitutions of square planar complexes; bond weakening as the cause of the *trans* effect implies a dissociative mechanism. Theoretical interpretations which explicitly recognise the formation of a trigonal bipyramidal five-coordinate intermediate have focussed upon the relative stabilities of the possible geometrical isomers; the three equatorial ligands comprise the entering group, the leaving group and the group *trans* to the leaving group in the square planar complex. Calculations of this kind will indicate which of the ligands should be the most probable leaving group, on thermodynamic grounds. Remember, however, that the rate of substitution depends on the activation barrier to formation of the five-coordinate intermediate; this will depend on a number of factors – steric repulsion, energy changes consequent upon the deformation of the square complex, possible participation by the solvent and entropy terms – other than the relative bond strengths in the five-coordinate intermediate. We are dealing with a kinetic contest between two (possibly more) pathways leading to different products. The winner will be the path which has the lowest activation barriers, and not necessarily the one leading to the more stable products via the more stable intermediate. This is illustrated schematically in Fig. 9.1, where we show the free energy/reaction coordinate profiles for two competing associative processes. The dashed curve is less favourable thermodynamically, but this pathway can be negotiated much more quickly than its rival. To summarise, it is perhaps asking too much to expect a simple, straightforward explanation for kinetic phenomena in terms of bonding theory. However, some empirical observations can be made. There is a fair correlation with electronegativity in the *trans*-directing series; ligands whose donor atoms are of relatively low electronegativity – C, H, P – have a strong *trans* effect, while the weakest effect is shown by N and O donor ligands. This may be further correlated with the softness of ligands as Lewis bases (Section 8.1); soft ligands tend to render hard ligands in *trans* positions labile. Why

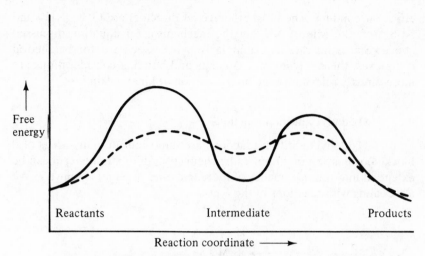

Fig. 9.1. Free energy/reaction coordinate profile for two competing associative pathways; the dashed line leads to less stable products via a less stable intermediate than the full line, but is faster because of smaller activation barriers.

trans and not *cis*? Two *trans* ligands in a square complex necessarily share the same central atom orbitals for bonding to a greater extent than two *cis* ligands; the reader may care to develop arguments along these lines, but we are now being drawn back into the fallacy of relying on thermodynamic arguments and the *trans* influence, rather than the *trans* effect.

Is there a *trans* effect in octahedral complexes? There is certainly experimental evidence for a *trans* influence similar to that found in square complexes, but a kinetic *trans* effect is less marked despite the fact that a dissociative mechanism applies in most cases. However, there is evidence in some types of octahedral complex for a *cis* effect; ligands having π donor characteristics render ligands *cis* to themselves labile and accelerate their departure as leaving groups. (In contrast, π acceptor ligands are among the most effective *trans* directors in square complexes.) The *cis* effect has been interpreted by considering π bonding between the *cis*-directing ligand and the central atom using a p orbital on the latter; this p orbital will be engaged in σ bonding with *cis* ligands. It generally happens that if an orbital on one atom is engaged in bonding to two other atoms, the strengthening of one bond causes weakening of the other. This explains the rendering of ligands *cis* to a π donor labile. Purely σ donors, like NH_3, exhibit no *cis* effect. Nor do π acceptor ligands like CO; overlap between the 4p orbitals of (say) cobalt with the π antibonding orbitals of CO – both of which are empty and at high energy – has no significant effect on bonding. The most effective *cis* directors are OH^-, NH_2^- and NCS^-; cf. the role of NH_2^- in the mechanism for aquation discussed above. Since the rate-determining step is dissociative for octahedral complexes, thermodynamic arguments and bonding considerations are more directly relevant in the interpretation of kinetic data.

9.6   Oxidative addition reactions

Oxidative addition reactions are important in many areas of d block coordination/organometallic chemistry, although the term can be extended into other branches of inorganic (and organic) chemistry. We are dealing with reactions of the type:

$$ML_n + X-Y \longrightarrow L_nM\begin{matrix} \nearrow X \\ \searrow Y \end{matrix}$$

where X—Y is a molecule which dissociates to give two separate ligands X and Y; these are formally to be regarded as anions, so that the oxidation number and coordination number of M both increase by two units. One way of looking at the situation is to imagine (without any mechanistic implications) that X—Y undergoes heterolytic cleavage to give X^+ and Y^-; X^+ then accepts two electrons from M to become X^-, and the two anions bind to the central atom, which has now had its oxidation state raised by two.

The most typical oxidative addition occurs with square coplanar complexes where the central atom has the d^8 configuration, especially Rh(I) or Ir(I). These conform to a 16-electron rule (Section 8.6); oxidative addition leads to an octahedral six-coordinate complex obeying the 18-electron rule, with a low-spin d^6 central atom, e.g.:

$$[Rh(PPh_3)_2(CO)Cl] + HCl \rightarrow [Rh(PPh_3)_2(CO)(H)Cl_2]$$
$$[Rh(PR_3)_3Cl] + CH_3I \rightarrow [Rh(PR_3)_3(CH_3)ClI]$$
$$[Ir(PR_3)_2(CO)Cl] + H_2 \rightarrow [Ir(PPh_3)_2(CO)(H)_2Cl]$$

The reverse process, reductive elimination, is also recognised as a common reaction in which the changes $d^6 \rightarrow d^8$ or $d^8 \rightarrow d^{10}$ occur.

The central atom in a square planar d^8 complex is sometimes described as *coordinatively unsaturated*. This term is best used to denote a tendency to take up additional ligands without change in oxidation state; it does not necessarily follow that a coordinatively-unsaturated species will be susceptible to oxidative addition. However, square d^8 complexes do have some tendency to take up a ligand to become trigonal bipyramidal, 18-electron species. Trigonal bipyramidal d^8 complexes formed by Fe(0),

Ru(0) and Os(0) are said to undergo oxidative addition in such reactions as:

$$[Os(CO)_3(PR_3)_2] + I_2 \rightarrow [Os(CO)_2(PR_3)_2I_2] + CO$$

Note, however, that one CO ligand is lost. The oxidation state of the Os increases by two units, from 0 to II; but the coordination number only increases by one. Both species obey the 18-electron rule. Oxidative addition is exhibited by some d^{10} systems, notably complexes of Pd(0) and Pt(0). A reaction such as:

$$\text{Pt(PPh}_3)_4 + \text{HgCl}_2 \longrightarrow \underset{\underset{\displaystyle Cl}{|}}{\overset{\overset{\displaystyle PPh_3}{|}}{Ph_3P - Pt - HgCl}} + 2PPh_3$$

may not seem like an addition; the oxidation number of the Pt increases from 0 to II, but the coordination number remains four (the coordination geometry, however, changes from tetrahedral to square planar). Its description as an oxidative addition can be justified if we recognise that the 18-electron species $Pt(PPh_3)_4$ suffers from steric overcrowding to the extent that it undergoes dissociation in solution to give the highly-reactive, coordinatively-unsaturated species $Pt(PPh_3)_3$ and $Pt(PPh_3)_2$ (16- and 14-electron species respectively). The *cis*-$Pt(PPh_3)_2$ fragment occurs in a great number of complexes, many of which can be prepared from $Pt(PPh_3)_n$, for example:

$$\text{Pt(PPh}_3)_n + \text{C}_2\text{H}_4 \longrightarrow (Ph_3P)_2Pt\overset{\diagup CH_2}{\underset{\diagdown CH_2}{\big|}}$$

$$\text{Pt(PPh}_3)_n + \text{(CF}_3)_2\text{CO} \longrightarrow (Ph_3P)_2Pt\overset{\diagup O}{\underset{\diagdown C(CF_3)_2}{\big|}}$$

These products could be regarded as Pt(0) complexes, i.e. as:

$$(Ph_3P)_2Pt \longleftarrow \overset{CH_2}{\underset{CH_2}{\|}} \qquad (Ph_3P)_2Pt \longleftarrow \overset{O}{\underset{C(CF_3)_2}{\|}}$$

where the ligands are bonded in a manner similar to that in Zeise's salt (Section 8.1) rather than as chelating dianions, and the term 'oxidative addition' is considered to be inappropriate. Whether these are to be classed as Pt(0) or Pt(II) complexes is a matter of taste.

The reader may wonder whether some of the ambiguities and descriptive problems in this section could be resolved by knowledge of the detailed reaction mechanisms of oxidative addition. Unfortunately, despite a great deal of work, the mechanisms are diverse, complex and not amenable to simple generalisations. The addition of H_2 to a square planar complex results in an octahedral product with *cis* hydrogens, suggesting a concerted mechanism in which the H–H bond breaks as the M–H bonds form. With HCl, CH_3I etc. the two new ligands may be *cis* or *trans*; the latter is favoured if the conditions encourage ionisation.

Oxidative addition reactions are important in synthetic coordination/ organometallic chemistry. Very unusual and unexpected ligands can be persuaded to bind to d block atoms. The $Pt(PPh_3)_n$ system is an important starting material for the preparation of square planar Pt(II) complexes, while square planar Rh(I) and Ir(I) complexes such as Vaska's compound $Ir(PPh_3)_2(CO)Cl$ provide routes to a great number of octahedral Rh(III) and Ir(III) complexes with 'soft' ligands like CO, PR_3 and H^-. Oxidative addition and reductive elimination are important in many catalytic applications of coordination/organometallic compounds (see Section 9.8).

A particularly important synthetic application is in the preparation of complexes with M–C σ bonds via cleavage of C–H bonds, by reactions of the type:

$$ML_n + RH \longrightarrow L_nM\begin{smallmatrix} R \\ \\ H \end{smallmatrix}$$

These occur most readily if the H atom involved is somewhat acidic, as in CH_3CN, CH_3NO_2 and other compounds where a strongly electron-withdrawing group is present, as well as acetylenes RC≡CH and aromatic compounds. A special type of C–H cleavage with formation of M–C bonds is called *cyclometallation*. This is often an intramolecular oxidative addition in which C–H bonds present in a ligand molecule (often an N- or P-donor) undergo cleavage, e.g.:

This reaction proceeds via an intramolecular oxidative addition, followed by reductive elimination of HCl. But many closely-related reactions apparently do not involve oxidative addition. Consider, for example, the 'orthomanganation' reaction:

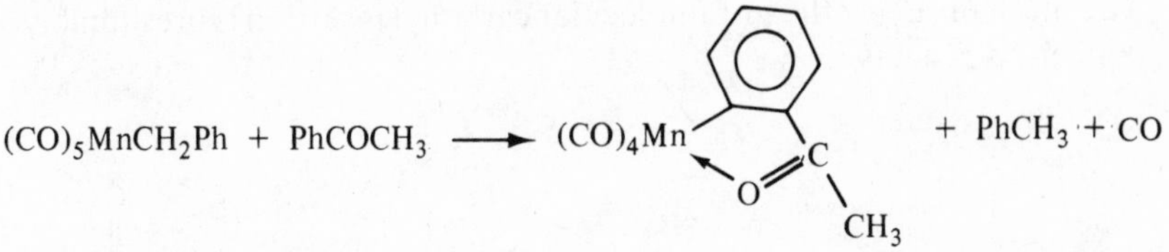

$$(CO)_5MnCH_2Ph + PhCOCH_3 \longrightarrow (CO)_4Mn \underset{O}{\overset{}{\diagdown}} \overset{}{\underset{}{\diagup}} C \diagdown CH_3 + PhCH_3 + CO$$

Both complexes are octahedral, 18-electron Mn(I) species. The formation of a seven-coordinate Mn(III) intermediate by oxidative addition is unlikely and the mechanism is probably a concerted process in which no Mn–H bond is ever formed.

9.7 Insertion reactions

These too have both synthetic and catalytic significance. Here we are dealing with reactions in which a molecule – such as CO – is inserted into the M–L bond in a coordination/organometallic compound, e.g.:

$$(CO)_5Mn—CH_3 + CO \rightarrow (CO)_5Mn—\overset{O}{\overset{\|}{C}}CH_3$$

The term is also apt for reactions of Main Group molecules which may not strictly qualify for admission to the realm of coordination chemistry; an example discovered in 1869 is:

$$SbCl_5 + 2HC≡CH \rightarrow SbCl_3(CH=CHCl)_2$$

The term insertion is usually restricted to reactions in which the central atom undergoes no change in oxidation state; in typical oxidative addition reactions, ML_n is inserted into X—Y, with oxidation of M.

In the case of the reaction of $Mn(CO)_5CH_3$ shown above, isotopic labelling studies (with ^{14}CO) show that the CO molecule inserted into the Mn–CH_3 bond is one of the five CO groups already coordinated, and not the 'external' CO shown in the stoichiometric formula. It appears that an equilibrium is set up between the $Mn(CO)_5CH_3$ and a five-coordinate, 16-electron (i.e. coordinatively-unsaturated) acyl complex:

$$(CO)_5Mn—CH_3 \rightleftharpoons (CO)_4Mn—\overset{O}{\overset{\|}{C}}—CH_3$$

The five-coordinate complex then reacts rapidly with external CO to give $(CO)_5MnCOCH_3$. Thermodynamically, the progress of the reaction to completion requires a ligand (not necessarily CO; PPh_3 will do) which can react quickly with the five-coordinate acyl complex to remove it from the system, so that more will be formed by rearrangement of $Mn(CO)_5CH_3$. The mechanism for the intramolecular carbonyl insertion is presumably a concerted process:

$$\underset{\overset{\displaystyle |}{(CO)_4Mn-CH_3}}{\overset{\displaystyle \overset{O}{\underset{|}{C}}}{}} \longrightarrow (CO)_4Mn \overset{O}{\underset{\diagdown}{\overset{C}{\diagup}}} CH_3 \longrightarrow (CO)_4Mn\overset{O}{\overset{\|}{C}}CH_3$$

Many other relatively small molecules can be inserted into M–L bonds, including SO_2, CO_2, CS_2 and O_2. Particularly important in industrial catalysis is the insertion of alkene molecules into M–H bonds (see Section 9.8):

$$L_nM—H + RHC=CH_2 \rightarrow L_nM—CRH—CH_3, L_nM—CH_2CH_2R$$

9.8   Catalysis by coordination/organometallic compounds in industry and biology

Our understanding of reaction mechanisms comes largely from kinetic studies. Catalysts affect reaction rates and the mechanisms of catalytic reactions are obviously of great interest. The use of coordination/organometallic compounds as catalysts in industrial processes has grown substantially in recent decades, and this has provided the stimulus (not to mention the funding) for much academic research in inorganic chemistry. During the same period, there has been an explosion of activity in bio-inorganic chemistry, or inorganic biochemistry. Although these two areas may seem to be poles apart, there are in fact some common principles and paradigms. We look first at some selected examples of industrial catalysis, and conclude with some remarks on the role of coordination chemistry in biology.

Wilkinson's catalyst $Rh(PPh_3)_3Cl$

This molecule would be expected to exhibit square planar coordination geometry, much favoured by central atoms having the d^8 configuration. However, the bulky triphenylphosphine ligands prefer tetrahedral coordination which keeps them further apart. The result of this conflict of interests is a compromise; the coordination geometry is best described as square planar with a pronounced distortion towards a tetrahedron. As with all compromises, neither party is entirely happy; the molecule is considerably strained, contributing to its reactivity. The compound catalyses the hydrogenation of $C=C$ bonds, an industrially-important process, at room temperature and atmospheric pressure. The reaction is homogeneous, i.e. the catalyst and the substrate alkene are in the same phase, which has some technical advantages.

Molecular hydrogen tends to react very slowly at ordinary temperatures, because any of its reactions must at some stage involve rupture of the strong H–H bond. However, Wilkinson's catalyst reacts rapidly with hydrogen:

$$
\begin{array}{ccc}
\underset{Cl}{\overset{Ph_3P}{\diagdown}}\!Rh\!\underset{PPh_3}{\overset{PPh_3}{\diagup}} & \xrightarrow{\ H_2\ } & \underset{Ph_3P}{\overset{Cl}{\diagdown}}\!\!\overset{\overset{PPh_3}{|}}{Rh}\!\!\underset{\underset{PPh_3}{|}}{\overset{H}{\diagup}}\!\!{\diagdown}H
\end{array}
$$

The product is an octahedral Rh(III) complex; this is an example of an oxidative addition reaction, in which the coordination and oxidation numbers of the central atom both increase by two units. In $Rh(PPh_3)_3(H)_2Cl$, the steric strain resulting from interligand repulsion is now sufficiently severe that a triphenylphosphine is easily lost to give the five-coordinate, trigonal bipyramidal complex $Rh(PPh_3)_2(H)_2Cl$. This is coordinatively unsaturated, since Rh(III) is normally found in octahedral six-coordination and such a five-coordinate complex will seize avidly upon any available ligand that is less sterically demanding than triphenyl-phosphine. Thus in the presence of an alkene, the complex $Rh(PPh_3)_2(H)_2Cl(alkene)$ is readily formed. Coordinated H^- now hydro-genates the alkene to alkane, leaving $Rh(PPh_3)_2Cl$; this is a reductive elimination, the reverse of oxidative addition. We now have another coordinatively-unsaturated molecule which readily reacts with H_2 to give $Rh(PPh_3)_2(H)_2Cl$ again, and the catalytic cycle continues.

The complex $Ru(PPh_3)_3Cl_2$ was mentioned in Section 8.3 as a rare example of five-coordinate Ru(II), the sixth coordination site being blocked by an H atom belonging to a phenyl group. This compound reacts with H_2 to give $Ru(PPh_3)_3(H)Cl$, which is also coordinately unsaturated but is trigonal bipyramidal rather than square pyramidal, and is a specific catalyst for the hydrogenation of 1-alkenes.

Hydroformylation

The hydroformylation reaction:

$$RCH=CH_2 + H_2 + CO \rightarrow RCH_2CH_2CHO$$

is of great industrial importance. The original process made use of cobalt salts as catalysts, and $Co(CO)_4H$ appears to be the active agent. This is now being superseded by better catalysts, especially $Rh(PPh_3)_3(H)CO$, a close relative of Wilkinson's catalyst (and indeed first prepared in Wilkinson's laboratory). The reaction scheme is shown in Fig. 9.2. Here we have a trigonal bipyramidal Rh(I) complex, which readily loses PPh_3 to give $Rh(PPh_3)_2(H)CO$. This picks up an alkene molecule, which the coordinated hydride converts into a σ bonded alkyl group. The four-coordinate Rh(I) can now accept a CO group, restoring it to five-coordination. The next step is an example of an insertion reaction, very

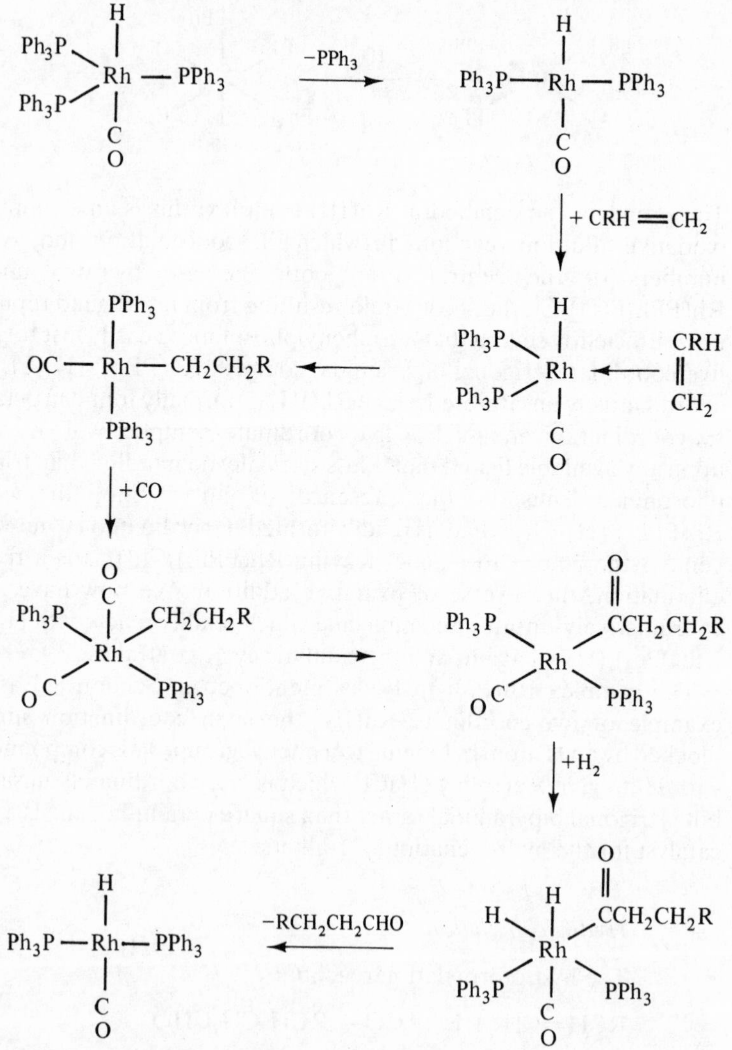

Fig. 9.2. Proposed mechanism for hydroformylation catalysed by $Rh(PPh_3)_3(H)CO$.

common in carbonyl chemistry (Section 9.7). A CO ligand inserts itself into the Rh–C bond so that we now have the makings of an aldehyde. Oxidative addition of hydrogen then occurs, followed by reductive elimination of the required aldehyde and the return of $Rh(PPh_3)_2(H)CO$, ready to begin another cycle.

The Wacker process

This is the oxidation of ethylene to acetaldehyde, catalysed by $PdCl_2$ and $CuCl_2$:

$$C_2H_4 + \tfrac{1}{2}O_2 \rightarrow CH_3CHO$$

The probable mechanism is shown in Fig. 9.3. Water adds to the

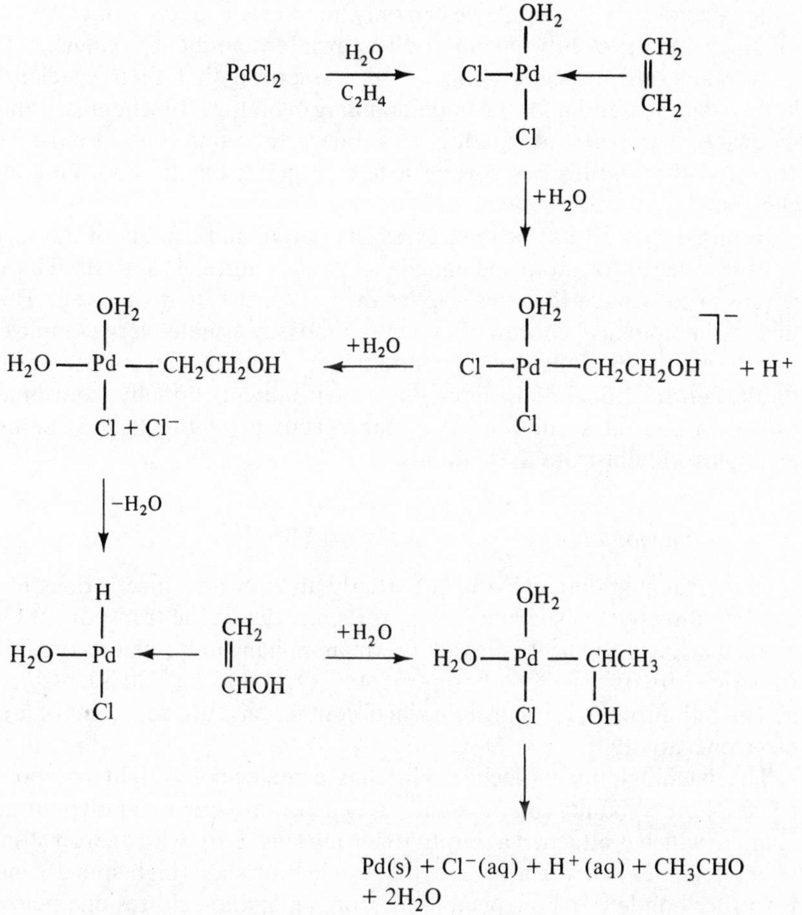

Fig. 9.3. Proposed mechanism for Wacker process for oxidation of ethylene to acetaldehyde.

coordinated ethylene. A hydride ion is then abstracted from the coordinated hydroxo-alkyl group, to become bonded to the Pd atom. The hydride then attacks the other C atom – in effect, H^- has been transferred from one C atom to the other – and CH_3CHO is eventually released. The Pd has been reduced to the elemental state (having oxidised the coordinated ligand) and is restored to the II state by the action of $CuCl_2$. The copper is reduced to the I state, and returns to the II state by the agency of O_2. The cycle can now continue.

Coordination compounds in biology

This is now a very large and rapidly-growing area of investigation. Coordination compounds of many elements have been identified in diverse life-forms, and complex formation is important in pharmacology and toxicology. In this book we can only hint at some of the principles by looking rather sketchily at some well-known bio-inorganic systems.

Inorganic chemists are prone to the assertion that their specialist knowledge is essential for the understanding of biology, biochemistry and medicine. There is some truth in this, but at the same time, it must be admitted that nature has sprung a few surprises on the coordination chemist.

In this section, it has been stressed that strain and conflict of interest between the central atom and ligands is often a feature of catalysts. There is reason to believe that this applies in biological systems as well. The ligands in biological coordination compounds are usually very complex, and the subtleties of their conformational requirements can impose strain on the central atom and its immediate environment. Equally, coordination to a central atom can affect the reactivity of the ligand. Some examples will illustrate these points.

Haemoglobin

Haemoglobin (Hb) is not strictly an enzyme, since it does not catalyse directly any specific reaction. Its function is the transport of O_2 from the lungs to the tissues of the mammalian body, where organic substrates are oxidised exothermically to CO_2 and water. Hb is classified as a metalloprotein, i.e. a protein which contains and utilises atoms of low electronegativity.

The haemoglobin molecule (Hb) has a molecular weight of about 64 000. Four subunits can be identified, each consisting of a polypeptide chain to which is attached a porphyrin group (Fig. 8.6), with an iron atom near its centre. The iron atom is in the II oxidation state (high-spin d^6) and is further bonded to a nitrogen atom from an amino-acid residue below the porphyrin ring. The conformation of the polypeptide chain prohibits completion of octahedral six-coordination by binding of a sixth donor atom (such as O or N) provided by the polypeptide. Since haemoglobin acts in (and is crystallised from) aqueous solution, one might expect a water molecule to occupy the sixth coordination site. However, this site is close to a *hydrophobic pocket*, a region within the molecule where a concentration of non-polar C–H bonds tends to repel water molecules. There is an H_2O molecule lurking close to the hydrophobic pocket, but it is too far away from the Fe atom to be regarded as anything better than semi-coordinated; a closer approach to the Fe atom to complete full coordination would cost it dearly in terms of hydrogen bonding to other groups. Thus the Fe atom can be described as coordinatively unsaturated. Any potential ligand which can find its way into the hydrophobic pocket will be avidly seized upon by the Fe atom in an associative process, with a minimal activation barrier. Molecules such as O_2, CO and PF_3 are unafraid of hydrophobic pockets and are readily bound. The high toxicities of CO and PF_3 are attributable to their irreversible coordination to Fe in Hb. Oxygen molecules (being somewhat poorer ligands) are reversibly coordinated and are given up where the O_2 concentration is low.

The mode of O_2 coordination was for some time a subject of controversy. It now seems settled that the ligand molecule is bonded via one O atom, with an Fe–O–O angle of *c.* 120°. This is consistent with the prediction of a rather crude VB description of the O_2 molecule, which allocates two lone pairs to each atom; the valence state of O is $(sp^2)^2(sp^2)^2(sp^2)^1(p)^1$. This description is inconsistent with the observed paramagnetism of O_2 (which can readily be explained in terms of MO theory). However, oxygenated haemoglobin $Hb.4O_2$ is diamagnetic and we may fairly describe the Fe–O bond in terms of a filled sp^2 hybrid donor orbital from the O atom.

Molecular oxygen will usually oxidise an Fe(II) complex to Fe(III), rather than coordinate reversibly. The mechanism for oxidations such as:

$$4Fe^{2+}(aq) + O_2(aq) + 4H^+(aq) \rightarrow 4Fe^{3+}(aq) + 2H_2O$$

always appears to involve the formation of a dimeric complex such as $[(H_2O)_5Fe—O—O—Fe(H_2O)_5]^{4+}$, which then undergoes hydrolysis to give H_2O_2 and an Fe(III) complex. In Hb, dimerisation is impossible because the Fe atoms are kept well apart by the polypeptide chains.

The Fe(II) atom in unoxygenated Hb is not quite in the centre of the porphyrin ring (which is somewhat buckled); it lies about 70 pm from the centre of the (approximate) plane defined by the four N atoms. Evidently high-spin Fe(II) is a little too large to fit perfectly into the ring. But – as noted in Section 4.2 – low-spin Fe(II) is significantly smaller in radius; the change in spin state from high-spin to low-spin Fe(II) which accompanies oxygen uptake leads to subtle structural changes stemming from the movement of the Fe atom towards the centre of the porphyrin ring. These are responsible for the observation of *cooperativity*; the coordination of one O_2 molecule to an Fe(II) atom increases the oxygen affinities of the remaining, unoxygenated Fe atoms. This ensures that, given a sufficiently high concentration of O_2, an Hb molecule will take up a full load and will deposit a full load whenever the O_2 concentration is low enough.

Carboxypeptidase A

This digestive enzyme catalyses the hydrolysis of peptides, i.e. the reaction:

$$
\begin{array}{ccc}
R\!-\!C\!=\!O & & R\!-\!COOH \\
\mid & \xrightarrow{\ H_2O\ } & \\
NH & & + \\
\mid & & \\
R' & & R'\!-\!NH_2
\end{array}
$$

Each molecule (molecular weight ~30 000) contains one zinc(II) atom, which is (approximately) tetrahedrally coordinated to two N atoms and one O atom from amino-acid residues plus a water molecule. The structures of both the enzyme and some enzyme–substrate complexes have been carefully studied and the detailed mechanism of the hydrolysis is now quite well understood. Without going into details, a crucial factor appears to be the pronounced distortion from regular tetrahedral coordination about the Zn(II), apparently imposed by the conformational requirements of the polypeptide chain. The conflict of interest between the needs of the Zn(II) atom – which, when four-coordinate, always assumes tetrahedral coordination – and the ligands induces an *entatic state*, a condition of strain and tension which enhances the reactivity at the active site. The Zn atom binds the substrate peptide via the O atom of the —CONH— peptide link, and the entatic state of the free enzyme facilitates formation of the enzyme–substrate complex.

The Zn atom can be replaced by Co without much loss of enzymatic activity; Co(II) shares with Zn(II) a liking for tetrahedral four-coordination. But if the Zn(II) is replaced by Ni(II) or Cu(II), the activity is virtually lost. Tetrahedral d^8 and (especially) d^9 complexes are subject to Jahn–Teller distortions, so that the imposition of a distortion from regular tetrahedral geometry does not lead to any strain; there is no entatic state.

Vitamin B_{12} coenzyme (5'-deoxyadenosylcobalamin)

This (as its name implies) functions as a co-catalyst in conjunction with various enzymes. It is involved, *inter alia*, in the catalysis of rearrangements via hydride transfer:

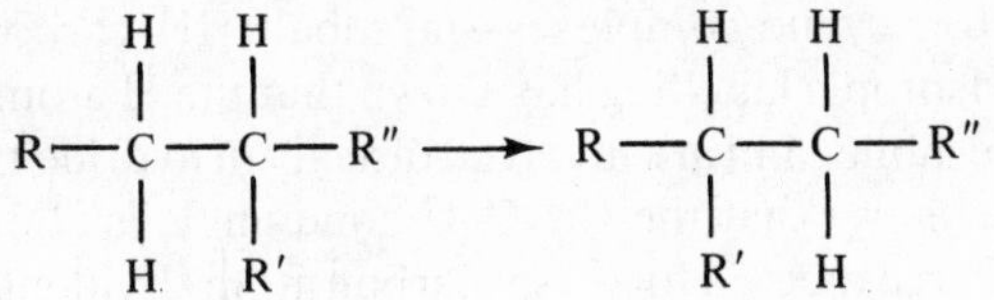

$$R-\overset{\overset{\displaystyle H}{|}}{\underset{\underset{\displaystyle H}{|}}{C}}-\overset{\overset{\displaystyle H}{|}}{\underset{\underset{\displaystyle R'}{|}}{C}}-R'' \longrightarrow R-\overset{\overset{\displaystyle H}{|}}{\underset{\underset{\displaystyle R'}{|}}{C}}-\overset{\overset{\displaystyle H}{|}}{\underset{\underset{\displaystyle H}{|}}{C}}-R''$$

e.g.:

$$CH_2OHCH_2OH \rightarrow CH_3CH(OH)_2 \rightarrow CH_3CHO + H_2O$$

It had been known since the mid-1940s that Vitamin B_{12} and most of its derivatives contain octahedrally-coordinated cobalt(III), one atom per molecule. Vitamin B_{12} as originally crystallised from aqueous potassium cyanide was found (not surprisingly) to have a CN^- ligand together with N-donor atoms (four from a planar macrocycle, corrin, and one from a heterocyclic group) attached to the Co(III) atom. The announcement in 1958 by Dorothy Hodgkin of the structure of the coenzyme – isolated without recourse to cyanide – caused a sensation among coordination chemists (Fig. 9.4). One of the ligands was a carbanion CH_2R^- (deoxy-adenosyl). This was not only the first organometallic compound to be isolated from a biological system; it was also the first organometallic Co(III) complex. For over a century, chemists had studied octahedral Co(III) complexes; several thousand had been prepared and charac-

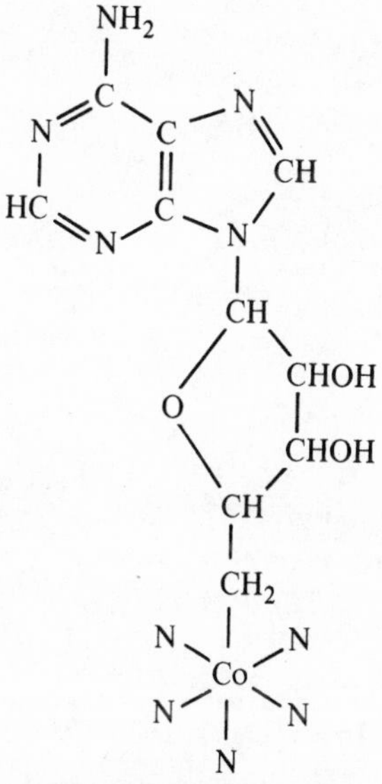

Fig. 9.4. Vitamin B_{12} coenzyme structure, showing bonding of deoxy-adenosyl carbanion to cobalt(III). Four of the five N atoms bonded to Co are furnished by the corrin ring and one is from a benzimidazole group.

terised. But – apart from cyano-complexes – no cobalt(III)–carbon bonds had been identified. Isotopic labelling has shown that the H atoms of the Co—CH$_2$R group are labile, and are involved directly in hydride transfer. The crystal structure shows that the Co–C–C bond angle is about 125°, compared with 109.5° as expected for an sp^3 carbon atom. Evidently there has been some rehybridisation of the C atom on coordination to the Co atom, which renders the C–H bonds labile.

To conclude this section, the reader should note the powerful analogies between industrial and biological catalysis by coordination and organometallic compounds. These two areas will certainly attract increasing activity and attention in the future, but should be seen as closely-related and mutually-supporting branches of inorganic chemistry.

9.9   Further reading

In addition to relevant chapters in inorganic texts, see the books listed in Section A.9 of the Appendix. An interesting account of donor/acceptor interactions is given by Gutmann, V. (1978). *The Donor–Acceptor Approach to Molecular Interactions*. New York: Plenum. For bio-inorganic chemistry, see the book by Hughes (1982) cited in Section A.12. For Sections 9.6 and 9.7, Cotton and Wilkinson's (1988) text (Section A.3) should be consulted.

10

The preparation of inorganic substances

10.1 Introductory remarks

It need hardly be said that the preparation of inorganic substances is important in the study of descriptive inorganic chemistry. However, the diversity of inorganic substances and preparative methods defies any attempt to set out a wholly satisfactory scheme of classification. Accordingly, this chapter concentrates on examples of inorganic preparations, chosen to illustrate some of the more important points. It would be inappropriate to place much emphasis on matters of practical technique; this is best studied in the laboratory rather than in the library. Success in synthetic inorganic chemistry depends as much on manipulative skill, experience, serendipity and good luck as on a thorough knowledge of thermodynamics, kinetics and bonding theory.

Many newly-prepared inorganic substances result from thinking along the lines of 'let's react A with B and see what happens'. Having characterised the product(s) of the reaction, further experiments may lead to improved methods of obtaining particular products. Although the substance may have been the completely unexpected product of a reaction, some degree of deliberate strategy – together, perhaps, with some trial-and-error experiment – is usually necessary in order to optimise the conditions. Ideally, the requirements for a convenient laboratory preparation should include:

(i)   readily-available starting materials;
(ii) smooth, rapid reaction leading to a high yield of the desired product;
(iii) facile separation and isolation of product from by-products, unreacted starting materials, solvent etc.

Many of the substances whose preparations are described in this chapter are available commercially, and some are manufactured on a very large scale. The preparative methods employed – economic considerations are, of course, dominant in the chemical industry – are largely of academic interest but the underlying principles are important to the student of inorganic chemistry. Even where a substance is commercially available, it may be necessary – for reasons of economy or urgency – to prepare a sample in the laboratory. The choice of method will depend on the materials and apparatus available, and is unlikely to be a scaled-down version of the industrial process.

Thermodynamic considerations are obviously dominant in planning a synthesis, and these, of course, include solubilities, boiling points etc. Kinetic and mechanistic considerations will determine the rate of the reaction(s) involved in the preparation, the extent of competing reactions (which may affect the yield of the desired product and the ease of its isolation) and, where appropriate, which of the possible isomeric forms of the product is favoured.

10.2   Elemental substances

Since an elemental substance is associated with the zero oxidation state, it follows that preparative methods will usually involve either reduction of the element E from a positive oxidation number, or oxidation from a negative oxidation number. The former will obviously be applicable where E is of low electronegativity, and the latter where E is of high electronegativity. In some cases, elemental substances are obtained by thermal decomposition of an E(0) compound, e.g. $Ni(CO)_4$.

Reductive preparation of elemental substances

Here we are dealing mainly with metallic elemental substances. The choice of reagents is governed by the usual thermodynamic and kinetic considerations, but since practically all elemental substances used in the laboratory are obtained from commercial sources, availability and cost are important considerations. The oxide or chloride of E is the usual starting material; the reduction may be accomplished by chemical or electrochemical means. The chemical reducing agent is usually an elemental substance. It is possible to set up a thermochemical series, analogous to the electrochemical series based upon redox potentials. The crucial quantity is $(-\Delta G_f^0/n)$ for an oxide or halide, where n is the oxidation number of E; for oxides such as E_2O_3 the stoichiometric unit is redefined as $EO_{n/2}$ to allow direct comparison with halides EX_n, so that the appropriate value of ΔG_f^0 for iron(III) oxide is half of that listed for Fe_2O_3. If an oxide $EO_{n/2}$ lies lower in the series than the oxide of another element E', then elemental E' should be able (under standard conditions, of course) to reduce $EO_{n/2}$ to yield elemental E. Values for a number of representative elements are given in Table 10.1.

Table 10.1 *Values of* $(-\Delta G^0_f/n)$ *for oxides and chlorides of selected elements (in kJ mol^{-1})*

Element	n	$EO_{n/2}$	ECl_n
H	1	118	95
Li	1	280	384
Cs	1	154	415
Ag	1	6	110
Be	2	290	223
Mg	2	285	296
Hg	2	29	89
B	3	199	130
Al	3	264	210
Fe	3	124	111
Ce	3	284	326
C	4	96	15
Si	4	214	155
Ti	4	222	184

For the chosen elements, the thermochemical series of oxides and chlorides is:

$$CsCl > LiCl > CeCl_3 > MgCl_2 > BeO > MgO > Ce_2O_3 >$$
$$Li_2O > Al_2O_3 > BeCl_2 > TiO_2 > SiO_2 > AlCl_3 > B_2O_3 >$$
$$TiCl_4 > Cs_2O > BCl_3 > Fe_2O_3 > H_2O > FeCl_3 > AgCl >$$
$$CO_2 > HCl > HgCl_2 > HgO > CCl_4 > Ag_2O$$

We can say that, for example, elemental magnesium will reduce aluminium(III) chloride, but hydrogen gas will not reduce SiO_2. Note that such predictions hold good only under standard conditions, viz. 25 °C/1 atm. Thus at room temperature carbon will not reduce iron(III) oxide, but it should be apparent from the production of $CO_2(g)$ that heating will drive the reaction forward, since the entropy change is positive. Carbon can in practice reduce most oxides where $(-\Delta G^0_f/n)$ is less than about 150 kJ mol^{-1}, which covers SnO_2 and ZnO as well as Fe_2O_3, but not Cr_2O_3 or MnO. Where stronger reducing agents than carbon are needed, gaseous hydrogen is often effective and may be commercially preferable to carbon in some parts of the world. The use of high pressures can effect reduction using hydrogen that would not be thermodynamically possible at one atmosphere. Elemental substances such as Al, Mg and Li are evidently good reducing agents for the preparation of metallic elemental substances E(s) where carbon or hydrogen fail. On an industrial scale, electrochemical methods are preferred for the preparation of elemental E in such cases, provided that an oxide/halide or complex oxide/complex halide of E can be found that forms a molten electrolyte at a reasonable temperature.

The reader will notice that the relative magnitudes of $(-\Delta G_f^0/n)$ for oxides and chlorides vary considerably over the chosen elements. Among the oxides, we can say that the largest values are found where E is small, of low electronegativity and in a high oxidation state. The combination of highly-disparate values for these variables accounts for the very comparable values found for E = Li, Be, Mg, Al and Ce, which are among the most 'thermopositive' elements, as far as the formation of oxides is concerned. Looking now at the chlorides, we see a rather different picture. For ECl_n, there is a tendency towards a large $(-\Delta G_f^0/n)$ compared with the corresponding oxide if E is large in size, and high in electronegativity, and if n is small. The best rationale of these observations – concealing a multitude of complex thermodynamic considerations – is to invoke the concept of hard and soft acids and bases (Section 8.1). The hard base O^{2-} will be preferred by hard acids: small cations E^{n+} where E is of relatively low electronegativity. The softer base Cl^- will tend to favour softer acids. Thus there is a large difference in behaviour between Li and Cs in Table 10.1; Cs^+ is a much softer acid than Li^+. Caesium metal will reduce any chloride under standard conditions, but is a relatively poor reducing agent for oxides.

The oxides formed with soft acids like Ag^+ and Hg^{2+} are of very low thermal stability, and the metal can be obtained from them by moderate heating. Gold, the most noble metal of all, does not form any stable oxide. Contrariwise, the bromides and iodides of elements whose cations are classed as hard acids can often be thermally decomposed at moderate temperatures. Such a procedure is rarely economic for industrial production, but can be useful for small-scale laboratory preparations, especially if the elemental substance is required in a state of high purity.

For many elements, a sulphide is the commonest natural source. It is usually most convenient to roast this in air to yield the oxide, which is then reduced. In some cases, the elemental substance can be obtained more directly by converting part of the sulphide ore to oxide, then heating the mixture of sulphide and oxide in the absence of air, e.g.:

$$Cu_2S + 2Cu_2O \rightarrow 6Cu + SO_2$$

Finally, some elemental substances can be isolated by precipitation from aqueous solution. For example, copper metal can be recovered from scrap or from low-grade ores (ultimately) by the reduction of $Cu^{2+}(aq)$ with iron metal.

Oxidative preparation of elemental substances

This is important only for the halogens. Elemental substances such as oxygen, nitrogen and sulphur are so abundant on Earth that their chemical preparation is of little interest nowadays. The heavier elements of Groups 15 and 16 – although they can exhibit negative oxidation states – are obtained as elemental substances by reduction of the oxides or complex oxides. Chlorine is always manufactured by the eléctrolytic oxidation of Cl^-, either in aqueous solution or in a fused salt such as NaCl; relatively few chemical oxidants can achieve this, and none is economically competitive with the electrochemical method. Bromine is usually prepared by oxidation of $Br^-(aq)$ with chlorine. Iodine is similarly prepared by oxidation of $I^-(aq)$, although $NaIO_3$ – which occurs in Chile saltpetre – is an important source. In the latter case, $IO_3^-(aq)$ is partially reduced to $I^-(aq)$ with bisulphite $SO_2(OH)^-$, and iodine is produced by the reaction:

$$5I^-(aq) + IO_3^-(aq) + 6H^+(aq) \rightarrow 3I_2(s) + 3H_2O$$

The preparation of elemental fluorine has been a continuing challenge to inorganic chemists for many decades. The element was recognised as early as 1812 – not long after the discovery of chlorine – but the elemental substance was not obtained until 1886. Moissan's original method – the electrolysis of KF dissolved in anhydrous HF – remained for a century the only practical preparation of fluorine from sources whose own preparations do not require elemental fluorine. There exists no chemical agent that will oxidise $F^-(aq)$ to fluorine, nor any that will oxidise a fluoride with evolution of fluorine. Some binary fluorides and complex fluorides decompose on mild heating with evolution of F_2, and provide convenient sources of fluorine for small-scale laboratory work. An example is $K_2[PbF_6]$:

$$K_2[PbF_6] \xrightarrow{150°} K_2PbF_4 + F_2$$

But $K_2[PbF_6]$ is prepared by the reverse reaction (performed at a lower temperature) and serves as a storage material for F_2 rather than as a genuine synthetic precursor. The search for a cheap, convenient preparative route to elemental fluorine by purely chemical means met with its first success in 1986, the centenary of Moissan's triumph. K. O. Christe took advantage of the fact that many complex halogeno-anions $EX_m^{(m-n)-}$ can be obtained as stable species in solution or in the solid state despite the fact that the binary halide EX_n cannot be isolated. Examples include $CeCl_6^{2-}$, $PdCl_6^{2-}$, CuF_6^{2-} and MnF_6^{2-}. The last of these provided the key to Christe's success. The solid $K_2[MnF_6]$ can be prepared by the reaction:

$$2KMnO_4 + 2KF + 10HF + 3H_2O_2 \rightarrow 2K_2[MnF_6] + 8H_2O + 3O_2$$

It is quite stable thermally: we expect the octahedral Mn(IV) complex (d^3) to enjoy a considerable degree of crystal field stabilisation. Christe reasoned that the unstable MnF_4 could be displaced by treating $K_2[MnF_6]$ with a stronger Lewis acid than MnF_4. Antimony(V) fluoride immediately sprang to mind (Section 9.2), and fluorine gas was obtained by the reaction:

$$K_2[MnF_6] + 2SbF_5 \rightarrow 2K[SbF_6] + MnF_3 + \tfrac{1}{2}F_2$$

This preparation is perhaps of more interest as an academic exercise than as a practical alternative to the electrochemical method.

10.3   Hydrides and halides

The most obvious route to a binary compound $E_xE'_y$ is the direct interaction between the elemental substances E and E'. This is often precluded by thermodynamic considerations; however, a great many compounds that are unstable with respect to the elemental substances are kinetically stable and can be prepared by other means. Even if the reaction between the elemental substances is possible, both from thermodynamic and kinetic viewpoints, an indirect synthesis may be preferable for practical reasons. We shall illustrate these principles by looking at some examples of synthetic routes to binary hydrides and halides.

These methods are, *mutatis mutandis*, applicable to the formation of E–H and E–X (X = halogen) bonds in compounds other than binary hydrides and halides.

Hydrides

The direct formation of a binary hydride EH_n from the elemental substances is usually favoured thermodynamically if E is much lower, or much higher, in electronegativity than H. Thus, for example, the free energies of formation ΔG_f^0 of LiH and HCl are respectively -68 and $-95\,kJ\,mol^{-1}$. In the case of HCl, it can be predicted from the relevant bond energies – rationalised in terms of Pauling's derivation of electronegativities – that the formation of HCl(g) from the elemental substances should be exothermic. Further, entropy considerations will favour the randomisation of atoms in molecules. For the reaction:

$$\tfrac{1}{2}H_2(g) + \tfrac{1}{2}Cl_2(g) \rightarrow HCl(g)$$

$T\Delta S^0$ at 298 K is $+3\,kJ\,mol^{-1}$. In the case of LiH, an ionic solid having the NaCl structure, a thermochemical analysis (see Chapter 5) can rationalise its thermodynamic stability relative to the elemental substances. However, ΔG_f^0 for LiH is much less negative than for LiF or LiCl (-616

and $-409\,kJ\,mol^{-1}$ respectively), despite the fact that the lattice energy of LiH is intermediate between the values for LiF and LiCl. This can be attributed to the higher dissociation energy of H_2 compared with Cl_2 and F_2 (Table 6.2) and to the less exothermic electron attachment energy of $H(g)$ relative to $F(g)$ and $Cl(g)$ (Table 4.4). Ionic hydrides EH_n are thermodynamically stable only where the elemental substance E has a low atomisation enthalpy, and where the resulting atomic substance $E(g)$ has a low ionisation enthalpy. These conditions are met only by the Group 1 elements and the heavier members of Group 2 (from Ca downwards, with Mg as a marginal case).

Lithium metal does not react at an appreciable rate with hydrogen gas at room temperature; the complete reaction requires temperatures in excess of 700 °C. The high dissociation energy of the H_2 molecule – which must be overcome before any hydride can be formed – renders gaseous hydrogen a rather sluggish reagent in the absence of heat or a catalyst which breaks the H–H bond. Hydrogen and chlorine do not react at room temperature, unless the mixture is exposed to light; photons of appropriate wavelength (<500 nm, towards the blue/violet end of the visible spectrum) are capable of cleaving the Cl–Cl bond, and the production of atomic chlorine leads us to consider the fast reaction:

$$Cl(g) + H_2(g) \rightarrow HCl(g) + H(g)$$

for which ΔG^0 at 298 K is $+2\,kJ\,mol^{-1}$. Although this quantity is positive, it is small enough in magnitude that any appreciable concentration of $Cl(g)$ will lead to the formation of a significant amount of $H(g)$, which undergoes the fast reaction:

$$H(g) + Cl_2(g) \rightarrow HCl(g) + Cl(g) \quad \Delta G^0 = -198\,kJ\,mol^{-1}$$

The production of more $Cl(g)$ allows the chain to continue.

The free energies of formation of hydrides EH_n where E is comparable in electronegativity with hydrogen are often positive. This applies to all hydrides of boron, all the hydrides of the Group 14 elements (except for a few of the lower alkanes), all the hydrides of the Group 15 and Group 16 elements (except NH_3, H_2O, H_2O_2 and H_2S) and HI. For example, the direct formation of SiH_4:

$$Si(s) + 2H_2(g) \rightarrow SiH_4(g) \quad \Delta G^0 = +57\,kJ\,mol^{-1}$$

seems to be distinctly improbable under any conceivable conditions. Nor are reactions such as the following any more promising:

$$SiO_2(s) + 4H_2(g) \rightarrow SiH_4(g) + 2H_2O(l)$$
$$\Delta G^0 = 494\,kJ\,mol^{-1}$$

$$SiCl_4(l) + 4H_2(g) \rightarrow SiH_4(g) + 4HCl(g)$$
$$\Delta G^0 = 297\,kJ\,mol^{-1}$$

Monosilane SiH_4 is indeed unstable with respect to decomposition to the elemental substances. It can, however, be isolated as a pure substance by reactions such as the following, where the number under each species is its free energy of formation ΔG_f^0 in kJ mol^{-1}:

$$SiO_2(s) + LiAlH_4(s) \rightarrow SiH_4(g) + LiAlO_2(s)$$
$$-857 \qquad -45 \qquad\qquad +57 \qquad -1131$$
$$\Delta G^0 = -172\,kJ\,mol^{-1}$$

$$SiCl_4(l) + LiAlH_4(s) \rightarrow SiH_4(g) + LiAlCl_4(s)$$
$$-620 \qquad -45 \qquad\qquad +57 \qquad -1030$$
$$\Delta G^0 = -308\,kJ\,mol^{-1}$$

It should be apparent that the real driving force for these reactions comes from the highly-negative free energies of formation of $LiAlO_2$ and $LiAlCl_4$, compared with SiO_2 and $SiCl_4$ respectively. This illustrates a valuable (though not infallible) principle, that the most stable arrangement of any collection of atoms will be that in which the least electronegative atoms are bonded to the most electronegative ones (cf. Pauling electronegativities and bond energies, Section 4.4). In each case, the most electronegative atoms – O and Cl respectively – finish up in combination with the least electronegative atoms, Li and Al. ($LiAlO_2$ is a mixed oxide having a wurtzite structure, with tetrahedral four-coordination of all atoms; $LiAlCl_4$ consists of Li^+ and $AlCl_4^-$ ions.) A great many synthetic routes owe their thermodynamic feasibility to the appearance on the right-hand side of the equation of species having very negative enthalpies/free energies of formation.

From the practical point of view, the preparation of SiH_4 is dominated by the need to exclude moisture and oxygen, either of which will lead to the formation of SiO_2. Given these precautions, SiH_4 is stable at room temperature despite the positive ΔG_f^0.

Another method of preparing 'unstable' hydrides EH_n is by the acidification of binary solids $E_x'E_y$ where E' is of lower electronegativity than E, e.g.:

$$Mg_2Si(s) + 4H^+(aq) \rightarrow SiH_4(g) + 2Mg^{2+}(aq)$$

The free energy of formation of $Mg^{2+}(aq)$ is $-455\,kJ\,mol^{-1}$ relative to the hydrated proton, and this drives the reaction. The familiar preparation of H_2S by acidification of a binary sulphide such as FeS comes into the same category.

Halides

Practically all known binary halides have negative free energies of formation, so that they can – in principle – be prepared directly from the elemental substances. Such direct preparation is indeed commonly employed. However, other methods are often necessary or preferable.

The preparation of binary fluorides themselves, as well as derivatives such as oxofluorides and organofluorine compounds, is not as straightforward as one might think. The most obvious fluorinating agent is, of course, gaseous fluorine, but this tends to react violently and indiscriminately, and specialised equipment is needed for its safe handling. Wet methods can sometimes be used for the preparation of an insoluble ionic fluoride (e.g., the addition of fluoride ion to an aqueous solution, or the treatment of a basic oxide with aqueous HF), but these methods are obviously inappropriate if the fluoride undergoes hydrolysis. They may produce hydrated fluorides, from which water cannot be removed without loss of HF and formation of a hydroxofluoride or oxofluoride. A range of fluorinating agents, from 'gentle' to 'fierce' in power, is available for the selective preparation of fluorides, oxofluorides etc. These agents can be regarded as sources of F atoms via endothermic decompositions, e.g.:

$$\Delta H^0 \, (\text{kJ mol}^{-1})$$

$LiF(s) \rightarrow Li(s) + F(g)$	695
$CsF(s) \rightarrow Cs(s) + F(g)$	633
$\frac{1}{2}ZnF_2(s) \rightarrow \frac{1}{2}Zn(s) + F(g)$	461
$AgF(s) \rightarrow Ag(s) + F(g)$	283
$AgF_2(s) \rightarrow AgF(s) + F(g)$	234
$TlF(s) \rightarrow Tl(s) + F(g)$	403
$\frac{1}{2}TlF_3(s) \rightarrow \frac{1}{2}TlF(s) + F(g)$	204
$CoF_3(s) \rightarrow CoF_2(s) + F(g)$	198
$\frac{1}{2}PbF_4(s) \rightarrow \frac{1}{2}PbF_2(s) + F(g)$	216
$ClF(g) \rightarrow \frac{1}{2}Cl_2(g) + F(g)$	133
$\frac{1}{2}ClF_3(g) \rightarrow \frac{1}{2}ClF(g) + F(g)$	133
$\frac{1}{2}BrF_3(l) \rightarrow \frac{1}{2}BrF(g) + F(g)$	182
$\frac{1}{2}BrF_5(l) \rightarrow \frac{1}{2}BrF_3(l) + F(g)$	158
$\frac{1}{2}XeF_2(s) \rightarrow \frac{1}{2}Xe(g) + F(g)$	132
$\frac{1}{2}F_2(g) \rightarrow F(g)$	79

Thus LiF and CsF are mild fluorinating agents, while ClF, ClF$_3$ and XeF$_2$ are among the strongest, apart from fluorine itself. In the case of ionic fluorides, the elemental metal is not, of course, a product of the reaction; we are usually dealing with a metathesis in which fluorine is exchanged for (say) chlorine. Thus the following data are relevant:

$$\Delta H^0 \, (\text{kJ mol}^{-1})$$

$LiF(s) + Cl(g) \rightarrow LiCl(s) + F(g)$	165
$\frac{1}{2}ZnF_2(s) + Cl(g) \rightarrow \frac{1}{2}ZnCl_2(s) + F(g)$	131
$CsF(s) + Cl(g) \rightarrow CsCl(s) + F(g)$	69
$TlF(s) + Cl(g) \rightarrow TlCl(s) + F(g)$	79
$AgF(s) + Cl(g) \rightarrow AgCl(s) + F(g)$	36

With a large cation, the difference in lattice energy between the fluoride and the chloride makes a relatively small endothermic contribution to the reaction. For silver (and, to a lesser extent, thallium), the lattice energy of the chloride is considerably larger than would be expected from the ionic model, the cation being strongly polarising. Thus the substitution of an E–Cl bond for an E–F bond by AgF will be exothermic if E–F is stronger than E–Cl by at least $36\,kJ\,mol^{-1}$; from Table 6.2, it will be apparent that this will nearly always be the case. As an easily-handled crystalline solid, AgF is a very useful fluorinating agent. For example, it reacts with NOCl to give NOF.

Elsewhere in this chapter, it has been suggested that atoms tend to rearrange themselves in the course of a chemical reaction in order to form bonds between the atoms of highest and of lowest electronegativity. This would suggest that CsF should be a very poor fluorinating agent; Cs and F being respectively the least and most electronegative atoms, CsF should appear as the product of any exothermic reaction if these elements are present in the reactants. The theoretical basis for the principle rests upon Pauling's equation (Section 4.4) in which covalent bond energies are enhanced by an amount proportional to the square of the electro-negativity difference. This is clearly inapplicable to ionic substances like CsF, where the lattice energy considerations outlined above are more relevant. Thus the reaction:

$$2CsF(s) + BeCl_2(s) \rightarrow 2CsCl + BeF_2(s)$$

is highly favourable ($\Delta G^0 = -311\,kJ\,mol^{-1}$; $\Delta H^0 = -315\,kJ\,mol^{-1}$). This might be rationalised by noting that the small Be atom can take better advantage than the large Cs atom of the opportunities for bonding afforded by F compared with Cl, notwithstanding the higher electro-negativity of Be compared with Cs. The concept of hard and soft acids and bases is often helpful in such cases; we might say that Be^{2+} is a harder acid than Cs^+, and therefore prefers the harder base F^-, while the softer acid/base pair Cs^+ and Cl^- come together (cf. Table 10.1 and the discussion in Section 10.2).

The fiercest fluorinating agents are appropriate for the oxidative preparation of a fluoride or oxofluoride, starting from a lower fluoride (or chloride, oxide etc.). Milder agents are preferred for non-oxidative preparations where, for example, Cl is displaced by F, e.g.:

$$U + 3ClF_3 \rightarrow UF_6 + 3ClF$$
$$AgCl + ClF_3 \rightarrow AgF_2 + \tfrac{1}{2}Cl_2 + ClF$$
$$AgF_2 + \tfrac{1}{2}XeF_2 \rightarrow AgF_3 + \tfrac{1}{2}Xe$$
$$2PCl_3 + 3ZnF_2 \rightarrow 2PF_3 + 3ZnCl_2$$

Reductive methods may be employed to prepare lower fluorides. For example, treatment of europium metal with even the mildest fluorinating agents yields only EuF$_3$. The difluoride can be prepared by reducing the trifluoride with hydrogen at a high temperature, or by heating the elemental substance with the trifluoride:

$$EuF_3 + \tfrac{1}{2}H_2 \rightarrow EuF_2 + HF$$
$$2EuF_3 + Eu \rightarrow 3EuF_2$$

Elemental chlorine, bromine and iodine are much more convenient reagents than fluorine for the preparation of compounds containing E–X bonds. Binary chlorides, bromides and iodides are often best made by direct interaction of the elemental substances.

However, other methods are commonly employed, sometimes because the elemental substance is expensive or inaccessible or because direct combination of the elemental substances produces the wrong oxidation state of E, or for other reasons of synthetic convenience. We will concentrate on preparative routes to chlorides. The principal methods may be summarised as follows:

(i) *Oxidative halogenation*
    This includes reactions between the elemental substances, e.g.:

$$Fe + \tfrac{3}{2}Cl_2 \longrightarrow FeCl_3$$

However, the hydrogen halide can also be used – either in the gas phase or in solution – as the oxidising/halogenating agent, e.g.:

$$Fe(s) + 2HCl(g) \rightarrow FeCl_2(s) + H_2(g)$$

Since HCl is a milder oxidising agent than Cl$_2$, the lower oxidation state of iron is obtained. This arises from the fact that HCl(g), whose free energy of formation is $-95\,kJ\,mol^{-1}$, appears on the left-hand side of the equation. Hence ΔG^0 for the formation of any chloride ECl$_n$ by treatment of E with gaseous HCl will be more positive by $95n\,kJ\,mol^{-1}$ than that for the direct reaction of the elemental substances.

Iron dissolves in aqueous HCl, and solids such as [Fe(H$_2$O)$_6$]Cl$_2$ and [Fe(H$_2$O)$_4$Cl$_2$] can be crystallised. The anhydrous chloride can be obtained by heating the hydrate in a stream of HCl, or by treatment with SOCl$_2$ which reacts with coordinated H$_2$O to give SO$_2$ and HCl. Such wet methods are obviously inapplicable to halides which undergo hydrolysis in water, e.g. BeCl$_2$, AlCl$_3$.

Higher halides can, of course, be made by oxidation of a lower halide with elemental halogen. For example, the sparingly-soluble TlCl can be easily obtained by precipitation from an aqueous solution of a Tl(I) salt. Treatment with chlorine under pressure (7 atm) at 160 °C yields TlCl$_3$. The higher-than-atmospheric pressure is required because TlCl$_3$ is thermally unstable, and begins to lose chlorine above 35 °C at atmospheric pressure. Higher temperatures than this are needed for the reaction between TlCl and Cl_2 to proceed at a reasonable rate, and the increased pressure suppresses the decomposition of $TlCl_3$ back to TlCl. High pressures are often needed to perform reactions which require high temperatures for kinetic reasons, but which involve a highly unfavourable entropy change; the preparation of carbonyl complexes is a good example, discussed in Section 10.4.

(ii) Reductive methods

Reductive procedures are required in cases where E in EX_n is in a relatively low oxidation state, and direct reaction between the elemental substances leads to a higher halide. 'Reproportionation' reactions are often used, e.g.:

$$2NdCl_3 + Nd \rightarrow 3NdCl_2$$

Other reducing agents than the elemental substance may be employed. Gaseous hydrogen is quite popular, e.g.:

$$2CrCl_3(s) + H_2(g) \rightarrow 2CrCl_2(s) + 2HCl$$

This reaction operates thermodynamically because the negative free energy of formation of $HCl(g)$ overcomes the effect of the less negative free energy of formation of $CrCl_2$ compared with $CrCl_3$.

Thermal decomposition of a higher halide is often the best route to lower halides, although this method can be rather messy, giving an impure product. For example, III is the most stable state for gold (apart, of course, from the very stable zero state in the relatively inert elemental substance). Gold(III) chloride can be obtained by direct interaction between the elemental substances, and decomposes to gold(I) chloride AuCl on heating. It is difficult, however, to obtain a pure product; too much heating leads to further decomposition to gold metal, while insufficient heating leaves some of the unchanged gold(III) chloride. But $AuCl_3$ – better written as Au_2Cl_6, because it consists of planar dimeric molecules in which each Au atom has the square planar coordination favoured by the d^8 configuration – is easily separated from AuCl by washing the product with ether, which readily dissolves Au_2Cl_6 but not the polymeric AuCl (which has the chain structure exemplified by HgO in Section 3.3).

(iii) Substitution methods

Here we consider the preparation of a halide EX_n by means of a substitution reaction in which the oxidation state of E remains unchanged.

Reaction of an oxide with hydrogen halide is a fairly obvious route. This is an acid/base reaction, and its thermochemistry will depend on the strengths of the oxide as a base and of HX as an acid. Consider, for example:

$$B_2O_3(s) + 6HCl(g) \rightarrow 2BCl_3(g) + 3H_2O(l)$$

$$\Delta G^0 = +279 \, kJ \, mol^{-1}$$
$$\Delta H^0 = +162 \, kJ \, mol^{-1}$$

Boron(III) oxide is an exceedingly weak base, so that the forward acid/base reaction does not progress far. The reverse reaction – the hydrolysis of the covalent, molecular boron(III) chloride – is, of course, highly favourable. However, compare this with the following:

$$Ce_2O_3(s) + 6HCl(g) \rightarrow 2CeCl_3(s) + 3H_2O(l)$$

$$\Delta G^0 = -389 \, kJ \, mol^{-1}$$
$$\Delta H^0 = -615 \, kJ \, mol^{-1}$$

Cerium(III) oxide is quite a strong base compared with boron(III) oxide (see Table 9.1), and the acid/base reaction is strongly exothermic. No appreciable hydrolysis occurs for $CeCl_3$, which dissolves in water to give $Ce^{3+}(aq)$. A detailed thermochemical analysis of the factors which lead to the dramatic difference in behaviour between B_2O_3 and Ce_2O_3 is not straightforward, but the considerations set out in Section 10.2 in the discussion of Table 10.1 are relevant.

Another substitutional method is the reaction between the oxide and the elemental halogen, with evolution of O_2. This is a useful method if it is thermodynamically feasible, because for many elements the oxide is the cheapest and most accessible starting material. Clearly the reaction:

$$EO_{n/2} + \tfrac{n}{2}Cl_2 \rightarrow ECl_n + \tfrac{n}{4}O_2$$

can have a negative ΔG^0 only if the free energy of formation of ECl_n is more negative than that of $EO_{n/2}$. This holds only in cases where E is fairly large and of low electronegativity, and n is relatively small (see Table 10.1 and the contrast between BCl_3 and $CeCl_3$ with respect to hydrolysis discussed above). Thus $TiCl_4$ cannot be prepared simply by treating TiO_2 with chlorine:

$$TiO_2(s) + 2Cl_2(g) \rightarrow TiCl_4(l) + O_2(g) \quad \Delta G^0 = +152 \, kJ \, mol^{-1}$$

But if the reaction is performed in the presence of a reducing agent which reacts exothermically with O_2, the reaction can proceed, e.g.:

$$TiO_2(s) + 2Cl_2(g) + C(s) \rightarrow TiCl_4(l) + CO_2(g)$$

$$\Delta G^0 = -231 \, kJ \, mol^{-1}$$

Halogen-exchange reactions provide important routes to halides. For a reaction such as:

$$ECl_n + E'Br_n \rightarrow EBr_n + E'Cl_n$$

the thermodynamics are likely to be favourable if E' forms stronger bonds to halogen than E, and derives more advantage than E from exchanging Br for Cl. Obvious considerations are the electronegativities and sizes of E and E'. For example, $FeBr_3$ can be prepared by the reaction:

$$FeCl_3 + BBr_3 \rightarrow FeBr_3 + BCl_3 \quad \Delta G^0 = -60 \, kJ \, mol^{-1}$$

Although many other syntheses of $FeBr_3$ could be devised, this has the advantage that the rigorously dry conditions necessary for the preparation of this highly deliquescent compound are maintained (BCl_3 and BBr_3 react rapidly with water), and the volatile BCl_3 and unreacted BBr_3 can be easily removed. An excess of BBr_3 is needed, because the product BCl_3 reacts with BBr_3 to give BBr_2Cl and $BBrCl_2$ (see also Section 9.5).

10.4   Coordination compounds

Provided that the necessary thermodynamic conditions are met, the preparation of a desired coordination compound may appear, *prima facie*, to be simply a matter of finding a suitable starting material containing the required central atom in the correct oxidation state, and flinging this together with the ligands. Obvious exceptions would include cases where the central atom in the required complex is in an unusual and inaccessible oxidation state, or where the ligand is unstable in the uncomplexed condition. Many coordination compounds can in fact be prepared by straightforward, direct interaction of this kind. For example, $[Ni(NH_3)_6]Cl_2$ can be obtained by the reaction of $NiCl_2$ with ammonia; note that this is a substitution rather than an addition reaction because the Ni in $NiCl_2$ is octahedrally bonded to 6Cl. Complexes of the type $[TiCl_4L]$ and $[TiCl_4L_2]$ can be prepared by direct reaction of molecular $TiCl_4$ (as the neat liquid, or in a non-polar solvent) with Lewis bases: these are simple addition reactions, and the products may appropriately be called adducts. Halogeno-complexes such as $[N(C_2H_5)_4]_2[NiCl_4]$ can be prepared by crystallisation from mixed solutions of the appropriate halides in a weakly-coordinating solvent.

However, more subtle procedures are often necessary. Preparative strategies may be conveniently classified according to whether the central atom M in the coordination compound has been oxidised or reduced, or whether no change in oxidation state has occurred and the reaction is a straightforward substitution, or addition of neutral ligands.

No change in oxidation state of central atom

We first consider the preparation of carbonyls $M_x(CO)_y$ – where M is, of course, in the zero oxidation state – by the direct interaction of the elemental substance M with CO gas. Of the numerous compounds in this category, only $Ni(CO)_4$ and $Fe(CO)_5$ can be conveniently prepared in this way:

$$Ni(s) + 4CO(g) \rightarrow Ni(CO)_4(l)$$
$$\Delta G^0 = -40\,kJ\,mol^{-1}; \Delta H^0 = -189\,kJ\,mol^{-1}$$

$$Fe(s) + 5CO(g) \rightarrow Fe(CO)_5(l)$$
$$\Delta G^0 = -20\,kJ\,mol^{-1}; \Delta H^0 = -219\,kJ\,mol^{-1}$$

The differences between ΔG^0 and ΔH^0 in these cases emphasise the highly unfavourable entropy changes involved. Nickel does react slowly with carbon monoxide at room temperature. Heating may be expected to accelerate the reaction; but the negative entropy change soon overturns the sign of ΔG as the temperature is increased. The optimum temperature for the preparation of $Ni(CO)_4$ is about 50 °C, at atmospheric pressure. The reaction between iron and carbon monoxide at room temperature and atmospheric pressure is too slow to be of any use. Raising the temperature to speed things up will reverse the sign of ΔG at about 55 °C; but this effect is counteracted by raising the pressure to about 300 atmospheres, in accordance with Le Chatelier's principle. A number of other carbonyls can be prepared by the direct reaction of the metal with CO at elevated temperatures and pressures, e.g. $Co_2(CO)_8$, $Mo(CO)_6$ and $W(CO)_6$; but these are better prepared by other means, described below.

Where the oxidation number of the central atom in both starting material and product is the same, but other than zero, we are usually dealing with a substitution reaction rather than addition. Even 'simple' compounds such as binary halides and oxoacid salts can be viewed as coordination compounds. Exceptions occur where the coordination number of the central atom is increased; for example, the reaction between $TiCl_4$ (which consists of tetrahedral molecules) and a neutral ligand L to give the octahedral complex $[TiCl_4L_2]$ is a straightforward addition, as already noted. But the preparation of an octahedral complex starting from a hydrated salt containing $[M(H_2O)_6]^{3+}$ ions will necessarily involve the replacement of water by other ligands. The kinetic and mechanistic considerations set out in Chapter 9 will obviously be relevant. As an example, we look in detail at the preparation of $[Cr(en)_3]Cl_3$, where en = ethylenediamine $NH_2CH_2CH_2NH_2$, an important starting material in Cr(III) coordination chemistry. The reaction:

$$Cr^{3+}(aq) + 3en(aq) \rightarrow Cr(en)_3^{3+}(aq)$$

is highly favourable thermodynamically (although, for reasons that will become apparent, the relevant functions have not been determined with certainty). One might therefore expect that $[Cr(en)_3]Cl_3$ could be obtained by dissolving $[Cr(H_2O)_6]Cl_3$ in water, adding ethylenediamine, and allowing the product to crystallise. In fact, all that happens is that the basic ligand raises the pH to the point where hydrated chromium(III) oxide is precipitated. Such precipitation can be prevented by adding HCl, so that the pH is kept low – although this will drive the equilibrium to the left. But the displacement of H_2O by ethylenediamine in the coordination sphere is very slow, a manifestation of the kinetic inertness of octahedral Cr(III) complexes (see Section 9.5). Up to about 1970, the usual procedure for the preparation of tris(ethylenediamine)chromium(III) salts was a tedious business. The sulphate was prepared by heating anhydrous chromium(III) sulphate with dry ethylenediamine overnight; these reagents had to be carefully dried in advance. The chloride was prepared by treating an aqueous solution of the sulphate with concentrated HCl.

A much improved method involves the treatment of $[Cr(H_2O)_6]Cl_3$ dissolved in methanol with ethylenediamine, in the presence of a small piece of zinc metal. The function of the zinc (virtually all of which can be recovered unchanged at the end of the reaction) is to produce a catalytic amount of Cr(II). The zinc dissolves very slowly:

$$Zn(s) + 2Cr(H_2O)_6^{3+}(m) \rightarrow Zn^{2+}(m) + 2Cr^{2+}(m) + 6H_2O$$

(the postscript (m) indicates solvation by methanol). The Cr(II) complex is kinetically labile, and the CH_3OH molecules are rapidly displaced by ethylenediamine in a thermodynamically-favourable process. We then have a moderately fast electron-transfer reaction:

$$Cr(H_2O)_6^{3+}(m) + Cr(en)_3^{2+}(m) \rightarrow Cr(en)_3^{3+}(m) + Cr^{2+}(m) + 6H_2O$$

The formation of another $Cr^{2+}(m)$ ion allows the chain reaction to continue, and a reasonable yield of $[Cr(en)_3]Cl_3.3H_2O$ (which is rather insoluble in methanol) is obtained in about an hour. For (presumably) kinetic reasons that remain obscure, the reaction does not work in water and the choice of methanol as solvent is largely an example of serendipity.

The kinetic and mechanistic considerations set out in Section 9.5 will, of course, be important in planning syntheses based on substitution reactions.

Oxidative methods

Here we look at preparative strategies for coordination compounds where the central atom is oxidised relative to its state in the starting material. Such a method is clearly indicated if the ligands in the required product tend to stabilise high oxidation states of the central atom.

The most obvious examples are in the preparation of octahedral Co(III) complexes. The most readily-available cobalt compounds are Co(II) salts; in the presence of suitable ligands – usually N-donors – these are oxidised to give Co(III) complexes by air or hydrogen peroxide. A few such easily-prepared complexes open up pathways to the vast number of known octahedral Co(III) complexes via substitution reactions. For example, $[Co(NH_3)_5(H_2O)]Cl_3$ is readily converted into $[Co(NH_3)_5X]Cl_2$ via anation reactions of the type discussed in Section 9.5, and salts containing the $[Co(NH_3)_4(CO_3)]^+$ ion (where the carbonate is bidentate, taking up two *cis* positions) are useful for the formation of *cis*-$[Co(NH_3)_4X_2]^+$.

Chromium(II) compounds are less accessible than Co(II) or Cr(III) compounds. However, their explicit preparation as routes to the inert Cr(III) complexes has some uses. We could prepare $[Cr(en)_3]Cl_3$ by first preparing $[Cr(en)_3]^{2+}$ from Cr^{2+}(aq), followed by treatment with a suitable oxidising agent. This in effect is what happens catalytically in the reaction described above.

Electrochemical methods are becoming increasingly popular in preparative coordination chemistry. Particularly interesting are those which start with an elemental substance (which is obtainable in very high purity, and any of which remains unreacted can be easily separated). For example, $[N(C_2H_5)_4][InI_2]$, which contains the interesting angular InI_2^- ion, can be prepared in an electrochemical cell from indium metal and tetraethylammonium iodide in acetonitrile. The oxidation of In(0) to In(I) is selectively and cleanly accomplished by means of a controlled electric potential in a convenient one-step process.

Oxidative addition reactions – discussed in Section 9.6 – provide valuable synthetic routes to octahedral d^6 systems, especially Rh(III) and Ir(III) complexes.

Reductive methods

Reductive strategies will be favoured for the preparation of complexes where the central atom is in a low oxidation state. Most carbonyl complexes $M_x(CO)_y$ are prepared by treating a halide such as $FeCl_3$ or $MnCl_2$ with a strong reducing agent such as metallic sodium or magnesium, in the presence of CO under a pressure of 100–300 atmospheres. The thermodynamics are dominated by the highly-negative free energy of formation of, e.g., $MgCl_2$, and by the use of high pressures to overcome the unfavourable entropy change. In some cases, carbon monoxide itself serves as a reducing agent, e.g.:

$$OsO_4 + 9CO \xrightarrow[\text{300 atm}]{\text{300 °C}} Os(CO)_5 + 4CO_2$$

The highly exothermic extraction of O atoms from OsO_4 by CO provides the thermodynamic driving force.

It might be thought that reductive methods would be suitable for the preparation of hydrido-complexes. In fact, most such compounds are prepared by substitution reactions, using a source of H^- such as $NaBH_4$; the formation of, e.g., $NaBCl_4$ as a product assists the thermodynamics (cf. the use of $LiAlH_4$ in the preparation of SiH_4 in Section 10.3 above).

10.5   Organometallic compounds

In this chapter, a distinction has been drawn between coordination and organometallic compounds because in the area of preparative methods the two overlapping sub-disciplines exhibit quite clearly the most distinctive differences in flavour. Organometallic chemists have to be very careful to exclude oxygen and water from their apparatus. Although many organometallics are (kinetically, at least) stable to oxidation and/or hydrolysis, their precursors in synthetic routes are often very sensitive to moisture and oxygen. In the case of an organometallic compound which can sensibly be viewed as a coordination compound, the organic ligands can be classified in three categories:

(i) Neutral molecules which are familiar as molecular substances in the uncomplexed state. Examples include benzene, acetylene and ethylene. All of these are thermodynamically unstable with respect to the elemental substances, let alone to a variety of products in the presence of oxygen and water. The reactivities of such molecules can be greatly enhanced by coordination. Moreover, the central atoms to which they are bound may be vulnerable to attack.

(ii) Carbanions such as CH_3^-, $C_5H_5^-$, C_2H^-. It is debatable whether these have any real existence in stable compounds or solutions, although they undoubtedly occur as reactive intermediates in many situations. Carbanions are powerful as reducing agents and as bases. We may surmise that the preparation of a compound in which CH_3^- is coordinated to a central atom or ion must involve the generation of the carbanion, and it must be protected from oxidation or hydrolysis before it can be affixed to the central atom.

(iii) Molecules which are very reactive in the free state and do not form isolable molecular substances. Examples include cyclobutadiene C_4H_4 and carbenes CX_2. There are fairly obvious ways of generating such molecules as unstable intermediates, but their lifetimes are so short that every care must be taken to protect them from harm until they can bind to the central atom. Carbene complexes are in fact more often made by reactions of coordinated ligands, e.g.:

$$L_n MCO \xrightarrow{\text{LiR}} L_n MC \overset{\displaystyle O^-}{\underset{\displaystyle R}{<}}$$

$$L_n MCF_3 \xrightarrow{\text{SbF}_5} [L_n MCF_2]^+ [SbF_6]^-$$

Homoleptic alkyls/aryls of the Main Group elements

The term 'homoleptic' is analogous to 'binary'; just as $SiCl_4$ is a binary chloride, $Si(CH_3)_4$ is a homoleptic alkyl in which Si is combined with just one type of group.

Alkyl/aryls ER_n and derivatives where E is of lower electronegativity than C tend to be unstable with respect to hydrolysis, oxidation by air and heating. It is not surprising that organometallic chemistry made relatively slow progress until the development of techniques for handling substances in air- and moisture-free apparatus; these date largely from the 1920s. Even the familiar Grignard reagents – in use since the early 1900s – require rigorously dry conditions for successful results.

An alkyl/aryl ER_n may, in a few extreme cases, be considered to contain R^- carbanions; alkyls/aryls of clearly covalent character behave chemically as sources of R^- if E is of lower electronegativity than C. There is a decrease in basicity along the isoelectronic series:

$$CH_3^- > NH_2^- > OH^- > F^-$$

Thus $CH_3^-(aq)$ cannot survive in aqueous solution, because the equilibrium:

$$CH_3^-(aq) + H_2O \rightleftharpoons CH_4(g) + OH^-(aq)$$

lies well to the right.

An alkyl such as KCH_3 – which is considered to have much ionic character – might therefore be expected to decompose on contact with water, just as ionic hydrides do. More covalent alkyls/aryls are susceptible to hydrolysis according to the scheme:

$$\begin{array}{c} E \relbar R \\ \curvearrowright \\ :\ddot{O}\relbar H \\ | \\ H \end{array} \longrightarrow \begin{array}{c} E \\ | \\ OH \end{array} + RH$$

Where the E—C polarity is small, there is little susceptibility to hydrolysis. This applies to BR_3, for example. Where the polarity is in the opposite direction, viz. $E^{\delta-}$—$C^{\delta+}$, there may be some tendency to the hydrolysis:

$$E—R + H_2O \rightarrow E—H + R—OH$$

With regard to the susceptibility of many organometallics to aerial oxidation, it must be remembered that most compounds containing C–H bonds are thermodynamically unstable with respect to oxidation by molecular oxygen to produce water and carbon dioxide. The mechanisms whereby (for example) many alkyls are spontaneously flammable in air are necessarily difficult to determine: the mechanisms involved in the combustion of hydrocarbons are still incompletely understood. Suffice it to say that the kinetic stability of ER_n to air seems to be low where E is of low electronegativity.

The thermal stabilities of Main Group organometallics are restricted by thermodynamic factors such as:

$$2CH_3(g) \rightarrow C_2H_4(g) + H_2(g) \quad \Delta H^0 = -239 \, kJ \, mol^{-1}$$
$$2CH_3(g) \rightarrow C_2H_6(g) \quad \quad \Delta H^0 = -376 \, kJ \, mol^{-1}$$

Thus the reaction:

$$E(CH_3)_2(s) \rightarrow E(s) + C_2H_6(g)$$

or:

$$E(CH_3)_2(s) \rightarrow EH_2(s) + C_2H_4(g)$$

is likely to be thermodynamically favourable at modest temperatures unless the E–C bonding is reasonably strong. Going down any Group, such bonding – whether it be ionic or covalent – is expected to weaken, and the thermal stabilities of organometallics tend to decline. Across a Period, E–C bond energies tend to increase, with increasing electronegativity and decreasing size of E.

Preparative routes to homoleptic alkyls/aryls must take account of the sensitivities of the products, as well as the usual thermodynamic and kinetic factors. The following will serve as examples.

Alkyllithium compounds LiR can be prepared by the reaction between lithium metal and an alkyl halide in an inert solvent such as benzene:

$$2Li + RCl \rightarrow LiR + LiCl$$

The highly-negative free energy of formation of LiCl ($-384 \, kJ \, mol^{-1}$) furnishes the driving force not only for this synthesis, but also for many of the reactions of lithium alkyls which make these among the most valuable reagents for organic and inorganic chemist alike.

Beryllium dialkyls BeR_2 are best made by one of the most general preparative methods for organometallics, by heating elemental beryllium with the corresponding mercury alkyl HgR_2:

$$Be + HgR_2 \rightarrow BeR_2 + Hg$$

The Hg–C bond is very weak, partly on account of the large size of the Hg atom and the low polarity of the bond (the mercury dialkyls are molecular substances, with linear C–Hg–C skeletons). Despite the low atomisation enthalpy of mercury compared with other metallic elemental substances, reactions of this kind are usually exothermic. The mercury dialkyls are relatively insensitive to air and moisture and are therefore convenient to store and use. They are prepared from Grignard reagents:

$$HgCl_2 + 2RMgBr \rightarrow HgR_2 + MgCl_2 + MgBr_2$$

Although the mercury dialkyls are not particularly stable (for example, the free energy of formation of liquid $Hg(CH_3)_2$ is $+140\,kJ\,mol^{-1}$) they can be prepared by a reaction in which the very stable $MgCl_2$ ($\Delta G_f^0 = -592\,kJ\,mol^{-1}$) and $MgBr_2$ ($\Delta G_f^0 = -504\,kJ\,mol^{-1}$) also appear as products. The required Grignard reagents – as some readers will know from personal experience – are prepared by treating magnesium metal with the alkyl/aryl halide RX in diethyl ether, under scrupulously dry and air-free conditions. The product is best formulated as $[MgRX(OEt_2)_2]$, although some disproportionation to MgR_2 and MgX_2 and some dimer formation take place. The formation of the strong Mg–X bond provides much of the thermodynamic driving force.

A related method of preparing organometallics is by 'metal exchange' reactions, e.g.:

$$3Al(C_2H_5)_3 + GaBr_3 \rightarrow Ga(C_2H_5)_3 + 3AlBr(C_2H_5)_2$$

The tendency of the Br atoms to become associated with Al atoms has been cited as evidence for the greater electronegativity of Ga compared with Al.

Compounds containing E–H bonds (where E is one of the Group 13 or 14 elements) can undergo insertion reactions with olefins and acetylenes, i.e.:

$$X_nE\!-\!H + RCH\!=\!CH_2 \rightarrow X_nE\!-\!CH_2CH_2R$$

These rarely produce homoleptic species, but boron trialkyls can be obtained from diborane:

$$B_2H_6 + 6RCH\!=\!CH_2 \rightarrow 2B(CH_2CH_2R)_3$$

Essentially similar methods can be employed to prepare non-homoleptic organometallic compounds of the Main Group elements, EX_mR_n etc. Some procedures unique to particular classes of compound are important, for example the mercuration of aromatic hydrocarbons by mercuric acetate:

$$C_6H_6 + Hg(CH_3COO)_2 \rightarrow C_6H_5HgO\overset{\displaystyle O}{\overset{\displaystyle \|}{C}}CH_3 + CH_3COOH$$

Organometallic compounds of the transition elements

We look first at the preparation of π complexes containing as ligands neutral organic molecules such as C_2H_4, C_6H_6 etc., or anions such as $C_5H_5^-$. These can often be obtained by substitution reactions. Thus Zeise's famous salt $K[PtCl_3(C_2H_4)]$ (see Section 8.1) was prepared in 1829 by passing ethylene through an ethanolic solution of $K_2[PtCl_4]$. Variations on this procedure, without any oxidation or reduction of the central atom, are employed to this day to prepare a host of alkene and alkyne complexes. Displacement of a CO ligand by an alkene, or of two CO ligands by a bidentate 1,3-dialkene, is common; irradiation by ultraviolet light is often necessary to break the M–C bond, leading to a coordinatively-unsaturated species.

As an example of a detailed preparative method, we look at the case of ferrocene $Fe(C_5H_5)_2$, historically one of the most important organometallic compounds. Ferrocene was first prepared accidentally in the course of an unsuccessful attempt to prepare dihydrofulvalene by the coupling of two C_5H_5 radicals:

The experimenters reasoned this could be achieved by the reaction of the Grignard reagent C_5H_5MgBr with $FeCl_3$:

$$2C_5H_5MgBr + 2FeCl_3 \rightarrow C_{10}H_{10} + MgCl_2 + MgBr_2 + 2FeCl_2$$

This is an oxidative coupling reaction, driven by the highly-negative free energies of formation of the crystalline dihalides. Instead, ferrocene $Fe(C_5H_5)_2$ was produced:

$$10C_5H_5MgBr + 4FeCl_3 \rightarrow 4Fe(C_5H_5)_2 + + 2C_5H_5Cl + 5MgCl_2 + 5MgBr_2$$

Having first isolated and characterised this important compound, researchers set about devising improved methods of synthesis, giving better yields etc. Perhaps the best method – doubtless performed in hundreds of undergraduate laboratories around the world – is the direct interaction between Fe^{2+} and $C_5H_5^-$. The latter is generated *in situ* by the deprotonation of cyclopentadiene C_5H_6, a very weak acid ($pK_a \sim 20$). Finely-ground potassium hydroxide powder can effect deprotonations which would be impossible in aqueous solution. If present in excess, the KOH can assist the thermodynamics of the process:

$$OH^- + RH \rightarrow H_2O + R^-$$

by absorbing the water formed in the exothermic reaction:

$$KOH(s) + H_2O(l) \rightarrow KOH.H_2O \quad \Delta G^0 = -29\,kJ\,mol^{-1}$$
$$\Delta H^0 = -38\,kJ\,mol^{-1}$$

Potassium hydroxide can deprotonate as weak an acid as $(C_6H_5)_3CH$ ($pK_a \sim 32$), and its desiccant properties ensure the necessary dry conditions.

The reaction must be performed in a dry, oxygen-free environment; although the product is not sensitive to air or moisture, the intermediate $C_5H_5^-$ is. Cyclopentadiene is added dropwise to a stirred slurry of KOH powder and 1,2-dimethoxyethane. A solution of $FeCl_2.4H_2O$ in dimethyl sulphoxide $(CH_3)_2SO$ is then added. The choice of solvents is determined by obvious considerations of solubility and miscibility, and chemical inertness towards the products and intermediates. Dimethyl sulphoxide is a fairly weakly-coordinating aprotic solvent, easily displaced by $C_5H_5^-$ from Fe(II). A mixture of aqueous HCl and crushed ice is then added; this neutralises the KOH (the ice absorbs the heat evolved) and causes the precipitation of ferrocene, which is (predictably) insoluble in water. The crude product is purified by sublimation.

The Grignard method is generally applicable to the preparation of π-allyl complexes, e.g.:

$$3C_3H_5MgCl + CrCl_3 \rightarrow Cr(\pi\text{-}C_3H_5)_3 + 3MgCl_2$$

The apparent simplicity of this reaction disguises the complications arising from the extreme sensitivity of the product to air and moisture; considerable care and skill are needed.

Reductive methods are often employed, for example in the preparation of dibenzenechromium(0) $Cr(C_6H_6)_2$:

$$3CrCl_3 + 2Al + AlCl_3 + 6C_6H_6 \rightarrow 3[Cr(C_6H_6)_2][AlCl_4]$$

The $Cr(C_6H_6)_2^+$ cation is further reduced with dithionite $S_2O_4^{2-}$ to give the required complex.

As a final example of the preparation of a π complex, we look at $Fe(\pi\text{-}C_4H_4)(CO)_3$. Cyclobutadiene defied the best efforts of synthetic organic chemists for many years. An obvious method would be to treat $C_4H_4Cl_2$ with an elemental substance such as Mg:

$$\begin{array}{ccc} HC-CHCl & & HC-CH \\ \| \quad | & + Mg \longrightarrow & \| \quad \| \\ HC-CHCl & & HC-CH \end{array} + MgCl_2$$

However, attempts along these lines, although they may well have led to the production of cyclobutadiene molecules as short-lived intermediates, gave dimers and oligomers of the desired product. Free cyclobutadiene does appear to be unstable, but a complex can be prepared by the reaction:

$$Fe_2(CO)_9 + C_4H_4Cl_2 \rightarrow Fe(C_4H_4)(CO)_3 + FeCl_2 + 6CO$$

Synthetic routes to compounds containing M–C σ bonds are fairly obvious. Substitution of, e.g., Cl by CH_3 can be effected by treatment with $LiCH_3$ or CH_3MgBr. A number of reaction types mentioned in Chapter 9 – oxidative addition, reductive elimination, insertion and cyclometallation (Sections 9.6 and 9.7) – have their uses in preparative routes to M–C bonds. The formation of organo-compounds of the lanthanides and actinides is an area of growing interest. Preparative methods are similar to those for other ER_n species where E is of relatively low electronegativity, e.g.:

$$LnCl_3 + 6LiCH_3 \rightarrow Li_3[Ln(CH_3)_6] + 3LiCl$$

The structures suggest the formation of Li–H–C three-centre bonds.

10.6   Preparation of compounds containing Si=Si and P=P bonds

As examples of inorganic syntheses which do not fall within the categories already covered in this chapter, but which epitomise much of the spirit of the subject, we look at compounds $R_2Si=SiR_2$ and RP=PR. For half a century, inorganic chemists confidently 'understood' the 'nonexistence' of such compounds and their students cheerfully accepted their 'explanations' (see Chapter 6). The successful preparation of many compounds in these categories by several groups in the 1980s bears testimony to a proper appreciation of thermodynamic and kinetic considerations in synthetic chemistry, and manipulative skill, not to mention faith and perseverance.

The familiar alkenes can be prepared by reactions such as:

$$RCHClCH_2R' \xrightarrow{KOH} RCH=CHR'$$
$$RCHClCHClR' \xrightarrow{Na} RCH=CHR'$$

These are driven thermodynamically by, in the first example, the energetically-favourable neutralisation of HCl by KOH, and in the second example by the highly exothermic formation of NaCl. Attempts to prepare the analogous silicon compounds – disilenes – by similar methods lead to the production of polymeric materials $(SiR_2)_n$. The molecule Si_2H_4 has been identified spectroscopically but turns out to be $HSi—SiH_3$ and not $H_2Si=SiH_2$; the latter appears to undergo spontaneous rearrangement to the resonance-stabilised silylsilylene:

$$HSi—\overset{+}{Si}H_3 \leftrightarrow \overset{+}{HSi}=\overset{-}{Si}H_3$$

The double-bonded structure $\overset{+}{HSi}=\overset{-}{Si}H_3$ involves a π bond formed by overlap of a filled 3p orbital on the two-coordinate Si atom with an empty 3d orbital on the four-coordinate silicon. The $p_\pi-p_\pi$ bond in $H_2Si=SiH_2$ is evidently rather weak.

Thus $H_2Si=SiH_2$ has gloomy prospects, being unstable to polymerisation (and much else besides) as a substance and to rearrangement as a molecule. But the alkenes are also unstable to polymerisation (and to decomposition to the elemental substances); they are kinetically stabilised by virtue of the large activation barriers imposed against change in hydridisation at a carbon atom. The larger and more stereochemically-flexible silicon atom is kinetically more vulnerable to such changes. However, it was thought possible to stabilise a disilene $R_2Si=SiR_2$ by use of very bulky R groups, so that polymerisation might be sterically inhibited; rearrangement to $RSi—SiR_3$ would also be discouraged by steric interference within the SiR_3 group. The first success was achieved in 1981 by the photochemical reaction:

$$2R_2Si(SiMe_3)_2 \xrightarrow[254\,nm]{h\nu} R_2Si=SiR_2 + 2Si_2Me_6$$

where R is mesityl (2,4,6-trimethylphenyl), abbreviated as mes. The product $Si_2(mes)_4$ is very sensitive to oxygen, but is thermally quite stable. It can also be prepared by perhaps a more obvious route:

$$SiCl_4 + 2LiR \rightarrow SiR_2Cl_2 + 2LiCl$$
$$2SiR_2Cl_2 + 4Li \rightarrow Si_2R_4 + 4LiCl$$

In each reaction, the thermodynamic driving force is provided by the highly-negative ΔG_f^0 of LiCl. In the second step, the silylene SiR_2 appears to be formed (it can be trapped in an inert matrix at low temperatures) and dimerises to give the disilene, but under the reaction conditions no further polymerisation occurs.

A slightly less sterically-hindered R group 2,6-dimethylphenyl can be used in the following sequence:

$$3SiR_2Cl_2 + 6Li^+C_{10}H_8^- \longrightarrow \underset{R_2Si—SiR_2}{SiR_2} + 6C_{10}H_8 + 6LiCl$$

$$\downarrow h\nu$$

$$R_2Si = SiR_2$$

The naphthalenide radical anion $C_{10}H_8^-$ (with the odd electron in a delocalised π antibonding MO) is a very powerful reducing agent, and effects the abstraction of $2Cl^-$ from SiR_2Cl_2 in the form of LiCl. The less bulky R groups compared with mesityl allow trimerisation of SiR_2, but this can be photolysed to give the required disilene. Similar strategies can be employed to give $R_2Ge=GeR_2$ and $R_2Sn=SnR_2$ as reasonably stable bulk substances.

Attempts to prepare $RP=PR$ compounds date back to 1877, when $PhP=PPh$ was claimed as the product of the reaction:

$$PPhCl_2 + PPhH_2 \xrightarrow[-2HCl]{base} 2PhP=PPh$$

The strategy was perfectly sound from the thermodynamic viewpoint, but the product is in fact a mixture of cyclic oligomers of the phosphene PPh:

Presumably PPh molecules are formed, but these do not stop at mere dimerisation. A molecule $RP=PR$ will presumably have as its most stable geometrical isomer the *trans* form, especially with bulky R groups:

Even with a large substituent, the P atom is evidently more exposed to attack via polymerisation than the Si atom in $R_2Si=SiR_2$, so that even bulkier groups are needed to stabilise diphosphenes $RP=PR$. The best R groups are 2,4,6-tri(t-butyl)phenyl and $C(SiMe_3)_3$. The strategy is usually the reductive coupling of $PRCl_2$ with a strong reducing agent such as magnesium or sodium naphthalenide. The preparation of unsymmetrical species $RP=PR'$ is best effected by the 1877 strategy, e.g.:

$$(SiMe_3)_3CPCl_2 + RPH_2 \xrightarrow{base} (SiMe_3)_3CP=PR$$

where R = 2,4,6-tri(t-butyl)phenyl.

Similar strategies are employed to prepare compounds containing $Si=C$, $P=C$, $As=As$ and other double bonds with $p_\pi-p_\pi$ overlap involving the heavier p block atoms.

10.7 Further reading

The following give valuable accounts of the principles and practice of preparative inorganic chemistry: Jolly, W. L. (1970). *The Synthesis and Characterisation of Inorganic Compounds.* Englewood Cliffs, N.J.: Prentice-Hall. Angelici, R. J. (1969). *Synthesis and Technique in Inorganic Chemistry.* Philadelphia: Saunders.

Appendix
The literature of descriptive
inorganic chemistry

Here are listed the most important sources of further information on the subject matter of this book. It is far from comprehensive, but most of the titles should be found in a good university library.

A.1 Comprehensive reference works

Bailar, J. C. *et al.* (eds.) (1973). *Comprehensive Inorganic Chemistry*. Oxford: Pergamon Press.

Emeleus, H. J. (ed.) (1972). *MTP International Review of Science: Inorganic Chemistry*. London: Butterworth.

Gmelin, L. (1924–). *Handbuch der Anorganischen Chemie*. Weinheim: Verlag Chemie.

Mellor, J. W. (1922–37: Supplements since 1957). *A Comprehensive Treatise on Inorganic and Theoretical Chemistry*. London: Longmans, Green.

Remy, H. (1956). *Treatise on Inorganic Chemistry*. Amsterdam: Elsevier.

Sidgwick, N. V. (1950). *The Chemical Elements and Their Compounds*. Oxford University Press.

Sneed, M. C., Maynard, J. L. and Brasted, R. C. (1953–61). *Comprehensive Inorganic Chemistry*. New York: Van Nostrand-Reinhold.

Wilkinson, G. (ed.) (1982). *Comprehensive Organometallic Chemistry*. Oxford: Pergamon Press.

A.2 Other reference sources

Sharp, D. W. A. (1983). *The Penguin Dictionary of Chemistry*. Harmondsworth, Middlesex: Penguin Books.

Weast, R. C. (ed.) (published annually). *The Handbook of Chemistry and Physics*. Boca Raton, Florida: CRC Press.

Encyclopaedia Britannica, 15th edn (1974). Chicago: Encyclopaedia Britannica Inc.

A.3 Inorganic chemistry texts

These vary considerably in the level of treatment, and in the extent to which descriptive inorganic chemistry is covered. No attempt here is made to assess their relative merits; you should browse through the ones which are available to you, and see which best suit your needs or whose presentation you find most attractive.

Cotton, F. A. and Wilkinson, G. (1976). *Basic Inorganic Chemistry.* New York: Wiley.

Cotton, F. A. and Wilkinson, G. (1988). *Advanced Inorganic Chemistry*, 5th edn. New York: Wiley.

Emeleus, H. J. and Sharpe, A. G. (1973). *Modern Aspects of Inorganic Chemistry.* London: Routledge & Kegan Paul.

Greenwood, N. N. and Earnshaw, A. (1984). *The Chemistry of the Elements.* Oxford: Pergamon Press.

Heslop, R. B. and Jones, K. (1976). *Inorganic Chemistry: a Guide to Advanced Study.* Amsterdam: Elsevier.

Huheey, J. E. (1983). *Inorganic Chemistry*, 3rd edn. New York: Harper & Row.

Jolly, W. L. (1976). *Principles of Inorganic Chemistry.* New York: McGraw-Hill.

Jolly, W. L. (1984). *Modern Inorganic Chemistry.* New York: McGraw-Hill.

Lagowski, J. J. (1973). *Modern Inorganic Chemistry.* New York: Dekker.

Mackay, K. M. and Mackay, R. A. (1989). *Introduction to Modern Inorganic Chemistry*, 4th edn. London: Blackie.

Moeller, T. (1982). *Inorganic Chemistry: a Modern Introduction.* New York: Wiley.

Phillips, C. S. G. and Williams, R. J. P. (1965). *Inorganic Chemistry.* Oxford University Press.

Porterfield, W. W. (1984). *Inorganic Chemistry.* Reading, Massachusetts: Addison-Wesley.

Purcell, K. F. and Kotz, J. C. (1977). *Inorganic Chemistry.* Philadelphia: Saunders.

Purcell, K. F. and Kotz, J. C. (1980). *Introduction to Inorganic Chemistry.* Philadelphia: Saunders.

Sanderson, R. T. (1967). *Inorganic Chemistry.* New York: Reinhold.

Sharpe, A. G. (1981). *Inorganic Chemistry.* London: Longman.

A.4 More specialised works

Cotton, S. A. and Hart, F. A. (1975). *The Heavy Transition Elements.* New York: Wiley.

Earnshaw, A. and Harington, T. J. (1973). *Chemistry of the Transition Metals.* Oxford University Press.

Emsley, J. (1971). *Inorganic Chemistry of the Nonmetals.* London: Methuen.

Kepert, D. L. (1972). *The Early Transition Elements.* New York: Academic Press.

Massey, A. G. (1972). *The Typical Elements.* London: Penguin Books.

Parish, R. V. (1977). *The Metallic Elements.* London: Longman.

Powell, P. and Timms, P. (1974). *The Chemistry of the Non-metals.* London: Chapman & Hall.

Steudel, R. (1977). *Chemistry of Nonmetals.* Berlin: de Gruyter.

A.5 Chemical periodicity

Puddephatt, R. J. and Monaghan, P. K. (1986). *The Periodic Table of the Elements*, 2nd edn. Oxford University Press.

Rich, R. (1965). *Periodic Correlations*. New York: Benjamin.

Sanderson, R. T. (1960). *Chemical Periodicity*. New York: Reinhold.

van Sprosen, J. (1969). *The Periodic System of the Chemical Elements*. Amsterdam: Elsevier.

A.6 Solid state inorganic chemistry

Adams, D. M. (1974). *Inorganic Solids*. New York: Wiley.

Greenwood, N. N. (1968). *Ionic Crystals*. London: Butterworth.

Ladd, M. F. C. (1979). *Structure and Bonding in Solid State Chemistry*. New York: Wiley.

Wells, A. F. (1984). *Structural Inorganic Chemistry*, 5th edn. Oxford University Press. (This is by far the most comprehensive and up-to-date work on the structures of inorganic substances.)

A.7 Bonding theory and molecular shapes

Cartmell, E. and Fowles, G. W. A. (1977). *Valency and Molecular Structure*. London: Butterworth.

Gillespie, R. J. (1972). *Molecular Geometry*. New York: Van Nostrand. (The best account of VSEPR theory.)

McWeeny, R. (1979). *Coulson's Valence*, 3rd edn. Oxford University Press.

McWeeny, R. (1982). *The Shapes and Structures of Molecules*, 2nd edn. Oxford University Press.

Murrell, J. N., Kettle, S. F. A. and Tedder, J. M. (1979). *The Chemical Bond*. New York: Wiley.

Pauling, L. (1960). *The Nature of the Chemical Bond*, 3rd edn. Ithaca, New York: Cornell University Press.

Wade, K. (1971). *Electron-deficient Compounds*. London: Nelson.

Williams, A. F. (1979). *A Theoretical Approach to Inorganic Chemistry*. Berlin: Springer.

A.8 Thermodynamics

Dasent, W. E. (1965). *Nonexistent Compounds*. New York: Dekker.

Dasent, W. E. (1982). *Inorganic Energetics*, 2nd edn. Cambridge University Press.

Johnson, D. A. (1982). *Some Thermodynamic Aspects of Inorganic Chemistry*, 2nd edn. Cambridge University Press.

A.9 Kinetics and mechanism

Basolo, F. and Pearson, R. G. (1967). *Mechanisms of Inorganic Reactions*, 2nd edn. New York: Wiley.

Tobe, M. L. (1972). *Inorganic Reaction Mechanisms*. London: Thames and Nelson.

A.10 Coordination compounds

Basolo, F. and Johnson, R. C. (1964). *Coordination Chemistry*. New York: Benjamin.
Bell, C. F. (1977). *Metal Chelation*. Oxford University Press.
Kettle, S. F. A. (1969). *Coordination Compounds*. London: Nelson.
Nicholls, D. (1974). *Complexes and First-Row Transition Elements*. London: Macmillan.
Orgel, L. E. (1966). *Introduction to Transition Metal Chemistry*, 2nd edn. New York: Wiley. (Excellent account of ligand field theory.)

A.11 Organometallic chemistry

Coates, G., Green, M. L. H. and Wade, K. (1967). *Organometallic Compounds*. London: Methuen.
Coates, G. E., Green, M. L. H., Powell, P. and Wade, K. (1968). *Principles of Organometallic Chemistry*. London: Methuen.
Haiduc, I. and Zuckerman, J. J. (1985). *Basic Organometallic Chemistry*. Berlin: de Gruyter.

A.12 Biological aspects

Hughes, M. N. (1982). *The Inorganic Chemistry of Biological Processes*, 2nd edn. London: Wiley. (The best introduction to a complex but exciting field.)

Index